AF846465

# Advances in Geographical and Environmental Sciences

**Series Editors**

Yukio Himiyama, Hokkaido University of Education, Asahikawa, Hokkaido, Japan

Subhash Anand, Department of Geography, University of Delhi, New Delhi, India

Advances in Geographical and Environmental Sciences synthesizes series diagnostigation and prognostication of earth environment, incorporating challenging interactive areas within ecological envelope of geosphere, biosphere, hydrosphere, atmosphere and cryosphere. It deals with land use land cover change (LUCC), urbanization, energy flux, land-ocean fluxes, climate, food security, ecohydrology, biodiversity, natural hazards and disasters, human health and their mutual interaction and feedback mechanism in order to contribute towards sustainable future. The geosciences methods range from traditional field techniques and conventional data collection, use of remote sensing and geographical information system, computer aided technique to advance geostatistical and dynamic modeling.

The series integrate past, present and future of geospheric attributes incorporating biophysical and human dimensions in spatio-temporal perspectives. The geosciences, encompassing land-ocean-atmosphere interaction is considered as a vital component in the context of environmental issues, especially in observation and prediction of air and water pollution, global warming and urban heat islands. It is important to communicate the advances in geosciences to increase resilience of society through capacity building for mitigating the impact of natural hazards and disasters. Sustainability of human society depends strongly on the earth environment, and thus the development of geosciences is critical for a better understanding of our living environment, and its sustainable development.

Geoscience also has the responsibility to not confine itself to addressing current problems but it is also developing a framework to address future issues. In order to build a 'Future Earth Model' for understanding and predicting the functioning of the whole climatic system, collaboration of experts in the traditional earth disciplines as well as in ecology, information technology, instrumentation and complex system is essential, through initiatives from human geoscientists. Thus human geoscience is emerging as key policy science for contributing towards sustainability/survivality science together with future earth initiative.

Advances in Geographical and Environmental Sciences series publishes books that contain novel approaches in tackling issues of human geoscience in its broadest sense-books in the series should focus on true progress in a particular area or region. The series includes monographs and edited volumes without any limitations in the page numbers.

Gaurav Tripathi · Achala Shakya · Shruti Kanga ·
L. T. Sasang Guite · Suraj Kumar Singh
Editors

# Sustainability and Health Informatics

## A Systems Approach to Address the Climate Action Induced Global Challenge

Springer

*Editors*
Gaurav Tripathi
School of Agricultural and Environmental Sciences
Shobhit University
Meerut, Uttar Pradesh, India

Achala Shakya
School of Computer Sciences
University of Petroleum and Energy Studies
Dehradun, Uttarakhand, India

Shruti Kanga
Department of Geography
Central University of Punjab
Bathinda, Punjab, India

L. T. Sasang Guite
Department of Geography
Central University of Punjab
Bathinda, Punjab, India

Suraj Kumar Singh
Centre for Sustainable Development
Suresh Gyan Vihar University
Jaipur, Rajasthan, India

ISSN 2198-3542 ISSN 2198-3550 (electronic)
Advances in Geographical and Environmental Sciences
ISBN 978-981-97-6705-2 ISBN 978-981-97-6706-9 (eBook)
https://doi.org/10.1007/978-981-97-6706-9

© The Editor(s) (if applicable) and The Author(s), under exclusive license to Springer Nature Singapore Pte Ltd. 2024

This work is subject to copyright. All rights are solely and exclusively licensed by the Publisher, whether the whole or part of the material is concerned, specifically the rights of translation, reprinting, reuse of illustrations, recitation, broadcasting, reproduction on microfilms or in any other physical way, and transmission or information storage and retrieval, electronic adaptation, computer software, or by similar or dissimilar methodology now known or hereafter developed.
The use of general descriptive names, registered names, trademarks, service marks, etc. in this publication does not imply, even in the absence of a specific statement, that such names are exempt from the relevant protective laws and regulations and therefore free for general use.
The publisher, the authors and the editors are safe to assume that the advice and information in this book are believed to be true and accurate at the date of publication. Neither the publisher nor the authors or the editors give a warranty, expressed or implied, with respect to the material contained herein or for any errors or omissions that may have been made. The publisher remains neutral with regard to jurisdictional claims in published maps and institutional affiliations.

This Springer imprint is published by the registered company Springer Nature Singapore Pte Ltd.
The registered company address is: 152 Beach Road, #21-01/04 Gateway East, Singapore 189721, Singapore

If disposing of this product, please recycle the paper.

# Preface

The world faces an unprecedented challenge—one that transcends borders, disciplines, and sectors. The nexus between climate change and human health demands innovative, multidisciplinary solutions. As the impacts of a changing climate become more pronounced, so does the urgency for a systems approach that integrates health informatics, sustainability, and technology. It is within this context that our proposed book, *Sustainability and Health Informatics: A Systems Approach to Address Climate Action Induced Global Challenges*, finds its purpose.

The intersection of health informatics and sustainability represents a powerful synergy. This edited volume is a comprehensive exploration of the complex, multifaceted challenges posed by climate change and how a systems approach at the global, national, and local levels can address these challenges. It delves into the ethical and social implications of harnessing health informatics and climate change solutions, encompassing data privacy, equity, and access to healthcare.

We aim to bridge the existing gap in the available literature by assembling the concepts, theories, and practical experiences of specialists and professionals in the field within a single volume. This book will serve as a showcase of successful health informatics and climate change solutions from around the world, revealing innovative approaches to data collection, analysis, and dissemination.

*Sustainability and Health Informatics: A Systems Approach to Address Climate Action Induced Global Challenges* brings together leading experts in the fields of sustainability and health informatics to explore how information systems and technology can address the formidable challenges posed by climate change to global health. It offers a comprehensive overview of the latest research, technologies, and applications of health informatics to promote sustainability and address climate change. At its core, this book adopts a systems approach that acknowledges the intricate interconnections between the environment, health, and technology, and it offers practical solutions for sustainable development and climate action.

The chapters in this book explore a wide range of topics, from environmental sustainability and its impact on public health to the use of health informatics in climate-smart agriculture, disaster response, and telemedicine. We delve into big data and its applications in the context of climate action and health informatics. Our exploration also includes the role of health informatics in advancing Sustainable Development Goals (SDGs), e-health, education and training, ethics, and multi-stakeholder collaboration for climate action. The contents of this book are not just a compilation of knowledge; they represent a collective effort to make a difference. Each chapter contributes to our understanding of how we can leverage health informatics to mitigate the impacts of climate change on human health and build more resilient and sustainable communities.

We extend our heartfelt appreciation to the dedicated contributors who have made this book possible. Your insights and expertise are invaluable. We hope this book is an essential resource for researchers, practitioners, and policymakers in health informatics, sustainability, and climate change. It is a resource that will empower individuals and communities to address the complex challenges of climate change and enhance healthcare outcomes through innovative health informatics solutions.

In summary, the interconnected challenges of climate change and human health demand a multidisciplinary and integrated approach, one that harnesses the power of health informatics. By doing so, we can develop innovative solutions that safeguard the health and well-being of present and future generations and pave the way for more sustainable and resilient societies.

Meerut, India — Gaurav Tripathi
Dehradun, India — Achala Shakya
Bathinda, India — Shruti Kanga
Bathinda, India — L. T. Sasang Guite
Jaipur, India — Suraj Kumar Singh

# Contents

# Chapter 1
# Climate Change, Health, and Health Informatics

**Suchitra Mitra and R. K. Saran**

## 1.1 Introduction

Climate change is one of the most alarming issues of present time, and its severe effects on human health are becoming increasingly clear. Extreme Weather Events (EWE), such as floods, droughts, and heat waves, are becoming more common and severe, significantly impacting our health (Cianconi et al. 2020). The effects of climate change on human health are not just limited to developed countries whereas it has been affecting developing countries as well for decades. Even in developed countries, climate change is already having a significant impact. For example, heat waves are becoming more common in the United States, leading to increased deaths from heat stroke. To acknowledge the health impacts of climate change, we must have accurate and timely information. This information will help us understand the risks that climate change poses to our health and develop effective strategies for mitigating these risks (Ebi et al. 2021). The relation between climate change and health will be discussed throughout the study. The present study will explore how climate change affects our health adversely, and discuss the importance of accurate and timely information (Ahmad et al. 2023). It will also highlight some of the strategies that are being used to mitigate the health effects of climate change. Climate change is a complex issue, and there is still much that we do not know about its effects on human health. However, the available evidence makes it clear that climate change seriously threatens our health (WHO 2015). The health impacts of climate change are not evenly distributed. People who are already marginalized, such as those living in poverty or areas with poor infrastructure, are mostly vulnerable to

S. Mitra (✉)
Indian Institute of Science Education and Research, Kolkata, India
e-mail: mitrasuchitra85@gmail.com

R. K. Saran
Department of Microbiology, Maharaja Ganga Singh University, Bikaner, India

© The Author(s), under exclusive license to Springer Nature Singapore Pte Ltd. 2024

G. Tripathi et al. (eds.), *Sustainability and Health Informatics*, Advances in Geographical and Environmental Sciences,
https://doi.org/10.1007/978-981-97-6706-9_1

the effects of climate change (Abbas et al. 2022). There are several things that we can do to mitigate the health impacts of climate change. These include reducing our emissions of greenhouse gases, adapting to the changes that are already happening, and investing in research better to understand the links between climate change and human health (Singh et al. 2022).

Casson et al. (2023), conducted a comprehensive and insightful study to analyse the perception of the health impacts of climate change among people in Canada. Canadians perceive the health effects of climate change, which is crucial for developing effective strategies to mitigate these impacts. The study emphasizes the importance of raising awareness and understanding public perceptions to address climate-related health risks. Anik et al. (2023) reviewed the effects of climate change on water resources in Bangladesh and the subsequent health risks that arise, Bangladesh being a low-lying country with a high population density, is particularly highly impacted by the climate change on its water resources. Rapid changes in rising sea levels, precipitation patterns, and increased frequency of EWE pose significant threats to the availability and quality of water. Moreover, the article highlights the direct and indirect health risks associated with these changes. Waterborne diseases, such as diarrhea and cholera, can become more prevalent due to compromised water sources and inadequate sanitation systems (Farooq et al. 2022). Additionally, the disruption of water availability can impact agriculture, leading to food insecurity and malnutrition, further exacerbating health risks. Kamel Boulos and Wilson (2023) highlighted the importance of geospatial applications, such as remote sensing, data analysis tools, and geospatial analysis, in understanding and mitigating the significant effects of climate change on human health (Singh et al. 2021a, b, c). These techniques enable researchers and policymakers to assess the spatial distribution of climate-related health risks, such as heatwaves, air pollution, and vector-borne diseases (Meraj et al. 2021; Singh et al. 2021a, b, c).

The objective of writing the book chapter "Climate Change, Health, and Health Informatics" is to help people understand how climate change affects our health and to show how technology and data can be used to protect our well-being in a changing environment. It aims to encourage different experts to work together and find practical ways to keep us healthy despite the challenges of climate change.

## 1.2 Impacts of Climate Change on Health

### *1.2.1 Changing Patterns of Disease*

#### Vector-Borne Diseases

Climate change is expected to impact the spread of vector-borne diseases significantly. Vectors are organisms that transmit diseases to humans, such as mosquitoes, ticks, and fleas. These organisms are sensitive to changes in temperature, rainfall, and humidity changes, and climate change is expected to change their geographic range, abundance, and behaviour (Chua et al. 2023).

As a result of climate change, we are likely to see an increase in the number of people who are at risk of contracting vector-borne diseases. For example, mosquitoes that carry malaria are now able to survive in areas that were previously too cold for them. This is leading to an increase in the number of people who are at risk of contracting malaria in these areas. Other vector-borne diseases that are expected to be affected by climate change include Zika virus, Chagas, West Nile virus, Leishmaniasis, Lyme disease, Malaria, Yellow fever, Chikungunya, and dengue fever. These diseases are already widespread in many regions worldwide, and climate change is expected to make them even more widespread (Palraj 2022).

## Waterborne Diseases

These are the diseases that are caused by the consumption of contaminated water. Various pathogens, including bacteria, viruses, and parasites can cause these diseases. Climate change has already had a significant impact on waterborne diseases (Kumar et al. 2022). Here are some of the relevant approaches in which climate change is affecting waterborne diseases:

- Increased water temperatures: As the Earth's climate warms, water temperatures are also rising. This can lead to the proliferation of bacteria and other pathogens that can cause waterborne diseases. For example, cholera is a waterborne disease that is caused by the bacterium Vibrio cholera (Sinha et al. 2022). Cholera is more likely to occur in warmer water, so as water temperatures rise, the risk of cholera outbreaks is also likely to increase.
- Changes in precipitation patterns: Climate change is also causing changes in precipitation patterns. In some areas, there is more rainfall, while in other areas, there is less rainfall. These changes in precipitation patterns can lead to flooding and drought, which can both increase the risk of waterborne diseases. For example, flooding can contaminate water supplies with sewage and other contaminants, which can lead to outbreaks of waterborne diseases. Drought can also lead to water scarcity, which can make it difficult to maintain clean water supplies, increasing the risk of waterborne diseases.
- Rising sea levels: Climate change is also causing sea levels to rise. This can lead to saltwater intrusion into freshwater sources, which can contaminate drinking water and increase the risk of waterborne diseases. For example, saltwater intrusion has been linked to an increase in the incidence of cholera in Bangladesh.
- Disruption of ecosystems: Climate change is also disrupting ecosystems, which can lead to the spread of waterborne diseases. For example, climate change is causing the ranges of some mosquitoes to expand, which can raise the risk of mosquito-borne diseases such as Chikungunya, Malaria, Dengue fever, and Zika virus.

The impacts of climate change on waterborne diseases are likely to become more severe in the future. As the Earth's climate continues to warm, water temperatures will rise, precipitation patterns will become more erratic, sea levels will rise, and ecosystems will be disrupted. All these factors will increase the risk of waterborne diseases.

Various things can be done to lesser the significant impacts of climate change on waterborne diseases including:

- Reducing Greenhouse Gas Emissions (GGE): This will lead to slow the pace of climate change and reduce the overall risk of waterborne diseases.
- Improving water management: This includes ensuring that drinking water is safe and that wastewater is properly treated.
- Protecting ecosystems: This will help to prevent the spread of waterborne diseases by disrupting the habitats of mosquitoes and other vectors.
- Building resilience: This includes educating people about waterborne diseases and how to prevent them and developing an early warning system that can detect and respond to outbreaks situations.

**Foodborne Diseases**

These are diseases that are caused by the consumption of contaminated food. Various pathogens, including bacteria, viruses, and parasites, can cause these diseases. Climate change is expected to impact the spread of foodborne diseases significantly. This is because climate change is likely to lead to changes in food production, processing, and distribution, and the prevalence and intensity of EWE (Sudershan et al. 2014).

For example, climate change is expected to increase the frequency and intensity of floods. This could lead to food crops and livestock contamination, increasing the risk of foodborne diseases. In addition, climate change is expected to lead to an increment in food temperature. This could lead to the growth of bacteria and other pathogens in food, which could also increase the risk of foodborne diseases.

The health impacts of foodborne diseases can be significant. These diseases can cause serious illness, disability, and even death. In addition, they can have a significant economic impact, as they can lead to lost productivity and increased healthcare costs (Adley and Ryan 2016). Some mitigation approaches for severe health impacts due to climate change are given below:

- Improving food safety standards. This includes measures such as increasing the frequency of food inspections, improving the sanitation of food processing facilities, and educating food handlers about proper food handling practices.
- Investing in food infrastructure to protect food from contamination. This includes measures such as building flood-resistant storage facilities and improving the quality of water used to irrigate crops.

- Educating people about the risks of foodborne diseases and how to protect themselves. This includes information about properly cooking food, storing food safely, and identifying and avoiding contaminated food.

Foodborne diseases are a major and frequent cause of death and disability in developing countries. Climate change is expected to make foodborne diseases more common in these countries. Several actions can be taken to alleviate the health impacts of changes in global climate on foodborne diseases, such as improving food safety standards and investing in food infrastructure.

Specific examples of how climate change affects foodborne diseases:

I. ***Salmonella***: It is a bacterium that can cause food poisoning. It is commonly found in poultry, eggs, and other animal products. Climate change is expected to raise the risk of *Salmonella* infection by increasing the temperature of food and making it easier for the bacteria to grow.
II. ***E. coli***: It is a bacterium that can cause food poisoning. It is commonly found in raw or undercooked meat, poultry, and dairy products. Climate change is expected to increase the risk of *E. coli* infection by increasing the frequency and intensity of floods, which can contaminate food with bacteria.
III. ***Listeria***: It is a bacterium that can cause food poisoning. It is commonly found in soft cheeses, deli meats, and other processed foods. Climate change is expected to increase the risk of listeria infection by increasing the temperature of food and making it easier for the bacteria to grow.

### Respiratory Diseases

Respiratory diseases are diseases caused by inhaling harmful substances, such as pollutants, allergens, and pathogens. These diseases can range from mild to severe, and they can affect people of all ages. Climate change is expected to have a life-changing impact on the spread of respiratory-borne diseases (Joshi et al. 2020). This is because climate change is expected to lead to changes in the air quality, as well as the intensity and frequency of EWE.

For example, climate change is expected to lead to an increase in the frequency and intensity of wildfires. This could lead to the release of harmful pollutants into the air, which could increase the risk of respiratory diseases. In addition, climate change is expected to lead to an increase in the temperature of the air. This could lead to the growth of mold and other allergens, which could also increase the risk of respiratory diseases (D'Amato et al. 2014).

The health impacts of respiratory diseases can be significant. These diseases can cause serious illness, disability, and even death. In addition, they can have a significant economic impact, as they can lead to lost productivity and increased healthcare costs.

Mitigation measures include:

- Improving air quality standards: This includes measures such as reducing emissions from vehicles and factories and planting trees to help filter the air.

- Educating people about the risks of respiratory diseases and how to protect themselves: This includes information about how to avoid exposure to pollutants, allergens, and pathogens.
- Developing new treatments and vaccines for respiratory-borne diseases: This will help to improve the care of people who are already sick, and it will help to prevent people from getting sick in the first place.

By taking these steps, we can help to reduce the burden of respiratory-borne diseases and protect the health of people around the world.

Respiratory diseases are a major cause of disability and death in both developed and developing countries. Climate change is expected to make respiratory diseases more common in these countries. A number of things can be done to mitigate the health impacts of climate change on respiratory diseases, such as improving air quality standards and educating people about the risks (D'amato et al. 2016).

Specific examples of how climate change is affecting respiratory diseases include:

- Asthma. It is a chronic respiratory disease that affects millions of people around the world. Climate change is expected to increase the risk of asthma by making the air more polluted and by increasing the recurrence and intensity of EWE.
- Chronic obstructive pulmonary disease (COPD). COPD includes chronic bronchitis and emphysema, is a group of lung diseases. Climate change is expected to increase the risk of COPD by making the air more polluted and by increasing the frequency and intensity of EWE.
- Pneumonia. It is a lung infection that can be caused by a variety of bacteria, viruses, and fungi. Climate change is expected to increase the risk of pneumonia by making the air more polluted and by increasing the intensity and frequency of EWE (Fig. 1.1).

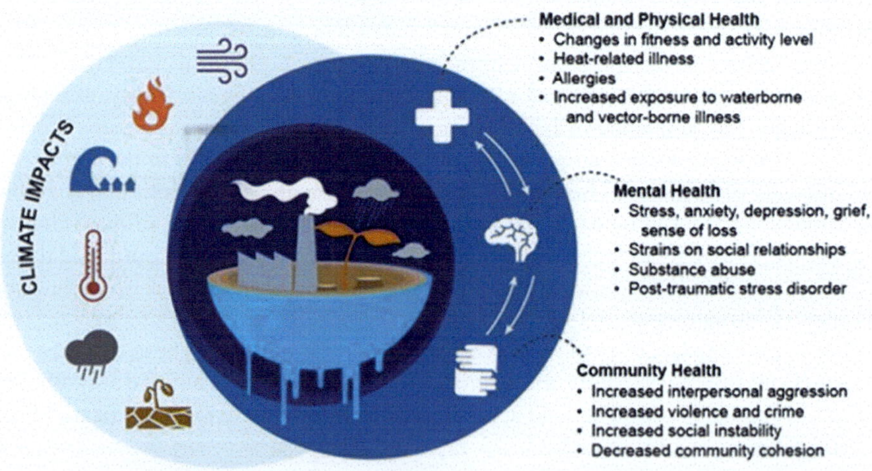


**Fig. 1.1** Health impacts of climate change (*source* modified from https://health2016.globalchange.gov/)

### *1.2.2 Extreme Weather Events (EWE)*

EWE are becoming more common and more severe due to climate change. These events can have a significant impact on human health, both directly and indirectly (Bergquist et al. 2019).

I. **Heatwaves**: Heatwaves are periods of prolonged hot weather that can be deadly. They are particularly dangerous for the elderly, the young, and those with chronic health conditions. Heatwaves can cause heat stroke, heat exhaustion, and other heat-related illnesses. They can also worsen existing health conditions, such as heart disease and respiratory disease.
II. **Hurricanes and Cyclones**: Hurricanes and cyclones are powerful storms that can cause extensive damage to property, buildings, and infrastructure. They can also lead to flooding, which can displace people and disrupt access to food and water. Hurricanes and cyclones can also cause injuries and deaths.
III. **Floods**: Floods are caused by an excessive amount of water in an area that is not normally flooded. They are caused by heavy rains, melting snow, or a combination of factors. Floods can not only cause damage to property and infrastructure, but also, can also lead to injuries and deaths.
IV. **Droughts**: Droughts are periods of prolonged dry weather that can lead to water shortages. They can also cause crop failures, which can lead to food insecurity and malnutrition. Droughts can also have a negative impact on mental health.
V. **Other EWE**: In addition to heatwaves, hurricanes, floods, and droughts, there are a number of other EWE that are becoming more common due to climate change. These include wildfires, extreme rainfall events, and extreme snowfall events.

#### The Health Impacts of EWE

The health impacts of EWE can be both direct and indirect. Direct health impacts include injuries, illnesses, and deaths caused by the event itself. Indirect health impacts include the effects of displacement, food insecurity, and water scarcity.

#### Mitigating the Health Impacts of EWE

Mitigation measures include:

- Improving early warning systems. These can help people to prepare priorly for and respond to EWE.
- Building resilient infrastructure. Resilient infrastructure can withstand the impacts of EWE.
- Educating people about the risks. People need to be aware of the risks of EWE and how to protect themselves.

### 1.2.3 *Environmental Pollution*

**Air Pollution**

Air pollution is a major environmental problem that significantly impacts human health and the environment. It is caused by releasing harmful atmospheric pollutants, such as gases, particulate matter, and volatile organic compounds. These pollutants can come from various sources, including transportation, industry, and power plants (Kaur and Pandey 2021). Air pollution is a major contributor to climate change. When pollutants such as carbon dioxide and methane are released into the atmosphere, they trap heat, which causes the planet to warm (Gulati et al. 2023). This warming can lead to several negative consequences, including rising sea levels, more ewe, and changes in agricultural yields (Manisalidis et al. 2020).

There are a number of things that can be done to reduce air pollution and mitigate climate change. These include:

- Switching to cleaner energy sources, such as solar and wind power.
- Boosting energy efficiency in homes and businesses.
- Reducing our reliance on cars and using public transportation, walking, or biking instead.
- Supporting policies that promote clean air and climate action.

Here are some specific examples of how air pollution can contribute to climate change:

When fossil fuels are combusted, they release carbon dioxide into the atmosphere. Carbon dioxide, being a greenhouse gas, traps heat in the atmosphere. This trapping of heat causes the planet to warm. Particulate matter (PM) is a type of air pollution that can be harmful to human health. PM can also contribute to climate change by absorbing sunlight and warming the atmosphere. Volatile organic compounds (VOCs) are another type of air pollution that can contribute to climate change. VOCs can react with other pollutants in the atmosphere to form ozone, which is a greenhouse gas (Singh et al. 2021).

**Water Pollution**

Water pollution is a major environmental problem that can significantly impact human health, ecosystems, and the climate. It is caused by releasing harmful pollutants into water bodies, such as ponds lakes, rivers, and oceans. The commencement of pollutants can be from a variety of sources, including industrial waste, agricultural runoff, and sewage. Water pollution can contribute to climate change in several ways. First, when pollutants such as organic matter and nutrients are released into water bodies, they can lead to the growth of algae and other microorganisms (Kotecha et al. 2024). This process, known as eutrophication, can deplete water oxygen levels, killing fish and other aquatic life. Decomposing algae can also release methane, a greenhouse gas that contributes to climate change (Lin et al. 2022).

Second, water pollution can increase the risk of flooding. When pollutants clog waterways, they can prevent water from flowing freely. This can lead to flooding, damaging property and infrastructure, and releasing greenhouse gases into the atmosphere. Third, water pollution can reduce the ability of forests to absorb carbon dioxide. When forests are polluted, they are less able to photosynthesize, which means they are less able to remove carbon dioxide from the atmosphere (Michalak 2016; Liu et al. 2023).

There are several things that can be done to reduce water pollution and mitigate climate change. These include:

- Reducing our reliance on pesticides and fertilizers.
- Treating sewage before it is released into waterways.
- Recycling and composting to reduce the amount of waste that ends up in landfills.
- Supporting policies that promote clean water and climate action.

Here are some specific examples of how water pollution can contribute to climate change:

- When agricultural runoff enters waterways, it can carry nutrients that promote algae growth. These algae can then decompose and release methane, a greenhouse gas.
- When sewage is released into waterways without treatment, it can carry bacteria that can kill fish and other aquatic life. This can deplete oxygen levels in the water, which can also contribute to climate change.
- When forests are polluted, they are less able to photosynthesize, which means they are less able to remove carbon dioxide from the atmosphere.

## Soil Contamination

Soil pollution is caused by the release of harmful pollutants into the soil, such as heavy metals, pesticides, and chemical fertilizers. The commencement of pollutants can be from a variety of sources, including industrial waste, agricultural runoff, and mining. Soil pollution can contribute to climate change in a number of ways. First, when pollutants such as heavy metals and pesticides are released into the soil, they can disrupt the soil's ability to store carbon. This can lead to an increase in GGE, as the soil releases more carbon dioxide into the atmosphere (Mocatta et al. 2022).

Second, soil pollution can reduce the ability of plants to grow. This can lead to a decrease in the amount of vegetation on the planet, which can also contribute to climate change. Plants help by absorbing carbon dioxide from the atmosphere and releasing oxygen, thus regulating the climate. Third, soil pollution can increase the risk of erosion. When the soil is polluted, it is less able to hold water and nutrients. This can lead to the soil being washed away by rain or wind, which can release greenhouse gases into the atmosphere.

Mitigation measures include:

- Reducing our use of pesticides and fertilizers.
- Recycling and composting to reduce the amount of waste that ends up in landfills.
- Supporting policies that promote sustainable agriculture and land management.

Here are some specific examples of how soil pollution can contribute to climate change:

- When heavy metals are released into the soil, they can disrupt the soil's ability to store carbon. This can lead to an increase in GGE, as the soil releases more carbon dioxide into the atmosphere.
- When pesticides are released into the soil, they can kill beneficial bacteria and fungi. These organisms help to break down organic matter and release carbon dioxide into the atmosphere.
- When soil is eroded, it releases carbon dioxide into the atmosphere. This is because the soil contains organic matter, which is a source of carbon.

## 1.3 Climate Change Communication and Health Information

### *1.3.1 Communicating the Health Risks*

Effective communication plays a crucial role in addressing the health risks associated with climate change. It is essential to raise public awareness, provide accurate information, and engage healthcare professionals in order to promote understanding, preparedness, and mitigation efforts. In this section, we will explore various strategies and approaches to communicate the health risks of climate change.

#### Public Awareness Campaigns

Public awareness campaigns serve as a vital tool for disseminating information about the health risks associated with climate change. These campaigns aim to educate and engage the general public, encouraging them to take action to protect their health and the environment. They utilize several media, such as social media, television, radio, and community events, to reach a wide audience. Public awareness campaigns often highlight the link between climate change and health, focusing on key issues like heat-related illnesses, vector-borne diseases, air pollution, and mental health impacts (Khatibi et al. 2021). By raising awareness and promoting behavior change, these campaigns empower individuals to make informed decisions and adopt sustainable practices that can mitigate health risks.

#### Health Education Programs

Health education programs are essential for equipping individuals and communities with the knowledge and skills necessary to understand and respond to the health risks of climate change. These programs can be implemented in schools, community centres, workplaces, and healthcare settings. Health education programs focus on providing evidence-based information about the specific health impacts of climate change, as well as strategies for prevention, adaptation, and resilience. They may cover topics such as recognizing heat-related illnesses, understanding air quality and its effects on respiratory health, managing allergies and asthma, and promoting mental well-being in the face of climate-related stressors. By empowering individuals with knowledge and resources, health education programs contribute to building a resilient and prepared society (Shaman and Knowlton 2018).

#### Engaging Healthcare Professionals

Healthcare professionals, and other medical personnel, play a crucial role in communicating the health risks of climate change. They are trusted sources of information and can effectively communicate the connection between climate change and public health to their patients and the wider community (Dupraz and Burnand 2021).

Engaging healthcare professionals involves providing them with the necessary training and resources to understand the health impacts of climate change, recognize related symptoms, and effectively counsel patients. This can be achieved through:

- Professional development programs, conferences, and workshops that focus on climate change and health.
- Integrating climate change and health topics into medical and public health curricula.
- Providing healthcare professionals with access to up-to-date information and resources on climate change and health.

By mobilizing healthcare professionals as advocates and educators, we can enhance public understanding, improve preventive measures, and facilitate appropriate responses to climate-related health risks.

### *1.3.2 Barriers to Effective Communication*

There are a number of barriers that can hinder effective communication about climate change. These include:

- Misinformation and disinformation: The misinformation and disinformation about climate change circulating online and in the media, can make it difficult for people to find accurate information and can lead to confusion and distrust.

- Psychological factors: People's psychological responses to climate change can also be a barrier to communication. For example, people may be in denial about climate change, or they may feel overwhelmed or helpless. These psychological factors can make it difficult for people to engage with information about climate change.
- Socioeconomic challenges: Socioeconomic factors can also be a barrier to communication about climate change. For example, people who are less educated or who have less access to information may be less likely to be aware of the risks of climate change. Additionally, people who are struggling to meet their basic needs may be less concerned about climate change.

### *1.3.3 Targeted Messaging and Outreach*

Effective communication about climate change requires tailoring messages to different audiences. This means using different communication channels, techniques, and languages to reach different people. For example, when communicating with the general public, it is important to use clear and concise language that is easy to understand. It is also important to focus on the benefits of taking action to mitigate climate change, such as improved air quality and reduced risk of disease (Pathak et al. 2021).

When communicating with healthcare professionals, it is important to use more technical language and to focus on the health impacts of climate change. It is also important to provide healthcare professionals with resources and training so that they can effectively communicate the health risks of climate change to their patients. When communicating with vulnerable populations, it is important to be sensitive to their cultural and linguistic needs. It is also important to focus on the ways in which climate change is already impacting these populations and how they can be affected in the future. By tailoring messages to different audiences, we can ensure that people are getting the information they need in a way that they can understand and act on.

Here are some additional tips for tailoring messages to different audiences:

- Consider the audience's level of knowledge about climate change.
- Use language that is appropriate for the audience's culture and language.
- Focus on the benefits of taking action to mitigate climate change.
- Use stories and examples that are relevant to the audience.
- Be patient and persistent. It takes time to change people's minds about climate change.
- By following these tips, we can help to ensure that our communication about climate change is effective and reaches the people who need it most.

### 1.3.4 Cultural and Linguistic Considerations

When communicating about climate change, it is important to be sensitive to cultural and linguistic differences. This means using language that is appropriate for the audience's culture and language and avoiding stereotypes or assumptions. For example, if you are communicating with a population that is predominantly Spanish-speaking, you should use Spanish-language materials and resources (Bayes et al. 2023). You should also avoid using stereotypes about Spanish-speaking people, such as the assumption that they are all poor or uneducated.

It is also important to be aware of the cultural context in which you are communicating. For example, in some cultures, it is considered disrespectful to talk about climate change in front of elders. In other cultures, it is considered important to involve traditional leaders in climate change discussions. By being sensitive to cultural and linguistic differences, we can ensure that our communication about climate change is effective and reaches the people who need it most.

### 1.3.5 Reaching Vulnerable Populations

Vulnerable populations are those that are disproportionately affected by climate change. This includes people who are poor, elderly or have disabilities. It also includes people who live in coastal areas, low-lying areas, or areas that are prone to EWE. It is important to reach vulnerable populations with information about climate change. This is because these populations are often the least able to adapt to the impacts of climate change (Kosanic et al. 2022).

There are a number of ways to reach vulnerable populations with information about climate change. These include:

- Working with community groups and organizations.
- Using local media outlets.
- Providing information in multiple languages.
- Using culturally appropriate materials.

## 1.4 Role of Health Organizations and Authorities

### 1.4.1 Surveillance and Monitoring

Surveillance and monitoring are essential tools for understanding and responding to climate change. They can help us to track the changing climate, identify the risks of climate change, and develop effective adaptation and mitigation strategies.

There are three main types of surveillance and monitoring:

- Early Warning Systems: EWS are designed to detect and warn of potential climate-related hazards, such as floods, droughts, and heat waves. This can help to save lives and property by giving people time to prepare for and respond to these hazards.
- Disease surveillance networks: Disease surveillance networks track the spread of diseases that are sensitive to climate change, such as malaria and dengue fever. This can help to identify outbreaks early and take action to control them.
- Environmental monitoring: Environmental monitoring tracks changes in the environment, such as air quality, water quality, and sea level rise. This can help us to understand the climate change impacts and to develop effective adaptation strategies.

Surveillance and monitoring are essential tools for responding to climate change. By tracking the changing climate, identifying the risks of climate change, and developing effective adaptation and mitigation strategies, we can help protect people and the environment from the impacts of climate change (Bowen et al. 2023).

Here are some specific examples of how surveillance and monitoring can be used to address climate change:

- Early warning system, EWS: It can be used to detect and warn of potential climate-related hazards, i.e., droughts, floods, and heat waves. This can help to save lives and property by giving people time to prepare for and respond to these hazards. For example, the Indian Meteorological Department's Early Warning System has been credited with saving millions of lives from floods and droughts.
- Disease surveillance networks: Disease surveillance networks can be used to track the spread of diseases that are sensitive to climate change, such as malaria and dengue fever. This can help to identify outbreaks early and take action to control them. For example, the GISRS of the World Health Organization (WHO) has been used to track the spread of influenza viruses and to develop vaccines to protect people from these viruses.
- Environmental monitoring: Environmental monitoring can be used to track changes in the environment, such as air quality, water quality, and sea level rise. This can help us to understand the impacts of climate change and to develop effective adaptation strategies. For example, National Centers for Environmental Information of National Oceanic and Atmospheric Administration (NOAA) collects data on a variety of environmental indicators, including air quality, water quality, and sea level rise. This data is used to track significant changes in the various parameters of environment and to develop climate change adaptation strategies.

### *1.4.2 Policy Advocacy*

Policy advocacy is the process of influencing public policy. It can be used to inform climate change policies, integrate health considerations into these policies, and strengthen health infrastructure.

There are a number of ways to engage in policy advocacy. These include:

- Advocating for policies.
- Building partnerships with other organizations to advocate for climate-friendly policies.
- By engaging in policy advocacy, we can help to ensure that climate change policies take into account the health of people and the environment.

### Informing Climate Change Policies

Climate change policies are the laws, regulations, and other measures that governments put in place to address climate change. These policies can have a significant effect on the health of people and the environment. It is important to inform climate change policies with health considerations. This means that during developing these policies, the health impacts of climate change should be considered as prime priorities (AchutaRao and Sagar 2023).

For example, when developing policies to reduce GGE, it is important to consider the impact of these policies on air quality and the spread of respiratory diseases.

### Integrating Health Considerations

In addition to informing climate change policies, it is also important to integrate health considerations into these policies. This means that the health of people and the environment should be considered throughout the development and implementation of climate change policies.

For example, when developing policies to adapt to the impacts of climate change, it is important to consider the impact of these policies on the health of people who are most vulnerable to climate change, such as the elderly and the poor.

### Strengthening Health Infrastructure

Finally, it is important to strengthen health infrastructure to address the health impacts of climate change. This means that governments should invest in health systems that are resilient to climate change and that can effectively respond to climate-related health emergencies.

For example, governments should invest in EWS to detect and warn of potential climate-related hazards, such as floods and droughts. By strengthening health infrastructure, we can help to protect people and the environment from the health impacts of climate change.

### 1.4.3 Collaboration and Partnerships

Climate change requires a global solution, as it is a global problem. International cooperation is essential to addressing climate change. This means working together with other countries to develop and implement policies that reduce GGE and help people adapt to the impacts of climate change (Smith et al. 2023).

There are a number of ways to engage in international cooperation on climate change. These include:

- Participating in international climate change negotiations.
- Adopting international climate change agreements.
- Providing financial and technical assistance to developing countries to help them address climate change.
- By engaging in international cooperation, we can help to build a more sustainable future for all.

**Interdisciplinary Approaches**

Climate change is a complex issue that requires a multidisciplinary approach. This means bringing together experts from a variety of fields, such as science, engineering, economics, and public policy, to address climate change. Interdisciplinary approaches are essential to understanding the causes and climate change impacts, developing effective solutions, and communicating the science of climate change to the public (Iwama et al. 2016).

There are a number of ways to promote interdisciplinary approaches to climate change. These include:

- Establishing interdisciplinary research centers and institutes.
- Funding interdisciplinary research projects.
- Providing training and education in interdisciplinary climate change studies.
- By promoting interdisciplinary approaches, we can help to build a more robust and effective response to climate change.

**Engaging Community Organizations**

Community organizations are essential to addressing climate change at the local level. These organizations can help to raise awareness of climate change, build resilience to climate change, and advocate for climate-friendly policies.

There are a number of ways to engage community organizations in addressing climate change. These include:

- Development and execution of plans for climate change adaptation in collaboration with various community groups.
- Providing technical as well as financial assistance to community organizations so they can combat climate change.

- Training community leaders on climate change.

By engaging community organizations, we can help to build a more sustainable future for all.

## 1.5 Harnessing Technology and Data for Health Information

### *1.5.1 Climate Models and Predictive Analytics*

Climate models are mathematical representations of climate system of the earth. They are used to predict how the climate will change in the future. Predictive analytics is the use of statistical methods to analyse data and predict future outcomes. Climate models and predictive analytics can be used to model the health impacts of climate change, forecast disease outbreaks, and identify high-risk areas (Etienne 2014).

#### Modelling Health Impacts

Climate models can be used to predict how climate change will affect the spread of diseases, the incidence of heat-related illnesses, and the availability of food and water. Predictive analytics can be used to identify populations that are most at risk from the health impacts of climate change.

#### Forecasting Disease Outbreaks

Climate models can be used to predict when and where disease outbreaks are likely to occur. Predictive analytics can be used to identify the factors that contribute to disease outbreaks and to develop interventions to prevent them.

#### Identifying High-Risk Areas

Climate models and predictive analytics can be used to identify areas that are most at risk from the impacts of climate change. This information can be used to develop adaptation strategies to protect people and the environment. By using climate models and predictive analytics, we can help to mitigate the health impacts of climate change and protect people and the environment.

Here are some specific examples of how climate models and predictive analytics have been used to address climate change:

- The Intergovernmental Panel on Climate Change (IPCC) uses climate models to project future climate change.
- The Centres for Disease Control and Prevention (CDC) uses predictive analytics to identify populations that are most at risk from the health impacts of climate change.
- WHO uses climate models to forecast disease outbreaks.

### *1.5.2 Digital Tools and Platforms*

Digital tools and platforms can be used to address climate change as follows:

- Educate people about climate change.
- Engage people in climate action.
- Track progress on climate change.
- Advocate for climate-friendly policies.

Various digital tools and platforms can be used to address climate change including:

- Mobile applications: Mobile applications can be used to provide people with information about climate change, to track their own carbon footprint, and to connect with other people who are interested in climate action.
- Online portals and websites: Online portals and websites can be used to provide people with information about climate change, to connect with other people who are interested in climate action, and to sign petitions and take other actions to support climate-friendly policies.
- Social media and influencers: social media and influencers can play a vital role to educate and aware people about climate change, and to engage people in climate action.

By using digital tools and platforms, we can help to reach a wider audience and to make a more significant impact on climate change.

### *1.5.3 Open Data and Citizen Science*

Open data is data that is freely available to anyone to use and reuse. Citizen science is a scientific method in which people volunteer their time to collect data. Such as:

- Collect data about the impacts of climate change.
- Develop new climate change mitigation and adaptation strategies.
- Educate people about climate change.
- Engage people in climate action.

There are a number of different ways to use open data and citizen science to address climate change. These include:

- Crowdsourced data collection: Crowdsourced data collection is a method of collecting data by asking people to contribute data. This can be done through online surveys, mobile apps, or other methods.
- Participatory monitoring: Participatory monitoring is a method of monitoring the environment by involving local communities. This can be done by training community members to collect data and by providing them with the tools and resources they need to do so.
- Citizen engagement initiatives: Citizen engagement initiatives are programs that are designed to engage people in climate action. These initiatives can take a variety of forms, such as community gardens, energy efficiency workshops, and climate strikes.

By using open data and citizen science, we can help to address climate change in a more informed and inclusive way.

Citizen engagement initiatives can be a valuable tool for addressing climate change. There are a number of different ways to use citizen engagement initiatives for climate change. For example, the Sierra Club organizes climate strikes. The Nature Conservancy organizes community gardens. The Alliance for Global Water Adaptation organizes energy efficiency workshops.

**Geospatial Distribution of Climate Data Including Surface Temperature, Precipitation, Relative Humidity (RH) and Solar Radiation (SR):**

See Figs. 1.2, 1.3, 1.4, and 1.5.

## 1.6 Discussion

### *1.6.1 Climate Change and Human Informatics: A Continent-Wise Analysis*

Climate change is a global challenge with diverse impacts, varying from continent to continent and country to country. Human informatics plays a critical role in understanding the regional intricacies of climate change and its consequences for the health and well-being of communities across the world. Let's delve into the continent-wise analysis of the major sources and reasons for surface changes, as well as the impacts of climate change on earth surface temperature, relative humidity, solar radiation, and precipitation, as represented in Figs. 1.2, 1.3, 1.4, and 1.5.

I. **North America**: In North America, the burning of fossil fuels from transportation and industrial activities is a significant contributor to climate change. This has led to more frequent and droughts, wildfires, severe heatwaves, and

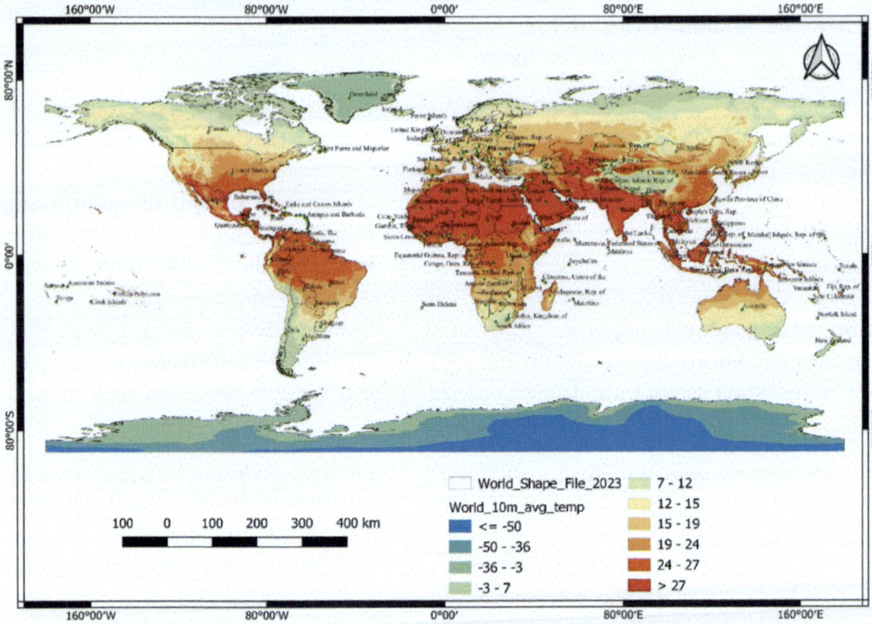


**Fig. 1.2** Geospatial distribution of surface temperature globally (Data obtained from https://climatedata.imf.org/pages/access-data and interpolated maps generated QGIS 3.32.2 'Lima')

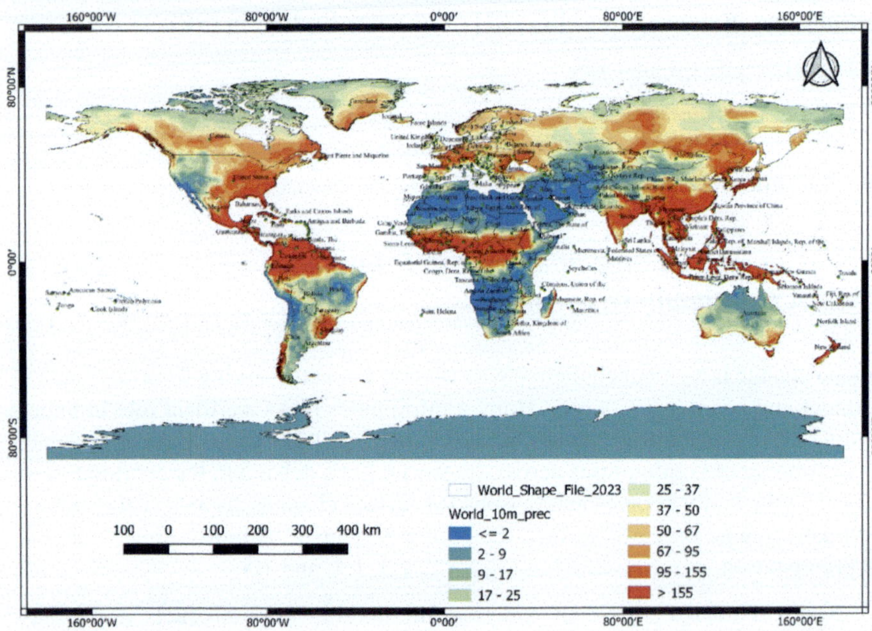


**Fig. 1.3** Geospatial distribution of precipitation globally (Data obtained from https://climatedata.imf.org/pages/access-data and interpolated maps generated QGIS 3.32.2 'Lima')

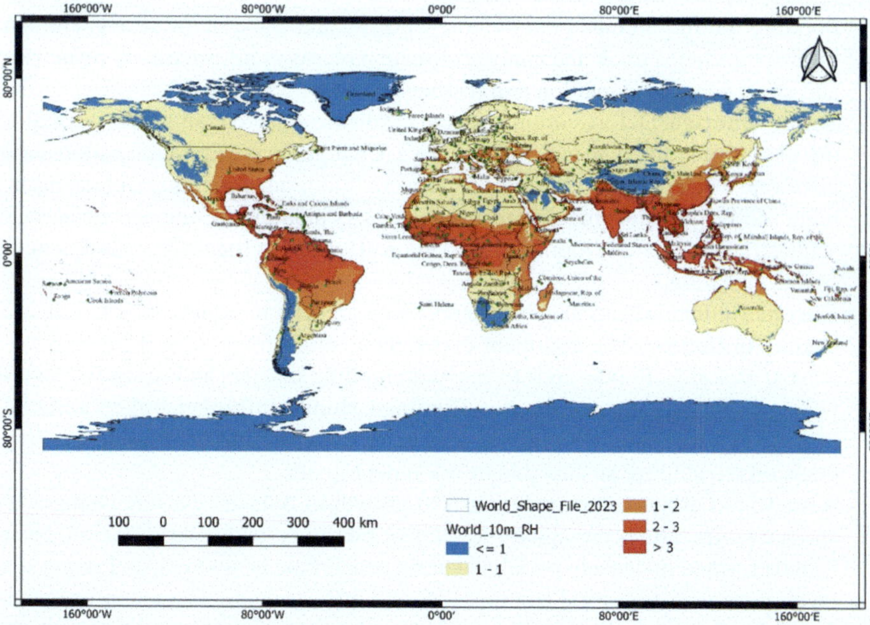


**Fig. 1.4** Geospatial distribution of relative humidity (RH) globally (Data obtained from https://climatedata.imf.org/pages/access-data and interpolated maps generated QGIS 3.32.2 'Lima')

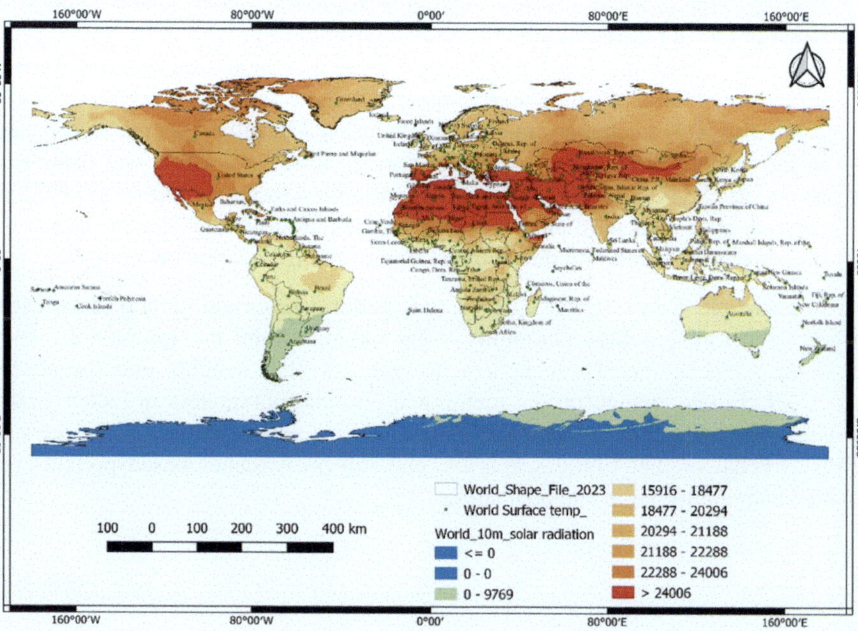


**Fig. 1.5** Geospatial distribution of solar radiation (SR) globally (Data obtained from https://climatedata.imf.org/pages/access-data and interpolated maps generated QGIS 3.32.2 'Lima')

flooding, putting human populations, wildlife, and ecosystems at risk. Human informatics helps track and analyse climate patterns to inform timely responses and strategies for adaptation and resilience.

II. **South America**: Deforestation, particularly in the Amazon rainforest, is a major driver of climate change in South America. Changes in precipitation patterns are also contributing to more frequent and severe floods and droughts. Human informatics aids in monitoring deforestation rates, evaluating the impacts on carbon sequestration, and developing conservation efforts to mitigate the consequences of deforestation.

III. **Europe:** Industrialization and transportation contribute significantly to climate change in Europe. The continent experiences more frequent and severe heatwaves, droughts, floods, and storms, with melting glaciers and rising sea levels posing threats to natural and cultural heritage. Human informatics plays a crucial role in modeling climate scenarios, assessing vulnerabilities, and guiding policy decisions to achieve sustainable development goals.

IV. **Asia**: Rapid urbanization and industrial growth in Asia are causing changes in surface temperature, precipitation, relative humidity, and solar radiation. Heat islands formed in densely populated areas exacerbate heatwaves, and rising sea levels pose risks to coastal communities and infrastructure. Human informatics facilitates the integration of climate data with health and disaster management systems to enhance preparedness and response measures.

V. **Africa**: Desertification and land degradation are major concerns in Africa, leading to increases in surface temperature due to the loss of vegetation that can help cool the air. More frequent and severe droughts, floods, and storms further exacerbate food and water insecurity. Human informatics supports EWS for drought and famine, aiding humanitarian efforts in vulnerable regions.

VI. **Australia**: Due to changes in climatic patterns, Australia is seeing an increase in the frequency and severity of heatwaves, storms, droughts and bushfires. Rising sea levels also threaten coastal communities and infrastructure. Human informatics assists in monitoring wildfire risks, optimizing resource allocation during emergencies, and developing climate change adaptation strategies.

The presented data in Figs. 1.2, 1.3, 1.4, and 1.5 represent annual mean earth surface temperature, relative humidity, solar radiation, and precipitation at 10 m across the entire globe. These figures provide essential insights into the severe impacts of climate change on the mentioned climatic parameters in recent years. Human informatics utilizes advanced data analytics and modelling techniques to identify trends, predict future scenarios, and inform evidence-based policies for climate change mitigation and adaptation.

### 1.6.2 Continent-Wise Summary of Climate Change Data Using Temperature, Precipitation, RH and SR

I. **Africa:**

*Temperature*: Africa's temperatures have slightly risen 1.2 °C since pre-industrial times. This warming is most pronounced in the Sahara Desert, where temperatures have increased by about 1.5 °C.

*Precipitation*: Precipitation patterns in Africa are becoming more erratic, with more intense rainfall events and longer droughts. This is leading to flooding in some areas and water shortages in others.

*RH*: RH levels in Africa are increasing, as the atmosphere can hold more water vapor as it warms. This can make heat waves more dangerous, as it can make it feel even hotter than the actual temperature.

*SR*: SR levels in Africa are increasing, as the Earth's atmosphere warms and traps more of this energy. This is leading to further warming.

II. **Asia:**

*Temperature*: Temperatures in Asia have increased by about 1.3 °C since the pre-industrial era. This warming is most pronounced in the Tibetan Plateau, where temperatures have increased by about 1.7 °C.

*Precipitation*: Precipitation patterns in Asia are becoming more erratic, with more intense rainfall events and longer droughts. This is leading to flooding in some areas and water shortages in others.

*RH*: RH levels in Asia are increasing, as the atmosphere can hold more water vapor as it warms. This can make heat waves more dangerous, as it can make it feel even hotter than the actual temperature.

*SR*: SR levels in Asia are increasing, as the Earth's atmosphere warms and traps more of this energy. This is leading to further warming.

III. **Europe:**

*Temperature*: Europe has witnessed an increase in surface temperature by about 1 °C since the pre-industrial era. This warming is most pronounced in the Arctic, where temperatures have increased by about 1.5 °C.

*Precipitation*: Precipitation patterns in Europe are becoming more erratic, with more intense rainfall events and longer droughts. This is leading to flooding in some areas and water shortages in others.

*RH*: RH levels in Europe are increasing, as the atmosphere can hold more water vapor as it warms. This can make heat waves more dangerous, as it can make it feel even hotter than the actual temperature.

*SR*: SR levels in Europe are increasing, as the Earth's atmosphere warms and traps more of this energy. This is leading to further warming.

IV. **North America**

*Temperature*: Temperatures in North America have increased by about 1 °C since the pre-industrial era. This warming is most pronounced in the Arctic, where temperatures have increased by about 1.5 °C.

*Precipitation*: Precipitation patterns in North America are becoming more erratic, with more intense rainfall events and longer droughts. This is leading to flooding in some areas and water shortages in others.

*RH*: RH levels in North America are increasing, as the atmosphere can hold more water vapor as it warms. This can make heat waves more dangerous, as it can make it feel even hotter than the actual temperature.

*SR*: SR levels in North America are increasing, as the Earth's atmosphere warms and traps more of this energy. This is leading to further warming.

V. **South America**

*Temperature*: Temperatures in South America have increased by about 0.9 °C since the pre-industrial era. This warming is most pronounced in the Andes Mountains, where temperatures have increased by about 1.2 °C.

*Precipitation*: Precipitation patterns in South America are becoming more erratic, with more intense rainfall events and longer droughts. This is leading to flooding in some areas and water shortages in others.

*RH*: RH levels in South America are increasing, as the atmosphere can hold more water vapor as it warms. This can make heat waves more dangerous, as it can make it feel even hotter than the actual temperature.

*SR*: SR levels in South America are increasing, as the Earth's atmosphere warms and traps more of this energy. This is leading to further warming.

Here is a summary of climate change data using temperature, precipitation, RH, and SR continent-wise:

*Temperature*: The average global temperature has risen since the pre-industrial era by roughly 1 °C. This warming is most pronounced in the Arctic, where temperatures have increased by about 2 °C.

*Precipitation*: Precipitation patterns are becoming more extreme, with more intense rainfall events and longer droughts. This is leading to flooding in some areas and water shortages in others.

*Relative Humidity (RH)*: RH is the amount of water vapor in the air. As the Earth warms, the atmosphere can hold more water vapor, leading to higher RH levels. This can make heat waves more dangerous, as it can make it feel even hotter than the actual temperature.

*Solar Radiation (SR)*: SR is the amount of energy from the sun that reaches the Earth's surface. As the Earth's atmosphere warms, it traps more of this energy, leading to further warming.

Here are some of the significant reasons for climate change:

- The burning of fossil fuels: It releases greenhouse gasses (GHGs) into the atmosphere, which trap heat and cause the Earth to warm.
- Deforestation: When decrease in number of trees leads to increased levels of carbon dioxide in the atmosphere and exacerbating the greenhouse effect.

- Agriculture: It releases GHGs such as methane and nitrous oxide into the atmosphere.
- Industrial processes: They release GHGs such as carbon dioxide and sulfur dioxide into the atmosphere.

The impacts and harmful effects of climate change around the globe are as follows:

- **Rising sea levels** are inundating coastal areas and displacing people.
- **More EWE** such as heat waves, droughts, and floods are becoming more common.
- **Changes in agricultural yields** are threatening food security.
- Climate change is **threatening biodiversity**, as many species are unable to adapt to the changing climate.

Following action can be taken to reduce GGE and mitigate the impacts of climate change:

- Switching to renewable energy sources: Renewable energy sources such as solar and wind power do not release greenhouse gases when they are used.
- Improving energy efficiency: Energy efficiency measures such as insulation and more efficient appliances can help to reduce energy consumption and emissions.
- Reducing deforestation: Reforestation and sustainable forest management can help to absorb carbon dioxide from the atmosphere.
- Changing agricultural practices: Changing agricultural practices such as no-till farming can help to reduce emissions and improve soil health.

## 1.7 Conclusion

In conclusion, the crucial role of health information in addressing climate change cannot be overstated. Health information serves as a powerful tool for raising awareness, promoting understanding, and driving action to mitigate the impacts of climate change on environment and living beings. By providing accurate and accessible information, we can empower individuals, communities, and healthcare professionals to take proactive measures in safeguarding their health and the environment. However, effective communication is just one part of the equation. Continuous monitoring of health impacts, coupled with clear and targeted communication strategies, is essential. By tracking and analysing data, researchers and responsible authorities can identify emerging trends, assess vulnerabilities, and evaluate the effectiveness of mitigation efforts. Furthermore, translating knowledge into tangible action is paramount. This involves empowering healthcare professionals, integrating climate change and health into education curricula, and fostering interdisciplinary collaborations. To create a resilient and prepared society, ongoing monitoring, communication, and action must be prioritized in our collective efforts to mitigate the health risks associated with climate change.

## References

Abbass K, Qasim MZ, Song H, Murshed M, Mahmood H, Younis I (2022) A review of the global climate change impacts, adaptation, and sustainable mitigation measures. Environ Sci Pollut Res 29(28):42539–42559

AchutaRao K, Sagar AD (2023) Climate science to inform adaptation policy: heat waves over India in the 1.5° C and 2° C warmer worlds. Clim Change 176(5):64

Adley CC, Ryan MP (2016) The nature and extent of foodborne disease. In: Antimicrobial food packaging. Academic Press, pp 1–10

Ahmad T, Gupta SK, Singh SK, Meraj G, Kumar P, Kanga S (2023) Unveiling nature's resilience: exploring vegetation dynamics during the COVID-19 era in Jharkhand, India, with the Google Earth Engine. Climate 11(9):187

Anik AH, Sultan MB, Alam M, Parvin F, Ali MM, Tareq SM (2023) The impact of climate change on water resources and associated health risks in Bangladesh: a review. Water Secur 18:100133

Bayes R, Bolsen T, Druckman JN (2023) A research agenda for climate change communication and public opinion: the role of scientific consensus messaging and beyond. Environ Commun 17(1):16–34

Bergquist M, Nilsson A, Schultz P (2019) Experiencing a severe weather event increases concern about climate change. Front Psychol 10:220

Bowen KJ, Ebi KL, Woodward A, McIver L, Tukuitonga C, Nayna Schwerdtle P (2023) Human health and climate change in the Pacific: a review of current knowledge. Clim Dev 1–15

Casson N, Cameron L, Mauro I, Friesen-Hughes K, Rocque R (2023) Perceptions of the health impacts of climate change among Canadians. BMC Public Health 23(1):1–13

Chua PL, Seposo XT, Hashizume M (2023) Heat exposure and the transmission of infectious diseases. In: Heat exposure and human health in the context of climate change. Elsevier, pp 189–221

Cianconi P, Betrò S, Janiri L (2020) The impact of climate change on mental health: a systematic descriptive review. Front Psychiatr 11:74

D'Amato G, Cecchi L, D'Amato M, Annesi-Maesano I (2014) Climate change and respiratory diseases

D'amato G, Pawankar R, Vitale C, Lanza M, Molino A, Stanziola A, D'amato M (2016) Climate change and air pollution: effects on respiratory allergy. Allergy Asthma Immunol Res 8(5):391–395

Dupraz J, Burnand B (2021) Role of health professionals regarding the impact of climate change on health—an exploratory review. Int J Environ Res Public Health 18(6):3222

Ebi KL, Vanos J, Baldwin JW, Bell JE, Hondula DM, Errett NA et al (2021) Extreme weather and climate change: population health and health system implications. Annu Rev Public Health 42(1):293–315

Etienne CF (2014) EHealth: harnessing technology on the road towards universal health coverage. Revista Panamericana de Salud Publica 35:317–319

Farooq M, Mushtaq F, Meraj G, Singh SK, Kanga S, Gupta A, Kumar P, Singh D, Avtar R (2022) Strategic slum upgrading and redevelopment action plan for Jammu City. Resources 11(12):120

Gulati B, Sharma R, Kanga S, Singh SK, Sajan B, Meraj G, Kumar P, Ramanathan AL (2023) Unraveling the relationship between stubble burning and air quality degradation in Punjab: a temporal and spatial analysis (2019–2022). J Clim Change 9(2):43–53

Iwama AY, Batistella M, Ferreira LDC, Alves DS, Ferreira LDC (2016) Risk, vulnerability and adaptation to climate change: an interdisciplinary approach. Ambiente Sociedade 19:93–116

Joshi M, Goraya H, Joshi A, Bartter T (2020) Climate change and respiratory diseases: a 2020 perspective. Curr Opin Pulm Med 26(2):119–127

Kamel Boulos MN, Wilson JP (2023) Geospatial techniques for monitoring and mitigating climate change and its effects on human health. Int J Health Geogr 22(1):1–7

Kaur R, Pandey P (2021) Air pollution, climate change, and human health in Indian cities: a brief review. Front Sustain Cities 3:705131

Khatibi FS, Dedekorkut-Howes A, Howes M, Torabi E (2021) Can public awareness, knowledge and engagement improve climate change adaptation policies? Discover Sustain 2:1–24
Kosanic A, Petzold J, Martín-López B, Razanajatovo M (2022) An inclusive future: disabled populations in the context of climate and environmental change. Curr Opin Environ Sustain 55:101159
Kotecha MJ, Bakori D, Agarwal S, Meraj G, Kanga S, Singh SK, Farooq M (2024) Geoinformation for integrated urban water resource management. In: Earth observation in urban monitoring. Elsevier, pp 93–111
Kumar P, Srivastava S, Banerjee A, Banerjee S (2022) Prevalence and predictors of water-borne diseases among elderly people in India: evidence from Longitudinal Ageing Study in India, 2017–18. BMC Public Health 22(1):993
Lin L, Yang H, Xu X (2022) Effects of water pollution on human health and disease heterogeneity: a review. Front Environ Sci 10:880246
Liu YR, van der Heijden MG, Riedo J, Sanz-Lazaro C, Eldridge DJ, Bastida F, Delgado-Baquerizo M (2023) Soil contamination in nearby natural areas mirrors that in urban greenspaces worldwide. Nat Commun 14(1):1706
Manisalidis I, Stavropoulou E, Stavropoulos A, Bezirtzoglou E (2020) Environmental and health impacts of air pollution: a review. Front Public Health 8:14
Meraj G, Kanga S, Kranjčić N, Đurin B, Singh SK (2021) Role of natural capital economics for sustainable management of earth resources. Earth 2(3):622–634
Michalak AM (2016) Study role of climate change in extreme threats to water quality. Nature 535(7612):349–350
Mocatta G, Allen K, Beyer K (2022) Towards a conceptual framework for place-responsive climate-health communication. J Clim Change Health 7:100176
Palraj R (2022) Vector-borne infections. In: A rational approach to clinical infectious diseases. Elsevier, pp 200–219
Pathak M, Roy J, Patel S, Some S, Vyas P, Das N, Shukla P (2021) Communicating climate change findings from IPCC reports: insights from outreach events in India. Clim Change 168:1–14
Shaman J, Knowlton K (2018) The need for climate and health education. Am J Public Health 108(S2):S66–S67
Singh C, Madhavan M, Arvind J, Bazaz A (2021a) Climate change adaptation in Indian cities: a review of existing actions and spaces for triple wins. Urban Clim 36:100783
Singh SK, Meraj G, Mondal N, Bera A, Verma MK, Tomar JS, Kanga S (2021b) Assessment of seasonal vegetation dynamics over parts of Thar desert using geospatial techniques. ANGRAU 105
Singh S, Singh H, Sharma V, Shrivastava V, Kumar P, Kanga S, Sahu N, Meraj G, Farooq M, Singh SK (2021c) Impact of forest fires on air quality in Wolgan valley, New South Wales, Australia—a mapping and monitoring study using Google Earth Engine. Forests 13(1):4
Singh H, Meraj G, Singh S, Shrivastava V, Sharma V, Farooq M, Kanga S, Singh SK, Kumar P (2022) Status of air pollution during COVID-19-induced lockdown in Delhi, India. Atmosphere 13(12):2090
Sinha KK, Gupta MK, Banerjee MK, Meraj G, Singh SK, Kanga S, Farooq M, Kumar P, Sahu N (2022) Neural network-based modeling of water quality in Jodhpur, India. Hydrology 9(5):92
Smith DJ, Mizelle E, Leslie SL, Li GX, Stone S, Stauffer P et al (2023) Intervention studies to reduce the impact of climate change on health in rural communities in the United States: a systematic review. Environ Res Health
Sudershan RV, Naveen Kumar R, Kashinath L, Bhaskar V, Polasa K (2014) Foodborne infections and intoxications in Hyderabad India. Epidemiol Res Int
WHO (World Health Organ.) (2015) Operational framework for building climate resilient health systems Rep., World Health Organ., Geneva

# Chapter 2
# Climate Change Impacts on Health in Remote Indigenous Community: A Case of the Great Andamanese Community in Andaman Islands, India

**Dipankar Dey**

## 2.1 Introduction

The Great Andamanese are indigenous community inhabiting the Andaman Islands, India. The Andaman Islands are an archipelago consisting of over 500 islands, with the Great Andamanese residing primarily on Strait Island, situated in the Southern part of the Andaman Islands. This community has a unique history and cultural heritage that sets them apart from other indigenous groups in the region. There have been distinctive characteristics of the Great Andamanese community are; Firstly, people believed to be one of the oldest human populations in the world. They have lived in relative isolation on the Andaman Islands for thousands of years (Thangaraj et al. 2005; Reich et al. 2009). The Great Andamanese language, which belongs to the Andamanese family of languages, is a group of languages with several dialects. Each dialect is specific to a particular tribe within the Great Andamanese community. Secondly, the Great Andamanese community has been hunter-gatherers, relying on the rich natural resources of the islands for their livelihood. They practice subsistence hunting, fishing, and gathering of forest products, sustaining themselves through their knowledge of the local environment. Thirdly, the Great Andamanese traditionally lived in a small, close-knit, and egalitarian community. They have a rich oral tradition, passing down stories, myths, and histories through generations (Narisetty et al. 2023).

However, the Colonial history of the Great Andamanese and their decline had started with the arrival of British colonizers in the nineteenth century, which had a significant impact on the Great Andamanese community (Mistra 2015). Due to the British occupation, the native population was uprooted from their ancestral lands and

D. Dey (✉)
Department of Political Science, Central University of Andhra Pradesh, Ananthapuramu, India
e-mail: dipankardey@cuap.edu.in

© The Author(s), under exclusive license to Springer Nature Singapore Pte Ltd. 2024

G. Tripathi et al. (eds.), *Sustainability and Health Informatics*, Advances in Geographical and Environmental Sciences,
https://doi.org/10.1007/978-981-97-6706-9_2

had their way of life altered by the advent of outside influences. Over the years, the Great Andamanese community has faced significant challenges, including diseases brought by outsiders, loss of land and resources due to colonization and development, and cultural assimilation (Chandi 2010). These factors have led to a sharp decline in their population, endangering their distinct languages and cultures.

In recent decades, the Indian Government has made efforts to protect and preserve the Great Andamanese community and their cultures (Srivastava et al. 2020). Some of these initiatives include setting up reserved areas and providing support for cultural revitalization programs. Despite conservation efforts, the Great Andamanese community continues to face modern challenges. Climate change, rising sea levels, and changing environmental conditions pose additional threats to their traditional ways of life and food security. The Great Andamanese community' unique cultural heritage and close relationship with their environment make them particularly potentially impacted by the effects of climate change and other external factors (Wu et al. 2016). Preserving their languages, traditions, and land rights is essential for the sustainable development and well-being of this indigenous community in the face of changing environmental and social conditions.

## 2.2 Objectives of the Study

- To comprehensively evaluate the specific climate change impacts experienced by the remote the Great Andamanese community, focusing on factors such as changing weather patterns, sea level rise, and altered ecosystems.
- To identify and analyze the vulnerabilities within the Great Andamanese community' health due to repercussions of climate change, including attainable harm to a person's mental and physical wellness as well as cultural customs.
- To investigate the ways in which traditional ecological knowledge held by the Great Andamanese community influences their perceptions, responses, and adaptations to climate change impacts on health.
- To explore the adaptation strategies employed by the Great Andamanese community to address the changing climate's impact on health, including both traditional and modern approaches.
- To develop context-specific recommendations for interventions that promote the health and well-being of the Great Andamanese community, considering their unique culture, location, and available resources.

## 2.3 Significance of Health Impacts of Climate Change on Remote Indigenous Community

Due to their geographic isolation, limited access to resources, and dependence on traditional subsistence techniques, remote indigenous communities are frequently among those who are most vulnerable to the consequences of climate change (Dolan and Walker 2006; Cunsolo Willox et al. 2015). Climate change can exacerbate existing health disparities, leading to a disproportionate impact on indigenous community. For instance, indigenous community like the Great Andamanese possess valuable traditional knowledge and practices related to healthcare, food, and environmental management. Understanding how climate change affects their health can help preserve and integrate this knowledge into broader adaptation strategies. Thus, Climate change-induced disruptions to ecosystems and traditional livelihoods can lead to a loss of cultural identity and negatively impact the mental well-being of indigenous peoples (Norton-Smith et al. 2016). Studying health impacts allows for a holistic understanding of the challenges they face in maintaining their cultural integrity and overall well-being.

Magni (2017) argues that integrating indigenous knowledge and perspectives into climate change adaptation plans and resilient approaches requires fostering more sustainability in the present context. By understanding the specific health risks faced by remote indigenous community, policymakers and researchers can develop targeted interventions to enhance their adaptive capacity. Indigenous community has the right to participate in decisions affecting their health and well-being. Thus, the health impacts of climate change should involve a participatory and culturally sensitive approach, respecting the autonomy and consent of this community.

Indigenous peoples often contribute minimal adverse harm activity towards environment but are heavily burdened by its effects. Additionally, the Great Andamanese and other indigenous communities' health problems due to climate change might draw attention to the need for global climate justice as well as for funding for efforts at adaptation and mitigation (Debnath et al. 2023). The wellbeing of this community depends on the wellbeing of its surroundings. In the current context of climate change, addressing the consequences of climate change on people's health helps to the preservation of biodiversity and ecological balance.

## 2.4 Methodology

The purpose of this study is to investigate the impacts of climate change on health in the isolated Great Andamanese population using a mixed-methods approach. The methodology integrates both qualitative and quantitative techniques to capture the multifaceted dimensions of this complex issue. Primary and secondary sources of data were collected from various government websites such as Andaman and Nicobar Administration, Directorate of Heath Services (A&N Islands), India Meteorological

Department, The Andaman Adim Janjati Vikas Samiti (AAJVS), Books, Journals, Open Sources and so on. The quantitative technique has been analysis based on Linear Estimation (regression) model to understand the population trend of the Great Andamanese and the qualitative technique is interpreted for the narrative analysis of the health condition of the Great Andamanese.

## 2.5 Identification of Climate Change Trends and Projections for the Region

The Andaman and Nicobar Islands, including the Great Andamanese community, have been experiencing several climate change trends and projections. The Andaman and Nicobar Islands have observed a rising trend in temperatures over the past few decades (Dash et al. 2007). The increase in average temperatures can result in more frequent and prolonged heat waves, which could have adverse effects on human health, especially for vulnerable populations like the Great Andamanese community. Due to these, there have been noticeable changes in rainfall patterns in the Andaman and Nicobar Islands. While some regions experience an increase in rainfall, others may face decreased precipitation. Changes in rainfall patterns can impact agriculture, water availability, and ecosystems, affecting the livelihoods of the local community. (Table 2.1).

Thus, due to the major influence of climate change, the islands are vulnerable to the effects of increasing sea levels. It contributes to coastal erosion, salty water

**Table 2.1** Comparative statistics of average (total) rainfall (mm) in Andaman and Nicobar Islands over the last three years (2021 to till June 2023)

| Comparative Statistics of average (total) rainfall (mm) of Andaman and Nicobar Islands over the last three years (2021 to till June 2023) | | | | | | | | |
|---|---|---|---|---|---|---|---|---|
| Month | 2021 | | 2022 | | 2023 | | % Variation in monthly rainfall of 2023 from 2022 | % Variation in cumulative rainfall of 2023 from 2022 |
| | Monthly | Cumulative | Monthly | Cumulative | Monthly | Cumulative | | |
| January | 54.57 | 54.57 | 52.85 | 52.85 | 40 | 40 | -24.31 | -24.31 |
| February | 51.98 | 106.55 | 171 | 223.85 | 31.28 | 71.28 | -81.71 | -68.16 |
| March | 44.11 | 150.66 | 132.28 | 356.13 | 12.33 | 83.61 | -90.68 | -76.52 |
| April | 146.53 | 297.19 | 47.02 | 403.15 | 24.68 | 108.29 | -47.51 | -73.14 |
| May | 380.58 | 677.78 | 540.30 | 943.45 | 133.52 | 241.81 | -75.29 | -74.37 |
| **June** | **289.42** | **967.19** | **271.95** | **1215.4** | **447.28** | **689.09** | 64.47 | -43.30 |
| July | 492.62 | 1459.81 | 491.57 | 1706.97 | | | | |
| August | 487.22 | 1947.03 | 363.07 | 2070.04 | | | | |
| September | 605.32 | 2552.35 | 373.76 | 2443.8 | | | | |
| October | 377.67 | 2930.02 | 325.58 | 2769.38 | | | | |
| November | 257.45 | 3187.47 | 181.07 | 2950.45 | | | | |
| December | 90.38 | 3277.85 | 187.73 | 3138.18 | | | | |

*Source* Directorate of Economics and Statistics, Andaman and Nicobar Administration. AverageRainfall3years.pdf (and.nic.in)

intrusion into freshwater sources, greater vulnerability to storm surges and floods, and other factors that put the lives of people who live in low-lying places at risk.

In addition, the Andaman and Nicobar Islands are prone to tropical cyclones, which can be exacerbated by climate change. It has been projected that cyclones may become more intense and frequent, putting coastal community at risk of damage to infrastructure and disruption to livelihoods.

Similarly, the Andaman Islands, nestled in the seismic heart of the Pacific Ring of Fire, have historically been a hotspot for tectonic activity, including earthquakes. This discussion delves into the complexities of the impact of earthquakes on these islands, focusing on their geophysical characteristics, societal consequences, and the strategies employed to mitigate their effects. The Andaman Islands are situated at the convergence of the Indian Plate and the Burmese Plate, leading to the subduction of the latter beneath the former (Acharyya et al. 1990). This tectonic interaction is a prime generator of earthquakes in the region. The subduction process can lead to both deep-seated and shallow earthquakes, each bearing distinct characteristics. Deep-seated earthquakes can be particularly devastating due to their potential to trigger tsunamis, as seen in the tragic events of 2004 (Chadha et al. 2005; Bahuguna et al. 2008). Shallow earthquakes, occurring closer to the Earth's surface, have the potential to cause significant ground shaking and structural damage (Chadha et al. 2005). Understanding the depth, magnitude, and location of these earthquakes is pivotal for accurate risk assessment and mitigation. (Table 2.2).

Additionally, increasing levels of carbon dioxide in the atmosphere are also affecting the oceans, leading to ocean acidification. Coral reefs and marine ecosystems in the Andaman and Nicobar Islands will face significant threats due to the changing ocean.

It is essential to monitor these climate change trends and projections continuously and take proactive measures to address their potential impacts. Adaptation and mitigation strategies, such as promoting sustainable practices, protecting coastal ecosystems, enhancing disaster preparedness, and integrating indigenous knowledge, can help community in the Andaman and Nicobar Islands, including the Great Andamanese, to cope with the challenges posed by climate change.

## 2.6 Understanding the Vulnerabilities of the Great Andamanese Community to Climate Change Impacts

The Great Andamanese are an indigenous community that inhabits the Andaman Islands in the Bay of Bengal. They consist of multiple distinct tribes, each with their own languages, cultures, and practices (Sarkar 2016; Gautam 2005; Sasikumar 2022). Historically, these communities have had limited contact with the outside world and have relied on traditional practices for their sustenance and well-being. But recently, they have developed good relations with the outside world and a modern way of life. Before modern interactions and interventions, the Great Andamanese community

**Table 2.2** Earthquake magnitudes 4.5 and above occurred in Andaman & Nicobar Islands in 2022 (on the Richter scale)

| Date | Time (Hrs) (IST) | Magnitude | Latitude | Longitude | Approx. geographical location |
|---|---|---|---|---|---|
| 4/1/2022 | 20:32 | 4.5 | 11.93 N | 93.68 E | Andaman Islands |
| 29/01/22 | 2:52 | 4.7 | 14.60 N | 93.18 E | Andaman Islands |
| 22/04/22 | 9:21 | 4.9 | 13.99 N | 92.99 E | Andaman Islands |
| 27–05-22 | 7:50 | 4.5 | 12.92 N | 93.44 E | Andaman Islands |
| 4/7/2022 | 11:05 | 4.6 | 10.44 N | 94.09 E | Andaman Islands |
| 4/7/2022 | 13:32 | 4.6 | 10.49 N | 93.92 E | Andaman Islands |
| 4/7/2022 | 13:55 | 4.7 | 10.30 N | 94.20 E | Andaman Islands |
| 4/7/2022 | 14:06 | 4.8 | 10.74 N | 94.41 E | Andaman Islands |
| 4/7/2022 | 14:37 | 5.6 | 10.53 N | 94.24 E | Andaman Islands |
| 4/7/2022 | 15:25 | 4.9 | 10.51 N | 94.28 E | Andaman Islands |
| 4/7/2022 | 15:39 | 4.5 | 10.30 N | 94.28 E | Andaman Islands |
| 4/7/2022 | 17:18 | 5 | 10.35 N | 94.21 E | Andaman Islands |
| 4/7/2022 | 17:50 | 4.6 | 10.29 N | 94.15 E | Andaman Islands |
| 4/7/2022 | 23:01 | 4.9 | 10.58 N | 94.24 E | Andaman Islands |
| 5/7/2022 | 2:54 | 5 | 10.57 N | 94.35 E | Andaman Islands |
| 5/7/2022 | 4:45 | 4.6 | 10.23 N | 94.24 E | Andaman Islands |
| 5/7/2022 | 5:57 | 5.1 | 10.92 N | 94.54 E | Andaman Islands |
| 5/7/2022 | 8:04 | 4.8 | 10.25 N | 94.07 E | Andaman Islands |
| 5/7/2022 | 11:36 | 4.9 | 10.49 N | 94.18 E | Andaman Islands |
| 5/7/2022 | 13:18 | 4.8 | 10.16 N | 94.16 E | Andaman Islands |
| 5/7/2022 | 15:14 | 4.7 | 10.33 N | 94.35 E | Andaman Islands |
| 5/7/2022 | 20:20 | 5 | 11.23 N | 91.81 E | Andaman Islands |
| 6/7/2022 | 13:31 | 5 | 10.52 N | 94.25 E | Andaman Islands |
| 9/7/2022 | 2:34 | 5.3 | 10.49 N | 94.18 E | Andaman Islands |
| 7/8/2022 | 21:14 | 4.8 | 10.35 N | 93.86 E | Andaman Islands |
| 20–10-22 | 12:17 | 5 | 11.59 N | 93.41 E | Andaman Islands |

*Source* Directorate of Economics and Statistics, Andaman and Nicobar Administration. AverageRainfall3years.pdf (and.nic.in)

had relatively good health in their traditional environment. However, with increased contact with the outside world, they faced various health challenges. Exposure to new diseases, lack of immunity to outside pathogens, and changes in their lifestyle and diet due to contact with mainstream society had a significant impact on their health.

The Great Andamanese community has historically practiced hunting, gathering, and fishing as their main means of subsistence (Abbi 2013). They have extensive knowledge of their environment, including plants and animals, which they use for food, medicine, and other resources. Their traditional practices were closely tied to the ecosystems of the Andaman Islands and allowed them to adapt and thrive in their natural surroundings.

However, the Great Andamanese community has faced numerous challenges due to colonization, displacement, and loss of their traditional lands. The encroachment of modern society, along with changes in lifestyle, has led to a decline in the practice of their traditional ways of life. This, in turn, has had an impact on their health and well-being as they transitioned to different diets and living conditions.

They were unable to defend their lands because they were plagued by illnesses. They were evicted from their home slowly but surely. Possession of the land disrupted their customary methods of subsistence (foraging), which led to pauperism. Furthermore, it destroyed their preexisting social structure as well as their modes of interpersonal communication and organization. Additionally, social groupings dispersed, disrupting the reciprocal relationships that support life.

The major reason they were relocated to Strait Island was to shield them from the many forms of exploitation they had previously experienced and to prevent their extinction as their number had dangerously declined to only 19 in 1961 (Mishra 2015; Srivastava et al. 2020).

Naturally, all steps thought necessary to safeguard them were taken, including designating the island as a tribal reserve and enforcing the Andaman and Nicobar Islands (Protection of Aboriginal Tribes) Regulations, 1956 (ANPAT), among other actions (Srivastava et al. 2020). These actions did provide the anticipated outcomes. The tribe's numbers started to rise. Some of the literate people among them were able to find work. Despite the law's protection, various types of exploitation—though small in scope and nature—occurred in the community that was relocated, including the provision of alcohol, the production of country booze, and other assaults.

The Great Andamanese had to relocate to Port Blair in the years after 2000, particularly after the 2004 tsunami (Thangaraj et al. 2005; Srivastava et al. 2020), as the island's essential utilities had been severely damaged, in addition to many of their homes. At the Strait Island, roughly a dozen concrete homes were also constructed in the years that followed the tsunami. They were situated on a hillside distant from the coast (the government had chosen this location in anticipation of a similar future occurrence). However, due to their distance from the coast, the islanders showed little interest in relocating to such homes. There were also concerns about the absence of drinking water facilities (Singh et al. 2021).

## 2.7 Climate Change Vulnerabilities of the Great Andamanese Community

A combination of geographical, social, and economic characteristics makes the Great Andamanese community especially susceptible to the effects of climate change. Some of the key climate change vulnerabilities they face in the contemporary time include:

1. Geographical Isolation: The Great Andamanese community lives on small, islands with low lying in the Bay of Bengal. It has also been observed that the significant hazards to their physical safety, homes, and infrastructure are posed by rising sea levels and a rise in the frequency of extreme weather events like hurricanes and cyclones.
2. Limited Access to Resources: This indigenous community relies heavily on natural resources for their livelihoods, including fishing, hunting, and gathering forest products. Climate change can disrupt these activities by affecting the availability and abundance of natural resources, leading to food insecurity and economic challenges.
3. Dependency on Traditional Agriculture: Traditional agriculture is a crucial aspect of their subsistence. Climate change-induced alterations in rainfall patterns and increased temperatures can impact crop yields and water availability, affecting food production and overall nutrition.
4. Health Impacts: As mentioned earlier, climate change can result in the spread of vector-borne diseases and waterborne illnesses, putting the Great Andamanese community at higher risk due to their limited access to healthcare facilities and medical resources (Gulati et al. 2023).
5. Loss of Biodiversity: Climate change has the potential to alter ecosystems and reduce biodiversity, which might result in the extinction or migratory movements of significant plant and animal species that are utilized by Great Andamanese. This can disrupt their traditional knowledge and cultural practices related to local flora and fauna.
6. Disruption of Cultural Practices: The Great Andamanese community' cultural identity and spirituality are deeply intertwined with their environment. Climate change impacts, such as loss of natural habitats or changes in wildlife migration patterns, can threaten their ability to practice traditional rituals and ceremonies.
7. Social and Economic Marginalization: Historically, this indigenous community has experienced marginalization, and climate change can exacerbate their socio-economic vulnerabilities. The less access to resources and limited integration into mainstream society can limit their ability to adapt to climate-related challenges.
8. Limited Institutional Support: The remote location of the Andaman Islands and the small size of the Great Andamanese community can result in limited access to government services and support during climate-related disasters or emergencies.

9. Loss of Land and Coastal Erosion: Rising sea levels and coastal erosion can lead to the loss of land, including sacred and culturally significant sites. This not only threatens their physical safety but also undermines their cultural heritage and traditional land rights.
10. Loss of Traditional Knowledge: Climate change impacts may disrupt the transmission of traditional knowledge from elders to younger generations. This loss of knowledge could diminish their ability to adapt to changing environmental conditions effectively.

Understanding and addressing the climate change vulnerabilities of the Great Andamanese community requires holistic and culturally sensitive approaches. Incorporating their traditional knowledge and practices into climate change adaptation strategies, supporting community-led initiatives, and ensuring equitable access to resources and healthcare are essential steps to enhance their resilience in the face of climate-related challenges.

## 2.8 Climate Change Impacts on Health in the Great Andamanese Community

The Great Andamanese were the subject of a recent physical anthropological research, which found that obesity, anemia, hypertension, and aberrant pulse rate ranges are all common in the group of Great Andamanese (Banerjee 2018). Surprisingly, teens and young adults are more at risk for these health issues. It may be efficiently controlled and reduced by embracing good eating habits, adhering to proper cleanliness, and engaging in regular exercise. It is mostly caused by changes in lifestyle and food practices in recent years (Oommen 2022). The possible preventative strategies include;

1. The Great Andamanese should be fed iron-rich plant and animal foods to reduce the incidence of anemia, especially in women. The local health department may also implement a program of iron supplementation. Additionally, they should be the target of comprehensive public awareness initiatives addressing the prevention of anemia.
2. The Great Andamanese's eating patterns have changed in recent years. There is an overall high calorie intake as a result of the consumption of fatty and junk meals. One of the main causes of the high incidence of obesity, hypertension, and irregular pulse rates is high calorie consumption combined with little activity. Future research may show that this is one of the main factors contributing to metabolic problems in young people. Therefore, the local government may organize intervention and sensitization programs on a regular basis.
3. Most Great Andamanese are observed to be addicted to chewing tobacco and gutka, whereas many appear to only drink alcohol occasionally. They must be made aware of the detrimental impact of these things on their health.

4. In order to assist avoid the onset of numerous health concerns, health awareness and sensitization programs regarding the significance of personal cleanliness and health should be routinely organized. Additionally, it must include how to maintain their home tidy.
5. It is necessary to extend the Strait Island dispensary and add some new specializations.
6. Anthropologists and medical professionals should collaborate on regular community health surveys and research.

The patterns of population are explaining the one of causes of ill health and climate change of Great Andamanese. The census data has been collected from 1951 to 2011 including the 2018 Directorate of Health, Andaman and Nicobar Administration released population data regarding the Great Andamanese. These population reports significantly described the vulnerability of remote community in the Andaman Islands. (Table 2.3).

Linear regression model estimation was done in order to track the growing number of the tribal population. (Table 2.4).

A linear regression analysis was performed using the demographic data to examine the growth trend of the tribes through time. It has explained that strong relationship between the Great Andamanese population pattern and census year, with the model explaining r-square values (0.795) clearly showing which factors account for the difference in Great Andamanese Population growth over time, which represents 79.5% of the population can rise over a year. The regression model explained the significance of relations between two variables ($f [1, 6] = 23.202$, $p < 0.005$).

At the same time, adverse change in climate forecasting considered factors like temperature, precipitation, and extreme events constantly impact the survival of

**Table 2.3** The Pattern of Growth of the Great Andamanese Population in the Andaman Islands

| Sl.no | Census | No. of population |
|---|---|---|
| 1 | 1951 | 23 |
| 2 | 1961 | 19 |
| 3 | 1971 | 24 |
| 4 | 1981 | 26 |
| 5 | 1991 | 45 |
| 6 | 2001 | 43 |
| 7 | 2011 | 44 |
| 8 | 2018 | 70 |

*Source* Basic Statistics, Andaman and Nicobar Administration

**Table 2.4** Linear regression model

| Indigenous tribe | $R^2$ | F | df1 | df2 | Sig |
|---|---|---|---|---|---|
| Great andamanese | 0.795 | 23.202 | 1 | 6 | 0.003 |

indigenous community. It has been argued that balanced ecosystem and good health can increase life expentency but the impact of climate change significantly influence the growth of population. In this regards, the Great Andamanese Community has been associated with the nature and biodiversity, which gradually became a severe influence on the survival of indigenous people.

## 2.9 Heath Status of the Great Andamanese

Indigenous communities are uniquely vulnerable to the health effects of climate change due to their dependence on the natural environment for sustenance, cultural practices, and traditional medicine (Chisholm Hatfield et al. 2018). Rising temperatures, changed precipitation patterns, and increasing frequency of extreme weather measures disrupt ecosystems, making it difficult for indigenous peoples to access clean water, nutritious food, and medicinal plants. These changes contribute to malnutrition, waterborne diseases, and mental health issues, which are exacerbated by limited healthcare access.

Additionally, indigenous communities often reside in remote or geographically marginalized areas, lacking infrastructure and resources to cope with climate-related health challenges (Maru et al. 2014). Socioeconomic disparities and historical injustices further amplify their vulnerability, limiting their ability to adapt to changing conditions. These factors create a vicious cycle where climate change worsens health outcomes, leading to increased susceptibility to future environmental changes.

## 2.10 Community Adaptation and Resilience Strategies

The Great Andamanese community has long developed intricate strategies for disaster risk reduction and climate change resilience, rooted in their traditional knowledge, resource management practices, and community solidarity. These approaches have enabled them to navigate the challenges posed by natural disasters and changing environmental conditions, allowing them to maintain their unique way of life while adapting to new circumstances (McMilla et al. 2022). It has explored the key components of their disaster risk reduction and climate change resilience strategies needed to respond to the climate change impact on the health of the Great Andamanese community. Some of the components are;

1. Traditional Ecological Knowledge (TEK): The Great Andamanese possess a deep understanding of their local ecosystems, including weather patterns, plant and animal behaviors, and coastal dynamics (Roy & Mukhopadhyay 2022). Passed down through generations, this knowledge equips them with the ability to predict and respond to environmental changes, enabling timely actions to minimize disaster impacts.

2. Adaptive Resource Management: The Great Andamanese community practice sustainable resource management techniques that have evolved over centuries. They understand the delicate balance between human activities and the environment, employing rotational cultivation, controlled burning, and selective harvesting to ensure ecosystem health and productivity. These practices enhance resilience by maintaining biodiversity and ecosystem services that contribute to disaster mitigation.
3. Community-based Early Warning Systems: Indigenous communities have developed intricate communication networks to disseminate information about approaching disasters. Traditional methods, such as observing animal behaviors, changes in wind patterns, and ocean currents, allow for early detection of potential threats. This knowledge is shared within the community, enabling collective actions to prepare and respond effectively.
4. Flexible Settlement Patterns: The Great Andamanese have historically chosen settlement locations that offer protection against natural hazards, such as tsunamis and storms. Their intimate knowledge of local geography and topography helps them identify safer areas for habitation, minimizing vulnerability to disasters.
5. Cultural Heritage Preservation: The Great Andamanese maintain strong cultural ties to their ancestral lands, which drives their commitment to preserving natural resources and ecosystems. This cultural connection fosters a sense of responsibility towards environmental stewardship, encouraging sustainable practices that enhance climate change resilience.
6. Local Governance and Decision-making: Indigenous governance systems promote community participation, consensus building, and decision-making processes that consider both short-term needs and long-term sustainability. These systems empower the Great Andamanese to make informed choices about resource use, disaster preparedness, and adaptation strategies.
7. Integration of Modern and Traditional Knowledge: The Great Andamanese recognize the value of integrating modern scientific knowledge with their traditional wisdom. By combining traditional ecological knowledge with scientific insights, they enhance their capacity to anticipate, adapt to, and recover from the impacts of climate change and disasters.

In regard to evolving climate change challenges and increasing disaster risks, the Great Andamanese community' resilience strategies offer valuable experience for wider disaster risk reduction and climate change adaptation efforts. It has also explained that collaborating with this community to identify their needs, respecting their knowledge systems, and incorporating their approaches into policy frameworks can contribute to more effective and culturally sensitive strategies that benefit both the Great Andamanese and the global community.

Though, recognizing the unique insights and expertise of the Great Andamanese people, community-led adaptation initiatives have emerged as promising avenues for enhancing disaster risk reduction and climate change resilience. These initiatives involve the active participation of community members in the design, implementation, and management of strategies aimed at mitigating the impacts of environmental changes (Rafiq et al. 2022). There are several benefits to making community-led adaptation such as; firstly, they contribute their deep understanding of local ecosystems, weather patterns, and resource management practices. This knowledge is integrated with scientific data to create more accurate and context-specific disaster risk assessments (Meraj et al. 2022; Farooq et al. 2022). Secondly, indigenous methods of observing natural signs and indicators have been harnessed to create early warning systems for impending disasters such as cyclones, tsunamis, and extreme weather events. This has proven crucial in enabling timely evacuation and preparation. Thirdly, the Great Andamanese community has a strong connection to their environment, relying on sustainable resource management practices. Initiatives that promote ecosystem-based adaptation, such as mangrove restoration and sustainable fishing techniques, help enhance the resilience of both the community and its ecosystems. Lastly, many adaptation initiatives also focus on preserving cultural practices that are inherently linked to environmental stewardship. By valuing and revitalizing these practices, the community ensures the transmission of knowledge and resilience from one generation to another.

***1. Recommendation of Effectiveness of Community-Led Initiatives for Integrating aboriginal Knowledge into Climate Change Adaptation Strategies***

- The effectiveness of community-led adaptation initiatives among the Great Andamanese community stems from their inherent relevance, ownership, and holistic nature.
- By leveraging their traditional knowledge and deep connection with the environment, these initiatives are better suited to address the unique challenges faced by this community.
- Furthermore, the involvement of community members in decision-making processes fosters a sense of agency and empowerment, which is vital for successful adaptation.

These initiatives have led to reduced vulnerabilities and increased resilience among the Great Andamanese community. Early warning systems have minimized casualties during disasters, while ecosystem-based approaches have maintained the health of vital resources. Additionally, the integration of cultural heritage preservation has not only enriched the community' identities but also reinforced their commitment to sustainable practices.

Integrating Indigenous knowledge into climate change adaptation strategies for the Great Andamanese community requires a collaborative and culturally sensitive approach. There are some recommendations to ensure the effective incorporation of traditional wisdom into adaptation efforts by engaging community members from the outset to identify their needs, concerns, and priorities related to climate change impacts.

***2. Recommendation of Strengthening Healthcare Services and Disaster Preparedness in Remote Community***

- The remote Great Andamanese community, nestled in the pristine landscapes of the Andaman Islands, is not only a repository of unique cultural heritage but also is vulnerable to the challenges of limited healthcare access and the increasing frequency of natural disasters.

- Recognizing the critical need for comprehensive healthcare services and robust disaster preparedness, a holistic approach is required to address the intricate interplay between these factors.
- In the heart of this remote community, the establishment of healthcare facilities assumes paramount importance. A well-designed healthcare infrastructure that combines modern medical expertise with Indigenous healing practices can significantly enhance the overall health and well-being of the Great Andamanese people.
- It has also been observed that establishing community-centric clinics staffed with healthcare professionals who respect and collaborate with Indigenous healers is essential. These clinics should offer preventive, curative, and maternal health services while also integrating traditional healing methods that have been passed down through generations.
- Similarly, launching health education initiatives tailored to the cultural context of the Great Andamanese community can empower individuals to take charge of their health. Workshops, awareness campaigns, and interactive sessions could cover topics ranging from hygiene practices to basic first aid skills.
- Also, mobile healthcare units equipped with medical supplies and staffed with healthcare workers could traverse challenging terrains to provide medical assistance, health check-ups, and immunizations directly to the residents.
- The Andaman Islands' geographic location renders the Great Andamanese community susceptible to natural calamities such as cyclones, tsunamis, and floods. An integrated disaster preparedness strategy that amalgamates traditional knowledge with contemporary techniques can be instrumental in minimizing loss and protecting lives.

## 2.11 Conclusion

The road ahead requires collective efforts and a deep commitment to safeguarding the health and well-being of the Great Andamanese community. It necessitates policy changes that prioritize Indigenous knowledge integration, allocation of resources for healthcare infrastructure, and capacity building that empowers community members to adapt to changing conditions. By preserving their cultural heritage, empowering their resilience, and fostering equitable partnerships, we can aspire to a future where the Great Andamanese community thrives despite the challenges posed by a changing climate.

In the face of climate change, the Great Andamanese community stands at a critical juncture where the delicate balance between their traditional way of life and the evolving environment is increasingly challenged (Subramanian et al. 2023). This exploration into the climate change impacts on health within this remote Indigenous community highlights the urgency of tailored and culturally sensitive interventions.

The unique vulnerability of the Great Andamanese to changing weather patterns, rising sea levels, and shifting ecosystems underscores the need for comprehensive adaptation strategies that prioritize their health and well-being. As witnessed in this study, the convergence of traditional ecological knowledge with modern scientific insights offers a promising pathway for building resilience (Smith et al. 2016).

The connections between the health of the Great Andamanese and the ecosystems they inhabit are intricately intertwined. Climate change-induced shifts, from altered food sources to increased vector-borne diseases, necessitate adaptive measures that not only safeguard physical health but also preserve cultural heritage. Therefore, any intervention must recognize the interplay between physical and mental health, community dynamics, and the preservation of Indigenous practices. While technological advancements play a role in monitoring and predicting climate impacts, the true strength of adaptation lies in the Great Andamanese community' deep-rooted wisdom. Their intimate understanding of local signs, resource management, and traditional healing practices offers an invaluable foundation for informed decision-making.

In conclusion, the imperative to protect the health and well-being of the Great Andamanese community in the wake of climate change requires collaborative efforts that respect and harness their intrinsic knowledge. The sustainability of their unique way of life, their deep connection to the land and sea, and the legacy they hold are at stake. By integrating Indigenous perspectives into policy, fostering partnerships that bridge traditional and modern systems, and empowering community-led initiatives, we can aspire to a future where the Great Andamanese community not only endures but thrives amidst the challenges of a changing world. This journey towards climate resilience is as much a commitment to their heritage as it is to the broader global endeavor of creating a sustainable and equitable future for all. The convergence of climate change and health impacts on the Great Andamanese community underscores the urgent need for holistic strategies that honor their traditional knowledge, address vulnerabilities, and build a resilient community capable of navigating an uncertain future. The case of this remote Indigenous community serves as a poignant reminder of the global imperative to protect the health of vulnerable populations as we collectively tackle the complexities of climate change.

## References

Abbi A (2013) A grammar of the great andamanese language: an ethnolinguistic study, vol 4. Brill

Acharyya SK, Ray KK, Sengupta S (1990) Tectonics of the ophiolite belt from Naga Hills and Andaman Islands, India. Proc Indian Acad Sci-Earth Planet Sci 99:187–199. https://doi.org/10.1007/BF02839389

Bahuguna A, Nayak S, Roy D (2008) Impact of the tsunami and earthquake of 26th December 2004 on the vital coastal ecosystems of the Andaman and Nicobar Islands assessed using RESOURCESAT AWiFS data. Int J Appl Earth Obs Geoinf 10(2):229–237. https://doi.org/10.1016/j.jag.2008.02.010

Banerjee A (2018) A Micro level study of socio-economic and health status of great Andamanese and Onges of Andaman Island. Orient Anthropol 18(1):19–29
Chadha RK, Latha G, Yeh H, Peterson C, Katada T (2005) The tsunami of the great Sumatra earthquake of M 9.0 on 26 December 2004–Impact on the east coast of India. Current Sci 1297–1301. https://www.jstor.org/stable/24110300
Chandi M (2010) Colonization and conflict resolution in the Andaman Islands: learning from Reconstruction of conflict between indigenous and non-indigenous islanders. The Jarawa Tribal Reserve Dossier, Cultural & Biological Diversities in the Andaman Islands
Chisholm Hatfield S, Marino E, Whyte KP, Dello KD, Mote PW (2018) Indian time: time, seasonality, and culture in traditional ecological knowledge of climate change. Ecol Process 7(1):1–11. https://doi.org/10.1186/s13717-018-0136-6
Cunsolo Willox A, Stephenson E, Allen J, Bourque F, Drossos A, Elgarøy S, Wexler L (2015) Examining relationships between climate change and mental health in the Circumpolar North. Region Environ Change 15:169–182. https://doi.org/10.1016/j.ijdrr.2021.102433
Dash SK, Jenamani RK, Kalsi SR, Panda SK (2007) Some evidence of climate change in twentieth-century India. Clim Change 85(3–4):299–321. https://doi.org/10.1007/s10584-007-9305-9
Debnath J, Sahariah D, Mazumdar M, Lahon D, Meraj G, Hashimoto S, Kumar P, Singh SK, Kanga S, Chand K, Saikia A (2023) Evaluating flood susceptibility in the Brahmaputra River Basin: an insight into Asia's Eastern Himalayan floodplains using machine learning and multi-criteria decision-making. In: Earth systems and environment, pp1–28
Dolan AH, Walker IJ (2006) Proceedings of the 8th international coastal symposium (ICS 2004). J Coastal Res 39(3):1316–1323. http://www.jstor.org/stable/25742967
Farooq M, Meraj G, Rishabh, Kanga S, Nathawat R, Singh SK, Ranga V (2022) Slum categorization for efficient development plan—a case study of Udhampur City, Jammu and Kashmir using remote sensing and GIS. In: Geospatial technology for landscape and environmental management: sustainable assessment and planning, pp 283–299
Gautam RK (2005) The primitive tribes of Andaman and Nicobar Islands. Primit Tribes Contemp India: Concept, Ethnogr Demogr 1:35–73
Gulati B, Sharma R, Kanga S, Singh SK, Sajan B, Meraj G, Kumar P, Ramanathan AL (2023) Unraveling the relationship between stubble burning and air quality degradation in punjab: a temporal and spatial analysis (2019–2022). J Clim Change 9(2):43–53
Leal Filho W (ed) (2015) Handbook of climate change adaptation. Springer, Berlin
Magni G (2017) Indigenous knowledge and implications for the sustainable development agenda. Eur J Educ 52(4):437–447. https://doi.org/10.1111/ejed.12238
Maru YT, Smith MS, Sparrow A, Pinho PF, Dube OP (2014) A linked vulnerability and resilience framework for adaptation pathways in remote disadvantaged communities. Glob Environ Chang 28:337–350. https://doi.org/10.1016/j.gloenvcha.2013.12.007
McMillan JM, Birkmann J, Tangwanichagapong S, Jamshed A (2022) Spatial planning and systems thinking tools for climate risk reduction: a case study of the Andaman Coast Thailand. Sustainability 14(13):8022
Meraj G, Farooq M, Singh SK, Islam MN, Kanga S (2022) Modeling the sediment retention and ecosystem provisioning services in the Kashmir valley, India, Western Himalayas. Modeling Earth Syst Environ 8(3):3859–3884
Mishra PP (2015) Great Andamanese of Strait Island, the Jarawas of South and Middle Andaman, the Onges of Little Andaman, the Sentinelese of the Sentinel Islands, and the Shompens of Great Nicobar. In: Steven D (ed) Native peoples of the world, vol 192. London
Narisetty NG, Tripathi G, Kanga S, Singh SK, Meraj G, Kumar P, Đurin B, Matijević H (2023) Integrated multi-model approach for assessing groundwater vulnerability in Rajasthan's semi-arid zone: incorporating DRASTIC and SINTACS variants. Hydrology 10(12):231
Norton-Smith K, Lynn K, Chief K, Cozzetto K, Donatuto J, Redsteer MH, Whyte KP (2016) Climate change and indigenous peoples: a synthesis of current impacts and experiences. United States Department of Agriculture, Forest Service, Pacific Northwest Research Station

Oommen MA (2022) An ecological reading of ritual and taboo in ethnographic accounts on Andamanese hunter-gatherers. In: Conservation through sustainable use: lessons from India
Rafiq M, Meraj G, Kesarkar AP, Farooq M, Singh SK, Kanga S (2022) Hazard mitigation and climate change in the Himalayas–policy and decision making. Disaster Management in the Complex Himalayan Terrains: Natural Hazard Management, Methodologies and Policy Implications. Springer International Publishing, Cham, pp 169–182
Reich D, Thangaraj K, Patterson N, Price AL, Singh L (2009) Reconstructing Indian population history. Nature 461:489–494. https://doi.org/10.1038/nature08365
Roy S, Mukhopadhyay M (2022) Transformation of a pristine livelihood in response to development—An overview of the Onge tribal community, Little Andaman Island, India. Indigenous People and Nature, pp 525–542
Sarkar J (2016) Conceptualizing tribe in Indian context. J Indian Anthropol Soc 51(3)
Sasikumar M (2022) The battle of Aberdeen: an indigenous resistance to colonialism. J Anthropol Surv India 71(2):281–292
Singh SK, Meraj G, Mondal N, Bera A, Verma MK, Tomar JS, Kanga S (2021) Assessment of seasonal vegetation dynamics over parts of thar desert using geospatial techniques. ANGRAU, p 105
Smith LT, Maxwell TK, Puke H, Temara P (2016) Indigenous knowledge, methodology and mayhem: what is the role of methodology in producing Indigenous insights? A discussion from mātauranga Māori
Srivastava VK, Kumar U, Sasikumar M, Pulamaghatta VN, Patel SK, Goyal PA (2020) Draft of the policies for great Andamanese and Sentinelese. J Anthropol Surv India 69(1):165–176. https://doi.org/10.1177/2277436X20912844
Subramanian A, Nagarajan AM, Vinod S, Chakraborty S, Sivagami K, Theodore T, Mangesh VL (2023) Long-term impacts of climate change on coastal and transitional eco-systems in India: an overview of its current status, future projections, solutions, and policies. RSC Adv 13(18):12204–12228
Thangaraj K, Chaubey G, Kivisild T, Reddy AG, Singh VK, Rasalkar AA, Singh L (2005) Reconstructing the origin of Andaman Islanders. Science 308:996. https://doi.org/10.1126/science.1109987
Wu Z, Keremane G, McKay J (2016) Sustainable urban water management through the lens of community – a photostory exercise in metropolitan Adelaide. Australasian J Environ Manag 23(2):141–156. https://doi.org/10.1080/13549839.2012.748733

# Chapter 3
# Environmental Sustainability and Its Impact on Public Health

**Ritambhara K. Upadhyay, Naval Kishore, Mukta Sharma, and Sandeep Sundriyal**

## 3.1 Introduction

According to Godfray (2015), environmental sustainability is the prudent management and preservation of natural resources to guarantee their availability for future generations. It entails actions that encourage resource efficiency, reduce adverse environmental effects, and work towards striking a balance between social, ecological, and economic demands. Conversely, the goal of public health is to safeguard and advance the health and welfare of the general public (Damschroder et al. 2009).

More people are realising how important the environment and public health are to long-term sustainability and human welfare. The degradation of ecosystems, climate change, pollution, and resource depletion have profound implications for human health. Environmental sustainability has emerged as a key concept in addressing these challenges and promoting positive health outcomes (Gruen et al. 2008; Holling 1973).

Given that environmental elements have a significant impact on human health and well-being, environmental sustainability is important in the context of public health (Ahmad et al. 2023). Encouraging sustainable policies and practices can help communities become healthier and more resilient while lowering health inequities and environmental risks (Kates et al. 2001). Achieving sustainable development and

R. K. Upadhyay (✉) · N. Kishore
Department of Geology, Panjab University, Chandigarh 160014, India
e-mail: ritambharaku@gmail.com

M. Sharma
Department of Civil Engineering, IKGPTU, Jalandhar, Punjab, India

S. Sundriyal
Department of Pharmacy, Birla Institute of Technology and Sciences Pilani, Pilani Campus, Rajasthan 333031, India
e-mail: sandeep.sundriyal@pilani.bits-pilani.ac.in

© The Author(s), under exclusive license to Springer Nature Singapore Pte Ltd. 2024

G. Tripathi et al. (eds.), *Sustainability and Health Informatics*, Advances in Geographical and Environmental Sciences,
https://doi.org/10.1007/978-981-97-6706-9_3

guaranteeing a healthy future for future generations requires the convergence of environmental sustainability and public health (Scheirer and Dearing 2011).

Considerable threats to public health arise from climate change, which is mostly caused by human activity. Among the direct and indirect effects of climate change include rising temperatures, harsh weather, altered disease patterns, food and water scarcity, and population displacement (Singh et al. 2022). In order to mitigate these health hazards, environmental sustainability initiatives—like cutting greenhouse gas emissions and adjusting to climate change—are essential (O'Loughlin et al. 1998; Tilman et al. 2011).

Unsustainable use of natural resources, including water, land, and energy, can lead to scarcity and inequitable distribution. This scarcity can have significant health consequences, particularly for vulnerable populations (Meraj et al. 2022). Environmental sustainability promotes efficient resource management, reducing the risk of resource-related health disparities (Hobley 2019; Verrier et al. 2022).

Healthy ecosystems provide essential services, such as clean air, water purification, nutrient cycling, and climate regulation, which are fundamental for human well-being (Singh et al. 2021a, b). Environmental sustainability aims to protect and restore these ecosystem services, ensuring their continued provision and promoting public health (Blasinsky et al. 2006; Turner et al. 2003).

Environmental sustainability is closely tied to social equity and justice. Marginalized communities often bear the brunt of environmental degradation and its associated health risks. By focusing on sustainable practices and policies, we can address social inequalities and create healthier and more equitable communities (Fisher et al. 2009).

The Sustainable Development Goals (SDGs) of the United Nations acknowledge the connections between public health and environmental sustainability. The integration of environmental sustainability is crucial to achieving improved public health outcomes, as highlighted by several goals including Good Health and Well-Being (SDG 3), Clean Water and Sanitation (SDG 6), Affordable and Clean Energy (SDG 7), Sustainable Cities and Communities (SDG 11), and Climate Action (SDG 13) (Liao et al. 2020; Hobley 2019; Verrier et al. 2022).

Given that environmental elements have a significant impact on human health and well-being, environmental sustainability is important in the context of public health (Meraj et al. 2021). Encouraging sustainable policies and practices can help communities become healthier and more resilient while lowering health inequities and environmental risks (Kates et al. 2001). Achieving sustainable development and guaranteeing a healthy future for future generations requires the convergence of environmental sustainability and public health (Scheirer and Dearing 2011) (Fig. 3.1).

**The objectives of the chapter are as follows:**

- Provide an overview of the interconnections between environmental sustainability and public health, highlighting the importance of integrating environmental considerations into public health policies and practices.
- Explore the various environmental factors that affect public health, including air pollution, water contamination, climate change, biodiversity loss, and chemical exposure.

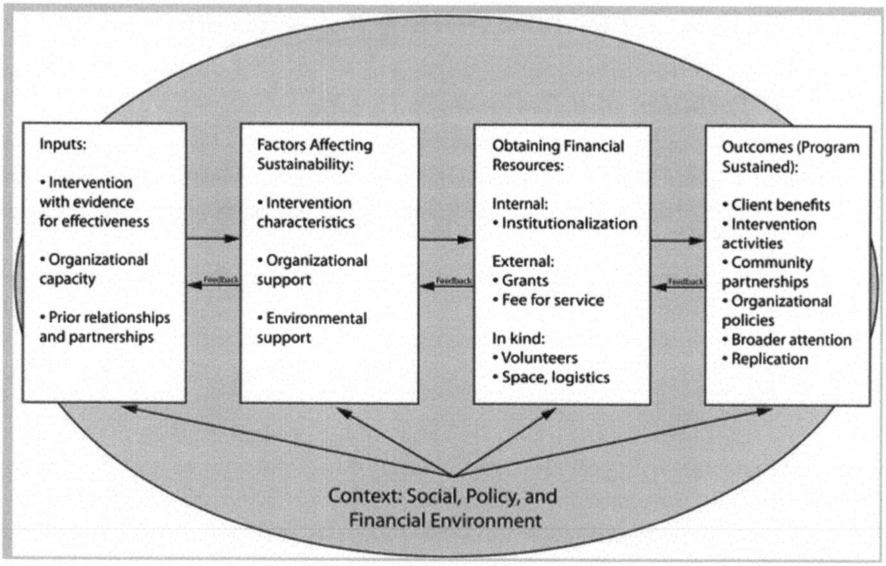


**Fig. 3.1** Conceptual framework for sustainability of public health programs (Scheirer and Dearing 2011)

- Examine the pathways through which these environmental factors impact human health, including direct exposure, environmental degradation, and social and economic disparities.
- Evaluate the potential benefits of environmental sustainability initiatives in improving public health outcomes, such as reducing disease burden, improving quality of life, and promoting equitable access to resources and opportunities.
- Identify successful case studies and examples of interventions that have integrated environmental sustainability and public health approaches, showcasing the positive impact on health outcomes. Identify challenges and barriers to achieving environmental sustainability and discuss strategies for overcoming them.
- Provide recommendations for promoting environmental sustainability and integrating it into public health policies, practices, and decision-making processes.
- Highlight the need for further research and collaboration to address emerging environmental health challenges and advance knowledge in the field.

## 3.2 Environmental Factors Affecting Public Health

### *3.2.1 Air Pollution and Respiratory Health*

Air pollution is a significant environmental factor that has detrimental effects on respiratory health. The presence of pollutants in the air, particularly fine particulate matter (PM2.5), nitrogen dioxide ($NO_2$), sulfur dioxide ($SO_2$), ozone ($O_3$), and volatile organic compounds (VOCs), can contribute to various respiratory conditions and diseases (O'Brien 2011).

Exposure to air pollutants can cause or exacerbate respiratory symptoms such as coughing, wheezing, shortness of breath, and chest tightness (Singh et al. 2021a, b). Air pollution can trigger asthma attacks in individuals with existing asthma and increase the risk of developing asthma in susceptible populations (Liao et al. 2020). Long-term exposure to air pollution, particularly PM2.5 and outdoor air pollution, is associated with the development and progression of Chronic Obstructive Pulmonary Disease (COPD), a chronic respiratory condition characterized by airflow limitation. Air pollution can weaken the immune system and increase susceptibility to respiratory infections such as bronchitis and pneumonia. Certain air pollutants, such as particulate matter and specific chemicals like benzene, are carcinogenic and linked to an increased risk of lung cancer. Air pollutants can induce inflammation in the respiratory tract, leading to irritation, swelling, and respiratory symptoms. Exposure to air pollution generates reactive oxygen species, causing oxidative stress in lung tissues and damaging cells. Prolonged exposure to air pollutants can lead to reduced lung function and impaired respiratory capacity. Air pollution has been found to cause epigenetic alterations in genes related to lung health, potentially leading to long-term respiratory health effects (Fisher et al. 2009).

Since their respiratory systems are still developing and they breathe more frequently than adults, children are especially susceptible to the effects of air pollution. Childhood exposure to contaminated air can have long-term effects on respiratory health. The negative impacts of air pollution on respiratory health are more likely to affect older persons, particularly those with pre-existing respiratory disorders. Exposure to air pollution can exacerbate symptoms and raise the likelihood of exacerbations in those with asthma, COPD, or other pre-existing respiratory illnesses (Damschroder et al. 2009).

Some of the mitigation strategies include reducing emissions, increasing awareness and education, improving indoor air quality, urban planning, etc. Implementing policies and practices that reduce emissions from industrial sources, vehicles, and household activities can significantly improve air quality and reduce respiratory health risks. Promoting public awareness of the health risks associated with air pollution and providing education on preventive measures can empower individuals to protect their respiratory health. Ensuring proper ventilation, using air purifiers, and minimizing indoor pollution sources can help reduce exposure to indoor air pollutants, benefiting respiratory health (Gruen et al. 2008). Designing cities with green spaces, pedestrian-friendly infrastructure, and sustainable transportation options can help reduce exposure to air pollution and improve respiratory health outcomes.

Addressing air pollution and its impact on respiratory health requires a multi-faceted approach involving governments, policymakers, industries, healthcare professionals, and individuals (Strange and Bayley 2008). By implementing effective measures to reduce air pollution and promote clean air, we can protect and improve respiratory health for individuals and communities.

### 3.2.2 Water Contamination and Waterborne Diseases

Water contamination is a significant concern that can lead to the spread of waterborne diseases and pose serious risks to public health. Contaminated water can contain various pathogens, chemicals, and pollutants that, when ingested or in contact with the body, can cause illnesses and diseases.

Water can be contaminated by microorganisms such as bacteria, viruses, and protozoa. Common sources include sewage and fecal matter, animal waste, and inadequate wastewater treatment systems. Water sources can be contaminated by various chemicals, including heavy metals, pesticides, industrial pollutants, and pharmaceuticals. Improper garbage disposal, industrial discharges, and agricultural runoff can all lead to the entry of these toxins into water bodies. Physical impurities that can turbidize water and degrade its quality include debris, suspended particles, and silt. Due to human activity, inappropriate land management techniques, and erosion, these toxins may find their way into water sources (Liao et al. 2020).

One of the main causes of diarrheal illnesses, such as cholera, typhoid fever, dysentery, and gastroenteritis, is contaminated water. These diseases are primarily transmitted through the ingestion of water contaminated with fecal matter-containing pathogens. Protozoa like *Cryptosporidium* and *Giardia* can cause waterborne infections, leading to diarrhea, stomach cramps, and other gastrointestinal symptoms. These pathogens can survive in water sources and resist conventional water treatment methods. Waterborne bacterial infections can result from exposure to bacteria such as *Escherichia coli*, *Salmonella*, and *Campylobacter*. These bacteria can contaminate water sources through sewage or fecal contamination, causing gastrointestinal illnesses and other systemic infections. Viruses like norovirus, hepatitis A, and rotavirus can be transmitted through contaminated water. These viruses can cause severe gastrointestinal symptoms, liver damage, and in some cases, even death (Damschroder et al. 2009).

Implementing effective water treatment processes, including filtration, disinfection (e.g., chlorination), and advanced treatment techniques (e.g., UV disinfection), is crucial to remove or inactivate pathogens and ensure safe drinking water. Promoting proper sanitation practices, such as improved sewage systems, proper waste disposal, and handwashing, can prevent fecal contamination and reduce the risk of waterborne diseases. Protecting water sources from contamination through measures such as watershed management, pollution control, and land-use regulations is essential for preventing waterborne diseases. Regular monitoring of water quality, including testing for microbial and chemical contaminants, helps identify potential risks and ensure the safety of water supplies. Raising public awareness about the importance of clean water, proper hygiene practices, and the risks associated with waterborne diseases can empower individuals to protect their health and take preventive measures (Kates et al. 2001).

Addressing water contamination and reducing the incidence of waterborne diseases requires a combination of infrastructure improvements, effective water treatment systems, sound sanitation practices, and community education. By implementing comprehensive strategies to ensure clean and safe water supplies, we can protect public health and reduce the burden of waterborne diseases on communities (Gruen et al. 2008).

### *3.2.3 Climate Change and Its Health Impacts*

Climate change is a global phenomenon that has far-reaching implications for human health. It affects various aspects of the natural and social environment, leading to a wide range of health impacts.

Dehydration, heat exhaustion, and heatstroke are among the heat-related ailments that can result from rising temperatures as well as an increase in the frequency and severity of heatwaves. Particularly at risk are vulnerable groups, such as the elderly, kids, and people with underlying medical concerns. Disease-carrying vectors like ticks and mosquitoes can fluctuate in number and behavior in response to climate change. This may result in the spread of diseases carried by vectors, such as West Nile virus, dengue fever, Zika virus, malaria, and Lyme disease. Water supplies can become contaminated by altered precipitation patterns, flooding, and extreme weather events, which can raise the risk of waterborne illnesses such as cholera, typhoid fever, and gastrointestinal infections (Gruen et al. 2008).

Climate change can worsen air quality, leading to increased levels of pollutants such as ground-level ozone and fine particulate matter. Prolonged exposure to these pollutants can exacerbate respiratory conditions (e.g., asthma) and cardiovascular diseases, leading to respiratory distress, reduced lung function, and increased cardiovascular mortality. Climate change impacts agricultural productivity, leading to shifts in crop yields, reduced food availability, and changes in nutritional content (Tilman et al. 2011). These factors can contribute to malnutrition, micronutrient deficiencies, and impaired growth and development, particularly in vulnerable populations. Water scarcity can be brought on by altered rainfall patterns and higher evaporation, which can impact the availability of clean drinking water and sanitary facilities. Lack of water can exacerbate illnesses linked to poor cleanliness, poor sanitation, and weakened public health. Natural catastrophes, relocation, and loss of livelihoods are examples of climate change-related events that can cause psychological distress, anxiety, despair, and post-traumatic stress disorder (PTSD). Eco-anxiety, persistent fear, and concern about the future of the planet and its impact on human well-being,

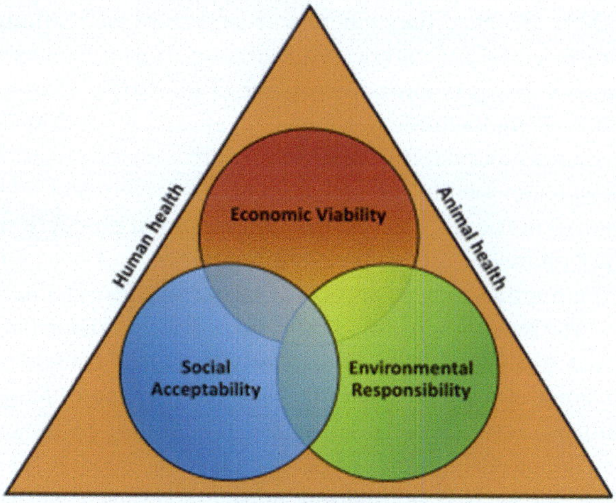


**Fig. 3.2** Pillars of sustainability and their interaction with one health (Jude 2017)

can be exacerbated by the existential danger posed by climate change and the ensuing environmental deterioration (Holling 1973).

Vulnerable populations, such as marginalized groups, indigenous people, and low-income communities, are disproportionately impacted by climate change. These groups are more vulnerable to risks and hazards associated with climate change, frequently have fewer resources, and have less access to healthcare.

Mitigation, or cutting back on greenhouse gas emissions, and adaptation, or increasing readiness and resilience, are two strategies for addressing the health effects of climate change. Key initiatives could include switching to clean and renewable energy sources, encouraging energy efficiency, lowering deforestation, and implementing sustainable farming methods (Holling 1973; Hobley 2019). improving heatwave management strategies, early warning systems, disaster preparedness, public health surveillance, and resilient infrastructure promotion. We can safeguard public health, lessen the burden of climate-related diseases, and create a more resilient and sustainable future by acting now to mitigate climate change and adapt to its effects (Jude 2017) (Fig. 3.2).

### *3.2.4 Deforestation and Biodiversity Loss*

Deforestation and biodiversity loss are closely linked environmental issues that have significant impacts on ecosystems, species, and ultimately human well-being. Deforestation refers to the permanent conversion of forested land into non-forest uses, such as agriculture, infrastructure development, logging, or urbanization. It is driven by various factors, including agricultural expansion, commercial logging, infrastructure development, and the extraction of natural resources (O'Brien 2011; Verrier et al.

2022). Because they can absorb and store vast volumes of carbon dioxide, forests are essential for carbon sequestration. It exacerbates climate change by increasing greenhouse gas emissions (Gaur et al. 2022). Many different kinds of plants and animals find a home in forests. It causes habitat destruction, upsets ecosystems, and reduces biodiversity. The loss of forest cover raises the possibility of soil erosion, which can lead to the sedimentation of rivers and other bodies of water, the depletion of nutrients, and decreased agricultural production. Indigenous tribes frequently rely on trees for both their traditional ways of life and means of subsistence. These communities may be uprooted and become less visible as a result (Verrier et al. 2022).

The reduction in the variety and richness of plant and animal species within ecosystems is referred to as biodiversity loss. It is caused by invasive species, pollution, overuse of resources, habitat destruction, climate change, and other human actions. It impairs ecosystem resilience and functioning by upsetting the complex interactions and relationships that exist within them. Biodiverse ecosystems provide valuable services such as pollination, nutrient cycling, water purification, and climate regulation. It diminishes the capacity of ecosystems to provide these essential services. It affects agricultural systems, including crop pollination and natural pest control, potentially leading to reduced food production and food security challenges (Tilman et al. 2011). Many plant and animal species have medicinal properties and contribute to traditional healing practices. It can result in the loss of potential medicinal resources (Holling 1973; Verrier et al. 2022).

The establishment and efficient administration of nature reserves, national parks, and protected areas aids in the preservation of significant habitats and hotspots for biodiversity. Supporting reforestation projects, sustainable agriculture, and agroforestry are examples of sustainable land use strategies that can help minimize deforestation and restore ecosystems (Kates et al. 2001; Verrier et al. 2022). It is possible to lessen the detrimental effects on biodiversity in sectors like mining, forestry, and fishing by implementing sustainable and ethical practices. Public support and involvement in conservation activities can be encouraged by increasing knowledge and fostering education about the value of biodiversity protection. Conservation of biodiversity can be aided by creating and implementing laws and rules that safeguard wildlife, ecosystems, and endangered species (Fisher et al. 2009).

Addressing deforestation and biodiversity loss requires a holistic approach that combines conservation efforts, sustainable land use practices, policy interventions, and public awareness. By preserving and restoring forests, protecting biodiversity-rich areas, and promoting sustainable resource management, we can safeguard ecosystems, maintain essential ecosystem services, and ensure a sustainable future for both nature and humanity.

### *3.2.5 Chemical Exposure and Toxicology*

Chemical exposure and toxicology play a significant role in understanding the potential health risks associated with exposure to various chemicals in the environment.

Chemicals can enter the environment through industrial processes, agricultural practices, consumer products, waste disposal, and other human activities. They can be present in air, water, soil, food, and everyday household items. They can enter the body through different routes, including inhalation (breathing in contaminated air), ingestion (swallowing contaminated food or water), dermal absorption (skin contact), and injection (e.g., through contaminated needles or medical procedures). Chemicals can include a wide range of substances, such as heavy metals, pesticides, industrial chemicals, volatile organic compounds (VOCs), endocrine disruptors, and pharmaceuticals (Holling 1973; Tilman et al. 2011).

Toxicology is the scientific study of the adverse effects of chemicals on living organisms and environmental health. It involves assessing the toxicity phenotypic changes dose–response relationships, mechanisms of action, and potential health risks associated with exposure to specific chemicals. Chemicals can have various health effects, including acute toxicity (immediate effects from high-dose exposures), chronic toxicity (long-term effects from repeated or continuous exposures), carcinogenicity (cancer-causing potential), mutagenicity (ability to cause genetic mutations), teratogenicity (ability to cause birth defects), and neurotoxicity (effects on the nervous system). Toxicologists and regulatory agencies identify potentially hazardous chemicals and evaluate their properties and potential health effects through laboratory studies, epidemiological research, and toxicological databases. The assessment of chemical exposure involves determining the extent, frequency, and duration of human exposure to a specific chemical, considering different exposure routes and exposure scenarios (Damschroder et al. 2009). Toxicologists evaluate the relationship between the dose of a chemical and the associated health effects to understand the threshold levels and potential risks at different exposure levels. Risk characterization integrates exposure assessment, dose–response assessment, and other factors to estimate the potential health risks associated with chemical exposure. chemical exposure. In addition, toxicologists are also involved in the mechanistic deconvolution of a chemical at the biochemical level and assist in understanding its effect at the molecular level. Such a detailed analysis often helps in inventing an antidote. This helps inform risk management decisions and regulatory actions.

Governments and regulatory agencies develop and enforce regulations, standards, and guidelines to limit chemical exposure, control emissions, and ensure product safety. Workplace regulations and safety measures aim to protect workers from hazardous chemicals through proper handling, personal protective equipment (PPE), and exposure control measures. Regular monitoring of environmental media (air, water, soil) and biomonitoring of individuals can help identify and mitigate chemical exposures in communities and inform appropriate interventions. Encouraging the use of safer alternatives and promoting the development of green chemistry can reduce the reliance on hazardous chemicals and minimize associated risks (Liao et al. 2020). Increasing public awareness about chemical exposure risks, proper handling of chemicals, and consumer choices can empower individuals to make informed decisions and protect themselves from potential harm (Baumol et al. 1998).

By conducting comprehensive toxicological assessments, implementing risk management strategies, and promoting preventive measures, we can mitigate the adverse effects of chemical exposure and safeguard human health and the environment.

## 3.3 Benefits of Environmental Sustainability for Public Health

### *3.3.1 Sustainable Urban Planning and Its Impact on Physical and Mental Health*

Sustainable urban planning refers to the design and development of cities and urban areas that prioritize environmental sustainability, social equity, and public health. It considers factors such as efficient land use, transportation systems, access to amenities, green spaces, and infrastructure that promote physical and mental well-being. It promotes walkability, cycling infrastructure, and efficient public transportation systems (Kates et al. 2001). By encouraging active modes of transportation, it increases physical activity levels and reduces sedentary behaviors, contributing to improved cardiovascular health, weight management, and overall fitness. It prioritizes the availability of fresh and nutritious food options through initiatives such as urban agriculture, farmers' markets, community gardens, and healthy food retail. Access to healthy food reduces the risk of diet-related diseases, such as obesity, diabetes, and cardiovascular disorders. Sustainable urban planning aims to minimize exposure to environmental pollutants by implementing measures such as green building standards, pollution control, and efficient waste management. This helps reduce the risk of respiratory illnesses, allergies, and other health conditions associated with environmental toxins. Designing urban areas with well-lit streets, pedestrian-friendly infrastructure, and crime prevention measures improves public safety. It promotes physical activity, outdoor recreation, and community engagement, reducing the risk of injuries and promoting overall well-being (Fisher et al. 2009).

Sustainable urban planning prioritizes the inclusion of parks, green spaces, and urban forests within cities. Access to nature and green environments has been linked to reduced stress, improved mental well-being, enhanced cognitive function, and reduced symptoms of depression and anxiety. Implementing measures to mitigate noise pollution, such as proper urban design, green buffers, and noise barriers, can significantly improve mental health outcomes. Reduced noise levels contribute to better sleep quality, reduced stress levels, and improved overall psychological well-being (O'Brien 2011). It fosters social connections and community engagement through the design of inclusive public spaces, community centers, and gathering places. Strong social networks and a sense of community belonging have positive effects on mental health, social support, and overall life satisfaction. Incorporating elements of biophilic design, such as natural lighting, views of nature, and indoor plants, in building and urban design has been shown to promote mental well-being, reduce stress, and enhance productivity.

Sustainable urban planning aims to address health disparities by considering the needs of all residents, including marginalized and vulnerable populations (Century et al. 2010). It focuses on equitable access to essential services, amenities, and infrastructure, reducing health inequities and promoting social justice. By incorporating principles of sustainable urban planning, cities can create environments that promote physical and mental health, enhance quality of life, and foster community well-being. The integration of health considerations in urban planning decisions is crucial for building resilient, inclusive, and sustainable cities that prioritize the health and well-being of their residents.

### *3.3.2 Clean Energy Transitions and Reduction of Air Pollution-Related Diseases*

Clean energy transitions refer to the shift from fossil fuel-based energy sources to cleaner and more sustainable alternatives, such as renewable energy and energy efficiency measures. This transition has significant implications for reducing air pollution-related diseases and improving public health (LaPelle et al. 2006; Hobley 2019). Air pollutants such as particulate matter (PM), sulphur dioxide ($SO_2$), nitrogen oxides ($NO_x$), and volatile organic compounds (VOCs) are mostly produced by burning fossil fuels to produce energy. Emissions of these pollutants are greatly decreased by clean energy transitions, such as the growing usage of renewable energy sources including solar, wind, hydro, and geothermal power. The broad use of electric vehicles (EVs) as a low-pollutant substitute for internal combustion engine vehicles can significantly lower emissions of pollutants including particulate matter and nitrogen oxides, which are primary causes of diseases linked to air pollution. Enhancing building, industrial, and transportation energy efficiency lowers the demand for energy overall, which in turn lowers the burning of fossil fuels. This results in lower emissions of air pollutants and improved air quality. Particulate matter, especially fine particulate matter (PM2.5), is associated with respiratory and cardiovascular diseases. Clean energy transitions help reduce the emission of particulate matter, leading to improved air quality and a lower risk of air pollution-related health issues. The transition to cleaner energy sources can reduce the emission of nitrogen oxides and volatile organic compounds, which are key contributors to ground-level ozone formation. Ozone is a respiratory irritant and can exacerbate respiratory conditions such as asthma. By reducing ozone formation, clean energy transitions contribute to improved respiratory health (Hobley 2019; Verrier et al. 2022).

By lowering exposure to air pollutants that can cause or exacerbate respiratory conditions including asthma, chronic obstructive pulmonary disease (COPD), and bronchitis, clean energy transitions have a favourable effect on respiratory health. A reduction in the risk of cardiovascular disorders, such as heart attacks, strokes, and other illnesses linked to prolonged exposure to air pollution, is possible with improved air quality brought about by clean energy transitions. Clean energy transitions can reduce death rates by reducing air pollution-related diseases, especially in vulnerable groups including children, the elderly, and people with pre-existing respiratory or cardiovascular disorders. By enhancing air quality in areas that are disproportionately affected by pollution, such as low-income neighbourhoods and marginalized populations, clean energy transitions can help address health disparities (Hobley 2019; Turner et al. 2003).

Clean energy transitions lower greenhouse gas emissions, which helps to mitigate climate change in addition to lowering ailments linked to air pollution. The argument for clean energy transitions is further strengthened by the health co-benefits of climate action, such as better air quality and lower health risks. Society can greatly reduce air pollution, enhance air quality, and lessen the burden of diseases linked to air pollution by switching to clean energy sources and implementing energy-efficient activities. These changes provide a method for communities to become healthier and more sustainable (Verrier et al. 2022).

### *3.3.3 Sustainable Agriculture and Nutrition*

The term "sustainable agriculture" describes the process of growing food in a way that is both profitable and socially and environmentally responsible. It seeks to ensure the production of healthy, safe food while reducing the detrimental effects of agriculture on the environment. It has significant implications for nutrition, both in terms of improving access to nutritious food and promoting healthier dietary choices. It encourages the cultivation of a diverse range of crops, including traditional and indigenous varieties. This promotes biodiversity, preserves genetic resources, and increases the availability of nutrient-rich foods. Practices such as organic farming, permaculture, and agroforestry emphasize natural soil fertility management, biodiversity conservation, and integrated pest management. These approaches enhance the nutritional quality of crops by minimizing the use of synthetic inputs and promoting natural nutrient cycling. It agriculture promotes organic farming practices, which minimize the use of synthetic fertilizers and pesticides. This reduces the risk of pesticide residues in food and increases access to pesticide-free produce (Holling 1973; Turner et al. 2003). By reducing the reliance on chemical inputs, sustainable agriculture contributes to safer food production and minimizes the potential health risks associated with pesticide exposure. It practices prioritize soil conservation and fertility. Practices such as cover cropping, crop rotation, and organic matter addition enhance soil structure, water-holding capacity, and nutrient availability, resulting in healthier crops and improved nutritional quality (Kates et al. 2001). It encourages efficient water management techniques, such as drip irrigation and rainwater harvesting (O'Loughlin et al. 1998). By minimizing water use and reducing water pollution from agricultural activities, helps protect water resources and ensures their availability for future food production. It promotes local and regional food systems, minimizing the distance between food production and consumption. This reduces the carbon footprint associated with transportation and supports the availability of fresh and nutritious food (Damschroder et al. 2009). Local and regional food systems enhance access to fresh fruits, vegetables, and other perishable goods, promoting a diet rich in vitamins, minerals, and antioxidants. It often prioritizes supporting small-scale farmers and local communities, creating opportunities for diversified and nutritious food production and promoting equitable access to nutritious food. It also integrates climate change adaptation and mitigation strategies. Practices such as agroforestry, conservation agriculture, and improved water management help build climate resilience, maintain agricultural productivity, and secure food availability in the face of climate challenges. By promoting climate-resilient agriculture, sustainable practices contribute to maintaining food security and ensuring a steady supply of nutritious food even in the face of climate-related disruptions (Durlak and DuPre 2008).

Sustainable agriculture plays a critical role in promoting nutrition by improving access to diverse and nutritious food, reducing chemical inputs, conserving natural resources, supporting local food systems, and building climate resilience. By embracing sustainable agricultural practices, societies can support both human health and environmental well-being.

### *3.3.4 Water Management and Waterborne Disease Prevention*

Water management and waterborne disease prevention are crucial for ensuring access to safe and clean water and reducing the risk of waterborne illnesses. Effective water management strategies and preventive measures are essential to protect public health. Proper water treatment processes, such as coagulation, sedimentation, filtration, and disinfection, are essential for removing or inactivating pathogens, chemicals, and other contaminants from water sources. Chlorination, ultraviolet (UV) disinfection, ozonation, and other disinfection techniques help destroy or inactivate disease-causing microorganisms, including bacteria, viruses, and protozoa, present in water supplies. Access to improved sanitation facilities, including proper toilets, sanitation systems, and wastewater management, is crucial for preventing the contamination of water sources with fecal matter and reducing the risk of waterborne diseases. Promoting good hygiene practices, such as handwashing with soap, safe food handling, and proper waste disposal, helps prevent the transmission of pathogens through contaminated water and reduces the risk of waterborne illnesses. Protecting and managing water catchment areas, such as forests, wetlands, and rivers, helps preserve water quality and minimize the contamination of water sources with pollutants and pathogens. Implementing measures to control and reduce pollution from industrial activities, agriculture, and urban runoff helps prevent the contamination of water bodies and safeguard water supplies. Proper land use planning and zoning regulations can help prevent activities that may contribute to water pollution, such as hazardous waste disposal, improper sewage systems, and chemical runoff. Monitoring the quality of drinking water sources, including routine testing for microbial and chemical contaminants, helps identify potential risks and ensure compliance with water quality standards (Blasinsky et al. 2006). Developing and implementing early warning systems for waterborne diseases, such as monitoring for specific pathogens or indicators, enables timely interventions and responses to prevent outbreaks. Tracking and investigating waterborne disease cases and outbreaks can help identify the sources of contamination, assess the extent of the problem, and implement appropriate control measures.

Promoting public awareness of the importance of safe water, sanitation, and hygiene practices is crucial for behavior change and the adoption of preventive measures. Engaging local communities in water management and promoting community ownership of water supply systems can improve water quality monitoring, maintenance, and sustainability.

By implementing effective water management practices, ensuring proper water treatment and disinfection, promoting sanitation and hygiene, protecting water sources, monitoring water quality, and raising public awareness, we can prevent waterborne diseases and ensure access to safe and clean water for all (Durlak and DuPre 2008). These efforts are essential for safeguarding public health and improving overall well-being.

### *3.3.5 Successful Initiatives in Sustainable Cities and Their Health Outcomes*

Integrated Sustainable Urban Development Curitiba, the capital city of Paraná state in Brazil, is known for its innovative and sustainable urban planning initiatives. The city has implemented a range of strategies that have had positive impacts on public health outcomes. Curitiba's bus rapid transit (BRT) system is recognized as one of the most successful in the world. The system provides affordable, accessible, and efficient public transportation, reducing reliance on private vehicles and decreasing air pollution levels (Macedo 2004). This has resulted in improved air quality and reduced respiratory diseases among residents. The city has prioritized the development and preservation of urban green spaces, including parks, gardens, and tree-lined streets. These green spaces provide opportunities for physical activity, stress reduction, and improved mental well-being among residents. It has implemented an innovative waste management system that encourages recycling and composting. The city's recycling program has achieved high participation rates, reducing waste sent to landfills and minimizing environmental contamination, which positively impacts public health. It has prioritized pedestrian-friendly urban design, with dedicated walkways, bike lanes, and pedestrian zones. This promotes active transportation, physical activity, and reduces the risk of traffic-related accidents, leading to improved physical and mental health outcomes.

Copenhagen, the capital city of Denmark, has embraced sustainable transportation and created a cycling-friendly culture. The city's initiatives have had several health benefits. Copenhagen has developed an extensive network of cycling infrastructure, including dedicated cycling lanes, bike-friendly streets, and bicycle parking facilities. This has encouraged residents to choose cycling as a preferred mode of transportation, leading to increased physical activity levels and improved cardiovascular health. By promoting cycling and reducing reliance on motorized vehicles, it has achieved significant reductions in air pollution. Improved air quality has positive effects on respiratory health, reducing the risk of respiratory diseases such as asthma and chronic obstructive pulmonary disease (COPD). The cycling culture has created a community-oriented and socially connected environment. Regular physical activity through cycling contributes to improved mental well-being, stress reduction, and overall quality of life. It has implemented measures to improve road safety for cyclists, including dedicated bike lanes, traffic calming measures, and cyclist-friendly intersections. This has reduced the risk of accidents and injuries among cyclists, resulting in improved road safety and public health outcomes.

These case studies highlight successful initiatives in sustainable cities that have demonstrated positive health outcomes, including improved air quality, increased physical activity, enhanced mental well-being, and reduced health risks. The lessons learned from these cities can inform and inspire other urban areas to prioritize sustainability and public health in their planning and development processes.

### *3.3.6 Renewable Energy Projects and Improved Air Quality*

Los Angeles, California, implemented a comprehensive Clean Air Action Plan (CAAP) aimed at reducing air pollution and improving air quality (Verrier et al. 2022). The plan included various renewable energy projects that contributed to improved air quality. The Los Angeles Department of Water and Power (LADWP) invested in renewable energy projects, including solar and wind power installations. By shifting from fossil fuel-based electricity generation to clean energy sources, the city reduced emissions of air pollutants, such as sulfur dioxide ($SO_2$), nitrogen oxides ($NO_x$), and particulate matter (PM), leading to improved air quality and public health outcomes. It introduced an ambitious plan to transition its transportation sector to electric vehicles (EVs). The increased adoption of EVs reduces tailpipe emissions and improves air quality, particularly in densely populated areas with high traffic volumes. The plan included the installation of EV charging infrastructure and incentives to promote EV adoption. Alongside renewable energy projects, it implemented energy efficiency measures in buildings and industries. By reducing energy consumption and reliance on fossil fuels, energy efficiency measures contribute to lower emissions and improved air quality. The CAAP's renewable energy projects and initiatives resulted in significant improvements in air quality in Los Angeles, including reductions in particulate matter (PM2.5) and ozone ($O_3$) levels. These improvements have been associated with reduced respiratory illnesses, cardiovascular diseases, and other health problems related to air pollution.

Güssing, a small town in Austria, transformed itself into a Renewable Energy Village by implementing various renewable energy projects. The transition to renewable energy had a positive impact on air quality. It utilized local biomass resources, such as wood chips and agricultural residues, to generate heat and electricity. Biomass-fired combined heat and power (CHP) plants replaced the use of fossil fuels, leading to a significant reduction in air pollution and greenhouse gas emissions (Verrier et al. 2022). It implemented a district heating system that utilized renewable energy sources. This centralized heating system, fueled by biomass and solar thermal energy, replaced individual fossil fuel-based heating systems, reducing emissions and improving local air quality. It promoted the use of electric vehicles (EVs) and established charging infrastructure. By transitioning from internal combustion engine vehicles to EVs, the town reduced emissions of air pollutants from transportation and improved air quality. The renewable energy projects in Güssing led to a substantial decrease in emissions of air pollutants, including carbon dioxide ($CO_2$), sulfur dioxide ($SO_2$), and nitrogen oxides ($NO_x$). This resulted in improved air quality and a healthier living environment for residents (Hobley 2019; Gaur et al. 2022).

These case studies demonstrate the positive impact of renewable energy projects on air quality. By transitioning to clean and sustainable energy sources, cities and communities can reduce air pollution, mitigate climate change, and improve public health outcomes.

### *3.3.7 Sustainable Food Production and Nutrition Programs*

The Green Revolution in India, which began in the 1960s, focused on improving agricultural productivity and food production. The initiative incorporated sustainable agricultural practices and nutrition programs, leading to positive outcomes. The introduction of high-yielding varieties (HYVs) of crops, along with improved irrigation and fertilizer management, significantly increased crop yields. This contributed to improved food production and availability, reducing the prevalence of hunger and malnutrition in India. It promoted crop diversification, encouraging farmers to cultivate a variety of crops. This led to a more diverse and nutritious food supply, improving dietary diversity and addressing micronutrient deficiencies. Alongside agricultural interventions, nutrition programs were implemented to address specific nutrient deficiencies. For example, the distribution of vitamin A supplements and fortified foods helped combat vitamin A deficiency, a major public health concern in India. It also emphasized the role of women in agriculture and rural development. Women's participation in agricultural activities and nutrition programs improved their access to resources, increased household food security, and enhanced nutritional outcomes for families. It demonstrated the potential of sustainable food production practices, coupled with targeted nutrition programs, to address food insecurity and improve nutrition outcomes.

The Purchase for Progress (P4P) program implemented by the United Nations World Food Programme (WFP) in Rwanda focused on enhancing smallholder farmers' capacity, improving agricultural practices, and supporting sustainable food production and nutrition outcomes. The P4P program provided training and resources to smallholder farmers, including improved seeds, fertilizer, and tools. Farmers learned sustainable farming techniques, such as agroforestry and soil conservation practices, which increased agricultural productivity and improved soil health. The program facilitated market access for smallholder farmers, connecting them with buyers and ensuring fair prices for their produce (Verrier et al. 2022). This encouraged sustainable production and increased farmers' income, improving their ability to access nutritious food for themselves and their families. It included nutrition education components that aimed to improve knowledge and practices related to nutrition and food safety. This contributed to improved dietary diversity, hygiene practices, and reduced foodborne illnesses. It focused on gender equality and women's empowerment in agriculture. Women were provided with training, resources, and support to actively participate in sustainable food production, which resulted in improved household nutrition outcomes. It demonstrated how interventions focusing on sustainable agriculture, market access, nutrition education, and gender empowerment can enhance food production, increase income, and improve nutrition outcomes for smallholder farmers and their communities (O'Loughlin et al. 1998).

These case studies highlight successful initiatives in sustainable food production and nutrition programs that have led to improved food security, increased agricultural productivity, and enhanced nutrition outcomes. By combining sustainable agricultural practices, market access, nutrition education, and gender empowerment, societies can work towards achieving sustainable food systems that promote health, well-being, and equitable access to nutritious food.

### 3.3.8 Eco-Friendly Water and Sanitation Systems

In West Bengal, India, groundwater sources were contaminated with high levels of naturally occurring arsenic, posing a significant health risk to the population. The implementation of an eco-friendly water treatment system addressed the issue. The eco-friendly water treatment system utilized a combination of low-cost and sustainable technologies to remove arsenic from groundwater (Godfray 2015). This included the use of iron oxide-coated sand filters, activated alumina filters, and household-level arsenic removal units. The project involved extensive community engagement and participation, empowering local communities to manage and maintain the water treatment facilities. This ensured the sustainability and long-term success of the project. Alongside the water treatment system, health and sanitation education programs were implemented to raise awareness about the health risks associated with arsenic-contaminated water and promote safe hygiene practices. The eco-friendly water treatment system in West Bengal significantly improved access to safe drinking water, reduced the health risks associated with arsenic contamination, and demonstrated the importance of community involvement in sustainable water management.

Kibera, one of the largest informal settlements in Kenya, Africa, faced significant challenges related to inadequate sanitation facilities and poor waste management. The Ecological Sanitation (EcoSan) project implemented eco-friendly sanitation systems. The EcoSan project introduced dry toilet systems, such as urine-diverting toilets and composting toilets, which required little to no water for flushing. These systems helped conserve water resources and reduced the pollution of water bodies. The project also emphasized the proper collection, separation, and treatment of human waste and organic waste. The composting of organic waste not only reduced environmental pollution but also produced nutrient-rich compost for use in agriculture. The project included hygiene promotion activities, such as education on handwashing, proper sanitation practices, and menstrual hygiene management, improving overall health and well-being in the community. The EcoSan project in Kibera showcased the potential of eco-friendly sanitation systems to address sanitation challenges in informal settlements, improve waste management practices, and promote better health and hygiene outcomes.

These case studies highlight successful initiatives in implementing eco-friendly water and sanitation systems that promote sustainability, improve access to clean water, and address sanitation challenges. By adopting environmentally friendly approaches to water treatment and sanitation, communities can protect water resources, improve public health, and contribute to a more sustainable future.

## 3.4 Challenges and Barriers

Socioeconomic and political challenges can pose significant obstacles to the implementation of sustainable practices and initiatives. Limited Resources and Funding: Implementing sustainable practices often requires financial resources and investments. However, limited financial resources and competing priorities can hinder the allocation of adequate funding for sustainable initiatives. This can impact the scale and effectiveness of sustainability projects and make it challenging to address pressing environmental and social issues. The political will to prioritize sustainability and enact supportive policies is crucial for driving change. However, political interests, short-term thinking, and a lack of awareness or understanding of sustainability issues can impede progress. It may be difficult to garner political support and create an enabling policy environment for sustainable practices (Century et al. 2010; Verrier et al. 2022). Socioeconomic inequalities and disparities can hinder the adoption of sustainable practices. Vulnerable communities often face challenges in accessing resources, education, and opportunities, making it harder for them to engage in sustainability initiatives. Addressing these disparities and ensuring that sustainability efforts are inclusive and equitable is essential. Resistance to change from established systems, industries, and vested interests can impede sustainable initiatives. Industries heavily dependent on fossil fuels or practices that harm the environment may resist transitioning to more sustainable alternatives due to concerns about economic impacts or resistance to new regulations (Godfray 2015; Turner et al. 2003). Limited awareness and understanding of sustainability issues can hinder progress. Lack of education and awareness about the benefits of sustainable practices, as well as the potential risks and consequences of unsustainable actions, can make it difficult to drive behavior change and gain public support for sustainability initiatives. Addressing global environmental challenges requires international cooperation and effective governance mechanisms. Differences in priorities, conflicts of interest, and challenges in reaching consensus among nations can impede coordinated efforts to tackle issues like climate change, biodiversity loss, and pollution. Transitioning traditional industries, such as agriculture, manufacturing, and energy, to more sustainable practices can face resistance due to concerns about job losses, economic impacts, and the availability of alternative livelihoods. Balancing the need for sustainable practices with socioeconomic considerations is crucial in managing this transition.

Overcoming these challenges requires collective action, stakeholder engagement, policy reforms, capacity building, public awareness campaigns, and supportive governance structures. It involves addressing social and economic inequalities, aligning political and policy agendas with sustainability goals, and fostering a global commitment to sustainable development (Century et al. 2010).

## 3.5 Policy and Governance Gaps

Policy and governance gaps refer to deficiencies or inadequacies in the development, implementation, and enforcement of policies and governance frameworks related to sustainable practices. These gaps can hinder progress in addressing environmental and social challenges. Here are some common policy and governance gaps: Insufficient or outdated policy frameworks may fail to address emerging sustainability issues or provide clear guidelines for sustainable practices. Gaps in policy frameworks can create ambiguity, lack of direction, and inconsistency in implementing sustainable initiatives (Blasinsky et al. 2006). Fragmentation and lack of coordination across different policy areas and government agencies can lead to disjointed efforts and inefficient use of resources. It is important to have integrated policies that consider the interconnectedness of environmental, social, and economic dimensions of sustainability. Even with robust policies in place, weak implementation and enforcement can undermine their effectiveness. Insufficient resources, inadequate monitoring systems, and lack of compliance mechanisms can limit the impact of sustainable policies and lead to non-compliance by individuals, businesses, or industries (Strange and Bayley 2008). Engaging stakeholders, including communities, civil society organizations, and businesses, is essential for effective policy development and implementation. However, gaps in stakeholder engagement processes can result in limited representation, lack of input, and reduced acceptance and ownership of sustainability initiatives (Century et al. 2010). Inconsistent or conflicting regulations and standards across regions or sectors can create uncertainty and hinder the adoption of sustainable practices. Harmonization and alignment of regulations and standards are necessary to provide clear guidelines and a level playing field for sustainable actions. Effective monitoring and reporting mechanisms are essential for tracking progress, identifying gaps, and ensuring accountability. Inadequate monitoring and reporting systems can result in a lack of data, making it difficult to assess the effectiveness of policies and governance framework. Insufficient financial resources and technical support can hinder the implementation of sustainable practices, particularly in developing countries or marginalized communities (Strange and Bayley 2008). Lack of access to funding, capacity-building initiatives, and technology transfer can create disparities in the adoption of sustainable solutions.

Addressing policy and governance gaps requires comprehensive and well-coordinated efforts. This includes strengthening policy frameworks, improving implementation and enforcement mechanisms, enhancing stakeholder engagement, promoting harmonization of regulations and standards, strengthening monitoring and reporting systems, and providing adequate financial and technical support (Durlak and DuPre 2008). Collaboration between governments, civil society organizations, businesses, and international institutions is crucial for bridging these gaps and advancing sustainable development.

## 3.6 Technological Limitations

Technological limitations refer to the constraints and challenges associated with the development, availability, and implementation of technologies in achieving sustainable goals. Many sustainable technologies, such as renewable energy systems or advanced waste treatment facilities, can have high upfront costs. Limited financial resources and affordability issues can hinder the adoption and widespread implementation of these technologies, particularly in developing countries or marginalized communities. Some sustainable technologies may face challenges in terms of scalability and efficiency. For example, certain renewable energy technologies may have limitations in generating electricity at a large scale or providing consistent power output. Advancements in technology are needed to improve scalability and enhance efficiency. The implementation of certain sustainable technologies may require supportive infrastructure, such as grid connections or distribution networks. Limited access to infrastructure, particularly in rural or remote areas, can impede the deployment of sustainable technologies and hinder their impact. The development and availability of sustainable technologies can vary across regions and sectors. Certain technologies may be more mature and readily available, while others are still in the early stages of development or require further research and innovation. Integrating sustainable technologies with existing infrastructure and systems can be challenging. Compatibility issues, technical barriers, and the need for retrofitting or upgrading existing systems can pose limitations and increase implementation costs. Some sustainable technologies rely on specific resources or materials that may be limited or geographically concentrated. This can lead to concerns about resource availability, extraction impacts, and potential conflicts over resource access. Successful implementation of sustainable technologies often requires changes in user behavior and public acceptance. Resistance to change, lack of awareness, or social and cultural barriers can impede the adoption of sustainable technologies and hinder their effectiveness.

Addressing technological limitations requires a combination of research and development efforts, investment in innovation, supportive policies, capacity building, and collaboration between academia, industry, and government. Advances in technology, including cost reductions, efficiency improvements, and increased accessibility, are essential for overcoming these limitations and accelerating progress toward sustainable development.

## 3.7 Behavioral and Cultural Barriers

Behavioral and cultural barriers refer to the social and psychological factors that can impede the adoption of sustainable behaviors and practices. These barriers can influence individuals, communities, and societies, affecting their willingness to change behaviors and embrace sustainable lifestyles. Limited awareness and understanding of sustainability issues can be a significant barrier. Many people may not be fully aware of the environmental, social, and economic impacts of their behaviors or the benefits of sustainable practices. Addressing this barrier requires education, outreach, and effective communication to increase awareness and knowledge. People often resist changing their behaviors or habits, especially when it requires significant adjustments to their daily routines or comfort zones. Resistance to change can stem from fear, perceived inconvenience, or the perception that sustainable practices are not beneficial or necessary. Overcoming this barrier involves addressing misconceptions, highlighting the benefits of sustainability, and emphasizing the urgency and importance of change (Blasinsky et al. 2006). Cultural and societal norms can shape individual behaviors and create barriers to sustainability. Peer pressure, social norms, and cultural values that prioritize consumption, materialism, or convenience can hinder the adoption of sustainable practices. Efforts to overcome this barrier require promoting new social norms and values that align with sustainability principles. The lack of access to sustainable alternatives and the perception that sustainable products or services are more expensive can be a significant barrier. Limited availability, affordability, or accessibility of sustainable options can discourage individuals from making sustainable choices. Addressing this barrier involves increasing the availability and affordability of sustainable alternatives, as well as promoting inclusive and equitable access to them. The absence of clear incentives or tangible rewards for sustainable behaviors can hinder their adoption. People are often driven by immediate benefits or rewards, and without sufficient incentives, sustainable behaviors may not be prioritized. Creating incentives, such as financial incentives, recognition, or social rewards, can help overcome this barrier. The emphasis on individualism and short-term thinking can hinder sustainable behaviors. People may prioritize their immediate needs and desires over long-term environmental or social considerations. Overcoming this barrier requires highlighting the long-term benefits of sustainability, emphasizing collective well-being, and promoting a sense of responsibility and interdependence. Psychological factors, such as cognitive biases, habits, and emotional attachments, can create barriers to sustainable behaviors. Biases like the status quo bias or the optimism bias can make it difficult for individuals to embrace change or take proactive action. Addressing psychological barriers involves strategies that promote self-reflection, behavior change techniques, and support systems that facilitate sustainable choices (Damschroder et al. 2009).

Addressing behavioral and cultural barriers requires a multi-faceted approach that includes education, awareness campaigns, social norms interventions, policy incentives, and community engagement (Strange and Bayley 2008). It is important to understand the specific barriers within a given context and tailor interventions to overcome them effectively. Promoting sustainable behaviors requires a shift in attitudes, values, and social norms, creating an enabling environment that encourages and rewards sustainable choices.

## 3.8 Recommendations for Promoting Environmental Sustainability and Public Health

Governments should integrate sustainability principles and goals into policies across sectors, including energy, transportation, agriculture, waste management, and urban planning. This integration should prioritize the protection of public health and the environment. It can establish incentives, subsidies, and tax breaks to promote sustainable practices and discourage unsustainable behaviors (Baumol et al. 1998). They should also implement and enforce regulations that address environmental pollution, resource depletion, and public health risks. Policymakers should adopt long-term planning approaches that consider the environmental, social, and economic impacts of development projects. This includes incorporating climate change adaptation and mitigation strategies, preserving natural resources, and promoting resilient infrastructure (Damschroder et al. 2009). Educational institutions should incorporate environmental education into their curriculum at all levels. This includes teaching students about sustainability concepts, environmental conservation, and the interconnections between the environment and public health. Governments, NGOs, and community organizations should run public awareness campaigns to promote sustainable behaviors, such as water and energy conservation, waste reduction, and responsible consumption. These campaigns should emphasize the benefits of sustainability for public health and well-being. Foster collaborations among government agencies, research institutions, NGOs, businesses, and community groups to address sustainability and public health challenges. Encourage knowledge sharing, joint research projects, and coordinated actions to develop holistic and effective solutions. Support interdisciplinary research initiatives that explore the nexus between environmental sustainability and public health. Encourage research collaborations to develop innovative technologies, policies, and interventions that promote both sustainability and public health outcomes. Involve communities in decision-making processes related to environmental sustainability and public health. Consultation and engagement allow communities to contribute their knowledge, needs, and concerns, ensuring that policies and programs reflect local realities. Provide resources, training, and support to empower individuals and communities to adopt sustainable practices and become advocates for environmental and public health. This can include workshops, skill-building programs, and the promotion of sustainable livelihoods. Support and facilitate community-led sustainability initiatives, such as community gardens, waste management projects, and renewable energy cooperatives. Encourage grassroots action and facilitate networks that enable knowledge sharing and collaboration among community members (O'Brien 2011).

By implementing these recommendations, governments, organizations, and communities can work together to promote environmental sustainability and public health (Durlak and DuPre 2008). Through policies, education, collaborations, and community engagement, a holistic approach can be taken to address pressing environmental challenges while improving public health outcomes.

## 3.9 Future Directions and Research Needs

- Conduct research on emerging contaminants, such as pharmaceuticals, microplastics, and endocrine-disrupting chemicals, to understand their potential health impacts and develop strategies for their mitigation and regulation.
- Investigate the health effects of climate change, including heatwaves, extreme weather events, vector-borne diseases, and air pollution.
- Develop adaptation and mitigation strategies to protect public health. Examine the health implications of urbanization, including the impact of urban environments on mental health, physical activity, and access to green spaces.
- Develop interventions to promote healthy urban living. Research occupational health risks, including exposure to hazardous substances and psychosocial stressors.
- Identify effective interventions and policies to protect workers' health and well-being.
- Explore and develop green technologies for renewable energy, sustainable agriculture, water treatment, waste management, and transportation. This includes the use of clean energy sources, precision agriculture, nature-based solutions, and smart infrastructure.

- Harness the power of digital technologies, such as remote sensing, artificial intelligence, and big data analytics, to monitor environmental health, predict disease outbreaks, and improve environmental management practices.
- Promote research on the circular economy, focusing on resource efficiency, waste reduction, and sustainable consumption and production patterns.
- Develop innovative solutions for recycling, upcycling, and waste-to-energy conversion. Investigate the health benefits of nature-based interventions, such as green infrastructure, urban forestry, and nature conservation.
- Assess their impact on air quality, mental health, and well-being.
- Conduct long-term studies to understand the cumulative health effects of environmental exposures over time.
- Monitor the health outcomes of populations exposed to specific environmental hazards and assess the effectiveness of interventions.
- Develop integrated monitoring systems that combine environmental data, health data, and social determinants of health. This will enable a comprehensive assessment of the links between environmental conditions and public health outcomes.
- Evaluate the effectiveness of environmental interventions and policies in improving public health. Assess their cost-effectiveness, scalability, and sustainability to inform decision-making and resource allocation.

By focusing on these future directions and research needs, we can advance our understanding of emerging environmental health challenges, develop innovative approaches and technologies, and ensure long-term monitoring and evaluation to protect public health and promote environmental sustainability (O'Brien 2011; Verrier et al. 2022). Continued research and collaboration are essential for addressing these challenges and improving our collective well-being.

## 3.10 Conclusion

Environmental sustainability is essential for safeguarding public health and well-being. Environmental degradation, pollution, and climate change have detrimental effects on human health, contributing to a range of diseases and health risks. Sustainable practices, such as renewable energy, sustainable agriculture, waste management, and clean water systems, can improve public health outcomes by reducing exposure to harmful pollutants, enhancing access to clean resources, and promoting overall well-being. Phytoremediation, eco-friendly water and sanitation systems, and sustainable food production are effective strategies for addressing environmental challenges while promoting public health. Socioeconomic and political challenges, and cultural barriers, technological limitations, and policy and governance gaps can hinder the integration of environmental sustainability and public health approaches.

The integration of environmental sustainability and public health approaches is crucial for addressing the complex challenges we face today. Policymakers need to prioritize health considerations in all sectors, recognizing that environmental sustainability is a prerequisite for ensuring long-term public health and well-being. Recognizing the interdependencies and feedback loops between environmental factors and public health outcomes can guide integrated strategies. Emphasizing prevention and precautionary principles can help mitigate environmental risks and reduce the burden of environmentally-related diseases. Taking proactive measures to protect ecosystems and reduce pollution is key. Environmental sustainability and public health initiatives must be pursued with equity and justice at their core. Disparities in access to clean resources and exposure to environmental hazards must be addressed to ensure equitable health outcomes for all.

By taking collective action, we can create a sustainable future that protects the environment, promotes public health, and ensures the well-being of current and future generations. The integration of environmental sustainability and public health is a shared responsibility and a pathway to a healthier and more sustainable world.

## References


Ahmad T, Gupta SK, Singh SK, Meraj G, Kumar P, Kanga S (2023) Unveiling Nature's Resilience: Exploring Vegetation Dynamics during the COVID-19 Era in Jharkhand, India, with the Google Earth Engine. Climate 11(9):187

Baumol WJ, Baumol WJ, Oates WE, Bawa VS, Bawa WS, Bradford DF, Baumol WJ (1988) The theory of environmental policy. Cambridge University Press

Blasinsky M, Goldman HH, Unutzer J (2006) Project IMPACT: a report on barriers and facilitators to sustainability. Adm Policy Ment Health 33(6):718–729

Century J, Rudnick M, Freeman CA (2010) Framework for measuring fidelity of implementation: a foundation for shared language and accumulation of knowledge. Am J Eval 31(2):199–218

Damschroder LJ, Aron DC, Keith RE, Kirsh SR, Alexander JA, Lowery JC (2009) Fostering implementation of health services research findings into practice: a consolidated framework for advancing implementation science. Implement Sci 4:50

Durlak JA, DuPre EP (2008) Implementation matters: a review of research on the influence of implementation on program outcomes and the factors affecting implementation. Am J Commun Psychol 41(3–4):327–350

Fisher B, Turner RK, Morling P (2009) Defining and classifying ecosystem services for decision making. Ecol Econ 68:643–653

Gaur A, Balyk O, Glynn J, Curtis J, Daly H (2022) Low energy demand scenario for feasible deep decarbonisation: whole energy systems modelling for Ireland. Renew Sustain Energy Transit 2(2022):100024

Godfray HCJ (2015) The debate over sustainable intensification. Food Secur 7:199–208

Gruen RL, Elliott JH, Nolan ML et al (2008) Sustainability science: an integrated approach for health-programme planning. Lancet 372(9649):1579–1589

Hobley A (2019) Will gas be gone in the United Kingdom (UK) by 2050? An impact assessment of urban heat decarbonisation and low emission vehicle uptake on future UK energy system scenarios. Renew Energy 142:695–705

Holling CS (1973) Resilience and stability of ecological systems. Annu Rev Ecol Systemat 4:1–23

Jude C (2017) Looking forward to a sustainable future—how do livestock productivity, health, efficiency and consumer perceptions interact? Cattle Pract 25:179–193

Kates RW, Clark WC, Corell R, Hall JM, Jaeger CC, Lowe I, McCarthy JJ, Schellnhuber HJ, Bolin B, Dickson NM et al (2001) Sustainability science. Science 292:641–642

LaPelle NR, Zapka J, Ockene JK (2006) Sustainability of public health programs: the example of tobacco treatment services in Massachusetts. Am J Public Health 96(8):1363–1369

Liao C, Agrawal A, Clark PE, Levin SA, Rubenstein DI (2020) Landscape sustainability science in the drylands: mobility, rangelands and livelihoods. Landsc Ecol 35:2433–2447

Macedo J (2004) Curitiba. Cities 21(Issue 6):537–549. ISSN 0264-2751

Meraj G, Kanga S, Kranjčić N, Đurin B, Singh SK (2021) Role of natural capital economics for sustainable management of earth resources. Earth 2(3):622–634

Meraj G, Farooq M, Singh SK, Islam MN, Kanga S (2022) Modeling the sediment retention and ecosystem provisioning services in the Kashmir valley, India, Western Himalayas. Model Earth Syst Environ 8(3):3859–3884

O'Brien K (2011) Global environmental change II: from adaptation to deliberate transformation. Prog Hum Geogr 36:667–676

O'Loughlin J, Renaud L, Richard L, Gomez LS, Paradis G (1998) Correlates of the sustainability of community-based heart health promotion interventions. Prev Med 27(5 pt 1):702–712

Scheirer MA, Dearing JW (2011) An agenda for research on the sustainability of public health programs. Am J Public Health 101(11):2059–2067

Singh S, Singh H, Sharma V, Shrivastava V, Kumar P, Kanga S, Sahu N, Meraj G, Farooq M, Singh SK (2021) Impact of forest fires on air quality in Wolgan valley, New South Wales, Australia—a mapping and monitoring study using Google Earth Engine. Forests 13(1):4

Singh SK, Meraj G, Mondal N, Bera A, Verma MK, Tomar JS, Kanga S (2021) Assessment of seasonal vegetation dynamics over parts of thar desert using geospatial techniques. ANGRAU 105

Singh H, Meraj G, Singh S, Shrivastava V, Sharma V, Farooq M, Kanga S, Singh SK, Kumar P (2022) Status of air pollution during COVID-19-induced lockdown in Delhi, India. Atmosphere 13(12):2090

Strange T, Bayley A (2008) Sustainable development: linking economy, society, environment. Organization for Economic Co-operation and Development, Paris, France

Tilman D, Balzer C, Hill J, Befort BL (2011) Global food demand and the sustainable intensification of agriculture. Proc Natl Acad Sci USA 108:20260–20264

Turner BL, Kasperson RE, Matson PA, McCarthy JJ, Corell RW, Christensen L, Eckley N, Kasperson JX, Luers A, Martello ML, Polsky C, Pulsipher A, Schiller A (2003) A framework for vulnerability analysis in sustainability science. Proc Natl Acad Sci USA 100:8074–8079

Verrier B, Li P-H, Pye S, Strachan N (2022) Incorporating social mechanisms in energy decarbonisation modelling. Environ Innov Soc Transit 45:154–169

# Chapter 4
# The Role of Health Informatics in Achieving Sustainable Development Goals

**Goldie Gabrani, Sunil Gupta, Lakshay Verma, and Sonali Vyas**

## 4.1 Introduction

The United Nations General Assembly adopted 17 global goals, known as the Sustainable Development Goals (SDGs), in 2015 with a goal to meet them by 2030. The objectives include eradicating poverty, conserving our planet, and promoting harmony and growth for everyone across all generations. Notably critical within these SDGs is health given its integral role toward sustainable progress; thus ensuring good health and wellbeing makes up the third SDG. Waiting where healthcare meets information technology lies Health Informatics—holding significant relevance concerning the achievement of this specific target along with other health-related SDGs too (Meraj et al. 2021). Intricacies surrounding roles played by Health informatics in relation to endorsing such developmental targets will be evaluated throughout this chapter consequent upon discussing associated challenges/ opportunities arising from utilizing data together with info systems aiding enhanced medical results plus advancing sustainability agendas overall. The reader can also anticipate assimilating ingenious strategies cultivated alongside methods devised through 'informatics' that allow for accomplishing the aforementioned Sustainable Developmental Goals, furthermore, anecdotes incorporated touching on diverse digital solutions (Robert et al. 2005).

G. Gabrani · L. Verma
School of Engineering, Manav Rachna University, Faridabad, India
e-mail: ggabrani@gmail.com

S. Gupta · S. Vyas (✉)
School of Computer Science, UPES University, Dehradun, India
e-mail: researcher.svyas@gmail.com

S. Gupta
e-mail: s.gupta@ddn.upes.ac.in

© The Author(s), under exclusive license to Springer Nature Singapore Pte Ltd. 2024 
G. Tripathi et al. (eds.), *Sustainability and Health Informatics*, Advances in Geographical and Environmental Sciences,
https://doi.org/10.1007/978-981-97-6706-9_4

Health informatics, a swiftly expanding area, incorporates the application of information technology to enhance healthcare provision, patient results, and overall population health. This field carries an extensive array of applications such as electronic medical records, telemedicine implementation, exchanging health data formally, systems for clinical decision support, and public supervision in terms of health (Eysenbach 2009). These technologies carry transformative potential with regards to how certain countries, particularly those classified under lower middle-income strata where access to necessary healthcare is notoriously scarce- offer their citizens quality care (Singh et al. 2022).

The Sustainable Development Goals (SDGs) provide a blueprint to tackle global health challenges while promoting sustainable progress. The third SDG aims for the promotion of healthy living and universal welfare across all ages, holding substantial significance in terms of health informatics. Achieving this target calls for an encompassing strategy that addresses social influences on health like poverty rates, educational attainment, and healthcare availability (Ahmad et al. 2023). Health Informatics could play a crucial role by introducing innovative solutions that transform the way healthcare services are provided, enhancing patient outcomes and reducing disparities correlating to overall societal wellbeing (Robert et al. 2005).

A significant hurdle to utilizing health informatics for long-term growth is the absence of data and information systems in a great deal of many low- to middle-wage countries. Monitoring health results, tracing disease occurrences, and crafting evidence-supported policies becomes troublesome without trustworthy data structures. Further complications arise from another issue—adopting an encompassing strategy while implementing such technology. Consequently, it's critical that solutions related with health informatics are built according to existing healthcare infrastructure catering specifically towards local population needs. (Mair et al. 2012).

While there are hurdles to overcome in the utilization of health informatics, like an absence of core competitiveness in the field's education and inhibitions about executing digital strategies within precarious circumstances, it can't be denied that digitized healthcare technologies offer myriad advantages. Health informatics could markedly transform how medical care is delivered in strife-ridden territories while simultaneously supporting the attainment of Sustainable Development Goals (SDGs).

Advancements in digital health like telemedicine, electronic medical documents (EMDs), wired fitness gadgets (wearables), mobile healthcare platforms(mHealth), and intuitive software applications propel the ability to surpass physical constraints while enabling access to internet services (Zheng et al. 2014). Also assisting are user-specific personal health records as well as non-health-related data along with environmental factors—all essential prerequisites for achieving global healthcare objectives efficiently. Furthermore, regulating EMD systems mixed with context-focused information storage tactics can facilitate a methodical approach toward assembling and handling precise data—this being vital when gauging SDG 3 targets accomplishments persistently.

Furthermore, digital health tools are able to enhance connection channels between various participants like patients, healthcare givers, and NGOs for the constant advancement of SDG 3. The synergy between medicine and technology harnesses potential benefits through innovative low-cost platforms in delicate regions to amplify both global and local abilities. Achieving "healthy lives alongside wellness at every age" prompts lower income nations toward showing some self-reliance strategies regarding capacity building. This independence should overcome any geographic constraints while keeping stability even amidst conflict.

## 4.2 Overview of Health Informatics

Health informatics refers to the combined study of design, development, and implementation of information technology (IT) alongside data science within healthcare. Its primary function is using tech tools for management and analysis purposes over health-related stats with an objective toward enhancing patient prognosis, increasing efficiency in medical service delivery along with curtailment of expenses. The scope extends far beyond just one segment comprising major activities such as creating electronic health records (EHRs), building clinical decision ancillary systems, promoting telemedicine plus developing health information exchange platforms or HIE structures (Jha et al. 2009) (Fig. 4.1).

The health informatics includes the following areas:

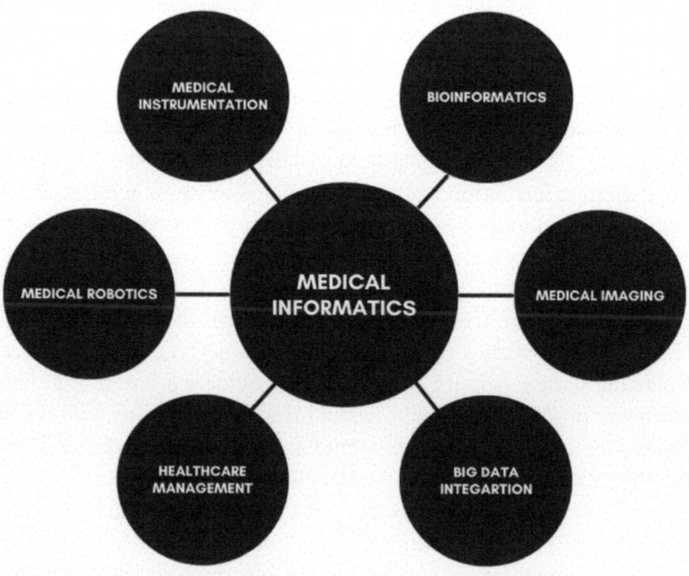


**Fig. 4.1** Medical Informatics types

### Clinical Informatics

This entails leveraging technology in aiding clinical judgments, fortifying patient security and boosting care quality. The scope of Clinical informatics involves creating Electronic Health Records (EHRs), digitized physician order entries known as CPOE systems, along with the development of support frameworks for crucial medical decisions (Jha et al. 2009).

### Public Health Informatics

IT's role in bolstering public health endeavors like disease monitoring, management of epidemics, and wellness propaganda is pivotal. Various tasks under Public Health Informatics encompass cultivating registry systems for diseases, syndromic surveillance configurations as well as the advancement of health information trading platforms (Eysenbach 2009).

### Consumer Health Informatics

This entails employing information technology to enhance patient involvement and self-care. The sphere of consumer health informatics embraces creating patient-focused platforms, crafting mobile health apps, and developing personal medical documentation.

### Research Informatics

IT's role in supporting research activities such as clinical trials, epidemiological studies, and health services investigation is a part of what is involved. This comprises the development of resources like data warehouses, databases specifically for research purposes, and tools designed to mine data.

The progression charted by Health informatics over the past several decades can be attributed to its substantial growth largely due to IT advancements alongside increasing demands for effective healthcare delivery methods. Focusing initially on managing patient information through automated systems with developments in computerized hospital details system (HIS) or laboratory insights scheme (LIS), it also streamlined administrative tasks including billing & scheduling during late 60 s and early 70 s era. Moving into subsequent periods around mid-80 s till end-90 s, there was a shift more toward clinically directed decision-making assistance schemes(CDSS). CDSS were developed with an objective that extended beyond providing relevant patient inputs bound within evidence-based regulations thus assisting healthcare professionals make better informed decisions while concurrently enhancing safety measures alerting personnel about potential risks inclusive yet not limited interactive drug reactions. (Borycki et al. 2017).

The dawn of the new millennium welcomed a substantial surge in embracing electronic health records (EHRs), substituting old-school paper-based medical documentation, and offering immediate access to patient information for healthcare professionals. EHRs eased up the transfer of crucial health details among various care providers, enhancing service coordination while minimizing medical mishaps (Jha et al. 2009). As we moved into 2010s, consumer-focused informatics got considerable attention where technology was leveraged to assist patients' active involvement and manage their own wellbeing—manifesting itself through innovative platforms like patient portals, mobile wellness apps, and individualized health logs; these offered consumers open-ended accessibility to their personal health reports as well as options for discussion with caregivers.

Contemporary health informatics consistently advances due to the progression in artificial intelligence, machine learning and comprehensive data examination. These innovations aid in creating precise models for disease identification and therapy along with recognizing populous health inclinations and upgrading healthcare administration. The advent of COVID-19 has swiftly sped up the utilization of virtual medicine consultation, including remote patient surveillance which is now a crucial factor while providing medical care yet lessening the affliction's widespread probability (Reeves et al. 2020). This transition into surge within this specialized domain stems from an increasing demand for more streamlined methods within healthcare services supported by upgrades done on information technology domains furthering smoother delivery systems. As we proceed forward it would be logical that there may arise inventive implementations concerning technology integrated into public welfare collectively altering our current patterns both involving offering as well as receiving professional medical assistance.

Health informatics is a tool that permits healthcare professionals to instantaneously access and examine patient information, aiding in their ability to make well-informed decisions about individual treatment plans. This potential can pave the way for enhanced patient results—less frequent readmissions into hospitals, improved management of long-term illnesses and increased compliance with prescribed medications are just some examples. Health informatics also simplifies health data exchange among different medical providers thereby enhancing collaborative care efforts as well as minimizing errors in treatments leading consequently toward better personal outcomes and an increasingly streamlined healthcare system overall.

Beyond this, administrative jobs such as billing processes or appointment scheduling can be automated using health informatics thus liberating more time for professionals to devote wholeheartedly on taking care of patients directly. It could additionally lessen both the duration needed and resources used up handling patient info which then optimizes how effectively followers' assistance may get offered inside our framework through achieving efficiency improvements across delivery bands.

Lastly yet importantly enough too; by enabling those who work within these setups—namely clinicians themselves—to—scrutinize wider population-based figures around disease patterns they're therefore able afterwards identifying worrisome emerging trends plus putting together proactive targeting strategies meant above all else preventing further spread along similar lately occurring incidences necessarily involved during tackling preventative medicine profiles specifically related tasks similarly involving active role participation programs put forth here comes courtesy "health Informatica". These tools really cannot fail regarding advantages when set against capacity, empowering people against otherwise understandably formidable challenges faced nowadays if suffering sadly under chronic condition(s) while hopefully encouraging others to never stop striving to aim high despite fate's severe hardships thrown upon oneself.

## 4.3 Understanding Sustainable Development Goals

In 2000, the United Nation adopted the Millennium Development Goals which addressed problems like hunger, disease, poverty, and other socioeconomic challenges faced by underdeveloped nations that later were used as foundation to shape the objectives for Sustainable Development Goals. Many of the MDGs' goals, like lowering rates of extreme poverty and infant mortality, increasing access to clean water and education, and battling illnesses like HIV/AIDS and malaria, were successfully attained. However, MDGs had flaws such as lack of attention to sustainable development, inadequate focus on socioeconomic concerns, and failure to address poverty and inequality (Sachs 2012).

The United Nations developed SDGs, which were then adopted in 2015 to address the shortcomings of MDGs with the goal of ending poverty, ensuring peace, and saving the planet by 2030. The SDGs are an extension of MDGs, adding a wide range of issues such as poverty, gender equality, clean water and sanitation, clean energy, economic growth, innovation, etc. (Sachs 2012).

The significance of the SDGs on a global level cannot be overstated as they offer a universal perspective and blueprint that allows nations, entities, and individuals to unite in their pursuit of crafting an environmentally balanced and just planet for all (Singh et al. 2021). An interplay between social concerns, economic factors, and environmental issues is acknowledged by these goals necessitating synergistic answers. Moreover, they highlight that there should be no one left behind—it's vital to include even those who're most vulnerable or marginalized within our growth initiatives. The review mechanism built into the SDGs includes benchmarks via various facets to ensure transparency while keeping progress checks intact over time. In essence then; what we have with this set of Sustainable Development Goals (SDGs) is essentially an innovative milestone-setting endeavor meant specifically toward worldwide development & sustaining ecological balance.

The goals of sustainable development (SDGs) are closely knit and mutually reliant, indicating that progress in a certain area aligns with advancements made elsewhere. To illustrate this point, the achievement of SDG 1—Eradicating Poverty has strong ties to accomplishing SDG 2—Eliminating Hunger as well as achieving Good Health & Well-Being (SGD3), alongside Quality Education(SDG4). This is due to poverty's contribution toward insufficient nutrition and poor health resulting in limited access to educational opportunities. In the same vein, attaining Gender Equality or SGD5 can have major impacts on Decent Work & Economic Growth(SGD8) along with Reduced Disparities underlined by SG10 because it

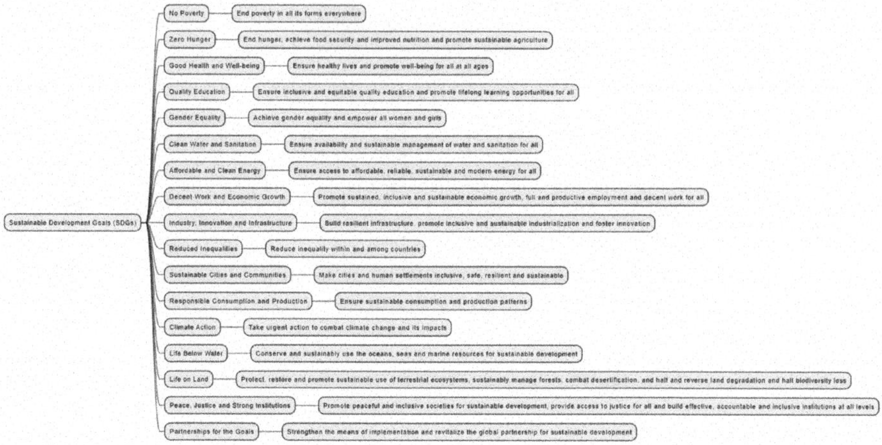


**Fig. 4.2** Sustainable development goals

directly addresses gender-discrimination which often leads women into constrained economic circumstances (Fig. 4.2).

The Sustainable Development Goals (SDGs) acknowledge the significance of ecological sustainability, particularly illustrated by SDG 13—Climate Action and SDG 14—Life Below Water which has a direct connection to both clean water, sanitation initiatives underlined in SDG 6; as well as affordable and neat energy solutions targeted at achieving goal number seven. Moreover, accomplishing sustainable cities & communities highlighted by SGD11 demands intricate links with industry improvements provided through innovation infrastructure endorsed by Goal nine while ensuring responsible consumption utilized according to outlined within TD12—promoting methods geared toward creating environmentally supportive products.

Fundamentally, the Sustainable Development Goals (SDGs) were designed to be cohesive and interdependent, underscoring that progression in one area reflects improvement across others. The SDGs furnish an operating plan for nations, institutions, and individuals alike striving toward building a world characterized by environmental stability powered with equality—a belief system grounded on noting the sophisticated associations between socio-economic-environmental matters.

## 4.4 The Intersection of SDGs and Health Informatics

"Good Health and Well-being", known as Sustainable Development Goal 3 (SDG 3), is a critical part of the United Nation's blueprint for sustainable development by 2030. This goal encapsulates worldwide efforts to guarantee everyone has access to a healthy life, while enhancing overall wellbeing at all stages of life. It covers an extensive range of health-related targets and key indicators dealing with both transmissible and non-transmissible diseases, maternal-child wellness, availability of crucial healthcare services along encouraging good mental health in general (Singh et al. 2021). Central to SDG 3 lies the basic human right—wellbeing. Its objectives include slashing mortality rates especially among children & women alongside combating disease proliferation. The global nexus acknowledges that assuring sound health without any exclusion not only fulfills moral obligations but also echoes economic sense; healthier populations improve productivity thus aiding societal advancement on-the-whole. To understand further about this diverse target areas specific within SDG 3 like its main aspects:

**Reducing Maternal Mortality (Target 3.1)**

Worldwide, the issue of maternal death continues to pose major health concerns especially in less affluent nations. The United Nations' Sustainable Development Goals (SDGs) highlight Target 3.1 with an aim towards significantly lowering statistics of global mother fatalities during childbirth. Reaching this goal provides important benefits for not just mothers but also impacts larger scopes involving healthcare standards and overall development goals positively. Employing a strategic approach through health informatics opens up opportunities for achieving better success rates on Target 3.1 by enhancing accessibility, quality, and results related to care during maternity.

Health informatics, blending information technology and data science with healthcare practices effectively addresses the issue of maternal mortality. Limited access to quality-oriented services for expectant mothers is a major stumbling block in minimizing this grim statistic globally. However, health informatics drastically improves accessibility by streamlining management of patient history details and other essential medical records related to maternity care—notably through Electronic Health Records (EHRs). EHRs enable secure storage along with easy retrieval across multiple checkpoints including prenatal clinics or birth centers as well as postnatal visits enabling coordinated response timely particularly during emergency situations associated with obstetrics. The instrumental role played by health informatics tools is critical resources empowering providers making right decisions impacting overall women's carrying pregnancies based upon a readily accessible knowledge base. Specifically, designed diagnostics tools backed up by predictive analytics enable risk profiling early enough allowing caregivers taking preventive actions ahead thereby fitting into SDG 3 objective aiming toward curbing down preventable loss due to pregnancy complications.

Telemedicine, which is a core element of health informatics, holds great importance in improving the provision of maternal healthcare services especially for those residing in remote and underprivileged regions. It facilitates pregnant mothers to reach out to medical professionals using video calls or mobile applications thus overcoming location-based challenges faced while seeking care. This unique advantage precisely mirrors SDG 3's objective (Target 3.8)—extending universal coverage of health services by making sure that irrespective of their geographical situation all women are able to access key maternity-related healthcare facilities along with expert support throughout their pregnancy till delivery.

Remote medical consultations backed by health informatics tools enable caregivers to closely watch over maternal health, offering immediate aid when needed. Such vigilant surveillance prompts an early identification of any complications and decreases potential risks associated with maternal death rates. It offers support towards patient's education and involvement, providing expectant mothers with vital knowledge about their treatment plan ensuring they can make informed choices regarding their personal care.

A vital component in cutting down the number of maternal deaths lies with prompt access to healthcare resources. The allocation of these significant resources can be assisted by health informatics, which offers an instantaneous flow of data regarding utilization and needs for maternity care. For instance, predictive analytics could help identify periods where there will likely be a high demand for such services so that facilities related to this field are well-prepared through proper staff scheduling and resource planning—bringing assurance about availability whenever mothers need it most; contributing directly toward meeting SDG 3's objective on lessening interactive maternal mortality rates (Goal 3.1).

**Ending Preventable Deaths of Newborns and Children (Target 3.2)**

Sustainable Development Goals (SDGs) have outlined an objective in Target 3.2, which is to put a stop to the unnecessary deaths of newborns and children below five years old. To make meaningful strides towards this aim, we need a comprehensive strategy that focuses on better healthcare access, enhancements in care quality standards, and harnessing innovative tech advancements effectively. It's within this scenario that health informatics comes into prominence as an efficient instrument with great promise for fast-tracking progress aimed at eradication of avoidable child mortality rates. Health Informatics domain consisting of usage data tracking systems or modern technology interfaces applied for medical purposes can counteract major issues while bringing about substantial contributions toward achieving SDG 3.2's aims.

Health informatics significantly aids in advancing healthcare accessibility, a key factor in decreasing child mortality rates. This is achieved through improved management of patient files and health-related data, ensuring the more efficient delivery of vital medical services. Digital Health Records (EHRs) provide secure storage for accessing patient's information which serves to streamline coordination among care providers. Pertaining specifically to SDG 3.2's goal, this enables a prompt exchange of critical details about newborns and young children enhancing their right timely treatment (Jha et al. 2009).

Moreover, the utilization of health informatics can improve monitoring and early identification of diseases which are fundamental components in lowering child mortality rates. With immediate data accumulation and interpretation, healthcare experts along with public health organizations gain an ability to observe disease emergence while pinpointing populations that may be at greater risk (Eysenbach 2009). This approach is especially crucial when handling children's afflictions more frequently seen like pneumonia, diarrhea or conditions preventable through vaccination. Swift recognition followed by apt responses toward outbreaks allows for numerous juvenile lives to be saved., Supporting essentially SDG 3.2's objective. (Yogesh et al. 2022).

The field of health informatics plays a critical role in enabling telemedicine and telehealth programs. These services broaden healthcare reach to isolated areas that are often underserved. For young ones such as newborns and children needing specialized care, prompt access to medical professionals could be vital for their survival. Telehealth hence facilitates specialist consultations regardless of geographical barriers which can make all the difference in reducing child mortality by providing timely assistance. The use of data analytics coupled with machine learning within health informatics creates significant benefits especially when employed for special predictive studies or projects like trends identification from child health data sets which aids early intervention measures helping prevent risks before they become serious threats. These capabilities also allow us to identify at-risk kids against malnutrition or infections sooner than typically possible—this way targeted preventative strategies corresponding specifically toward these dangers get enabled effortlessly resonating strongly with SDG 3.2's ambition aiming halting premature deaths among children Additionally, digital tools including mobile-health apps supported via Health Informatic structures fosters community-wide wellness campaigns ensuring parents/caregivers remain engaged tracking key development milestones besides accessing relevant details regarding their charge's wellbeing By empowering families through knowledge combined alongside efficient technological utility assisting them make precise choices about managing minor wards' healthy lifestyle thus supports significantly achieving sustainable goals defined under category SDG-3 emphasized on minimizing childhood fatality figures.

Using health informatics within the realm of immunization programs can enhance vaccine coverage, thus becoming a key factor in lowering childhood mortality. It accomplishes this by monitoring how vaccines are distributed and taken up as well as sending reminders to parents or guardians ensuring children get their vital shots on time. This doesn't just ward off diseases that could be avoided with vaccination but also aids in achieving Sustainable Development Goal (SDG) 3.2 through minimizing child death risks. On the other hand, protection and safety aspects related to data when dealing with private patient information cannot be ignored; it is crucial for maintaining confidentiality and integrity - an absolute necessity for gaining trust toward these systems leading to them being successfully utilized. Also, transitioning toward digitized medical resolutions compared to classic paper-based methodologies may call drastic shifts in work procedures culture. Lastly-but-certainly importantly-efficiency, future-proof relevance, and continuous updates tailored precisely according to the evolving needs of the technology landscape are imperatively essential regarding long-term implications.

**Combating Communicable Diseases (Target 3.3)**

Tackling infectious diseases (Target 3.3) is a vital part of Sustainable Development Goal 3 (SDG 3), striving for universal health and well-being. The efficient prevention, surveillance, and control of infectious diseases can strongly impact public health, societal welfare, and global economic growth. In this framework, the application of healthcare informatics stands out as an innovative strategy that could markedly enhance initiatives to fight off such diseases further propelling progress in SDGs.

Health informatics serves by tapping into its inherent power to spot signs early on, trace disease patterns consistently across vast geographies; and react timely while initiating decisive steps toward controlling further spread of these infectious illnesses. We live at times when globalization is taking center stage paired with fast-paced urban growth—situations that aid rapid pathogen mobility leading to swift cross-border transmission of pathogens causing contagion like never before seen. This highlights why such robust systems are particularly crucial during our times. Health surveillance stands out amongst the top benefits offered via advanced tech-aided solutions or simply called "health informatics". The whole process includes collecting real-time data from various sources linked through hi-tech platforms facilitating prompt analysis and reporting features thereby enabling public wellbeing officers to achieve improved accuracy & speed which directly reflects upon tracking any sparking point/disease's shroud time/place-wise thus making computerized/app-based tools an incredibly reliable option offering efficient linkages allowing shared access to all collected useful info spanning care-giving outfits also transcending geographic snags so warranted prevention can be implemented promptly whenever & wherever something hazardous gets kick-started pretty much aligning it firmly alongside SDG3 aspirations (Eysenbach 2009).

In the realm of health informatics, complex predictive models are built to predict potential disease outbreaks. These tools utilize various data types such as demographic details, environmental conditions, and epidemiological information for precise predictions. This early alert system assists healthcare systems in better preparation and response towards epidemics thus minimizing their negative impact on communities under the umbrella of SDG 3 goals by preventing avoidable deaths. Another important role played by health informatics is employing a data-driven approach for decision making when dealing with communicable diseases. Leveraging machine learning algorithms along with valuable insights from big-data analytics can allow authorities to allocate resources more efficiently whilst identifying threatened populations at risk during times like an influenza outbreak where areas with higher infection rates need targeted overseen vaccination programs and treatment allocation; further enabling progress against Target 3 of SDG goal number 30, i.e., cutting down burden imposed due to communicable diseases. Telemedicine has shown significant improvement in providing access to medical services especially benefiting rural population victims suffering from infectious illnesses so they consult doctors, remotely minimizing transmission rate inside med-centers resulting in timely care. The mutual sharing of data, proven protocols, and research results enables nations to jointly confront worldwide health dangers with greater efficacy. This method of collective action is well demonstrated in programs such as the Global Polio Eradication Initiative which applies health informatics for gathering information, scrutiny, and planning immunization drives.

## Non-Communicable Diseases (Target 3.4)

The prevalence of Non-Communicable Diseases (NCDs) as a major global health issue, responsible for many diseases and high death rates worldwide, is steadily increasing. The Sustainable Development Goals' target 3.4 advocates lessening premature deaths from NCDs by one-third through inhibition and healing by the year 2030. To accomplish this lofty objective, medical informatics has arisen as an impactful approach with vast capability to overhaul methods used in preventing, treating, and managing NCDs. While cardiovascular ailments, cancer diabetes control the majority of chronic illnesses that affect individuals throughout their lives—these persistent conditions demand efficient healthcare plans due to their lengthy nature. Navigating this challenge using health informatics; the application of data science alongside IT within healthcare may be instrumental thanks to its versatile role.

Health informatics greatly assists in achieving Target 3.4 by offering early detection and analyzing risk factors. The application of machine learning algorithms and data analytics helps to examine extensive datasets such as electronic health records or lifestyle patterns, which can single out individuals who are more prone to contracting NCDs. By identifying these elements before time, healthcare practitioners can design proactive strategies including changes in the patient's lifestyle that help avert the likelihood of an onset of NCDs. In addition, personalizing treatment plans for those already suffering from NCD becomes easier via technology-driven approach intrinsic to health informatics processing important information like genetic predispositions on an individual basis so customized regimen is offered according to each patient's specific needs.

Primarily a subset of health informatics, telemedicine improves the availability and convenience of medical services. This is especially beneficial for individuals dealing with Non-communicable diseases (NCDs) who might struggle with transportation or reside in distant locations. Telehealth sessions offer patients the opportunity to gain professional healthcare advice remotely, keep track of their ailments, and receive consistent care without necessitating regular face-to-face appointments; this factor contributes significantly toward NCD maintenance because persistent follow-ups as well as scrutiny are key elements essential for disease management by slowing its progression hence minimizing complications. Moreover, an additional role played by health informatics lies within devising strategies at population level aimed at prevention along with control efforts concerning NCDs. The accumulation followed by investigation involving data related to prevalence frequency linked to these conditions including risk factors attached plus usage patterns associated with healthcare gives policymakers a clear view aiding them while taking steps also facilitating efficient distribution regarding resources pivotal during the identification process concerned high risk communities enhanced plans targeting certain demographics vastly contribute reducing adversity due non-contagious specific neighborhoods is central properly correlating broader aspiration ensuring none being left behind under sustainable development goals ambit In addition scientific research which entails scrutinizing huge amount information uncover existing trending patterns unveil undiscovered threats alongside application new measures progressing exponentially resulting faster invention procedure medicine model area coupled preventive solutions.

Despite its potential, health informatics in the scope of Target 3.4 is faced with various hurdles that require attention. These include technical roadblocks like maintaining data privacy and security, dealing with interoperability problems, and the necessity for unified guidelines on sharing information. In addition to this, there's an essential need for spreading digital knowledge and reducing digital disparity so as to ensure less privileged groups also reap benefits from health informatics.

**Substance Abuse (Target 3.5)**

The issue of substance misuse has global health implications that significantly impact public wellbeing, which makes it a focal point in the Sustainable Development Goal—"Good Health and Well-being" (SDG 3). SDG 3's Target 3.5 pertains specifically to the detrimental use of alcohol along with other substances, advocating for robust prevention treatments and their enhancement. In this context, employing health informatics offers an optimistic pathway toward dealing with substance abuse more effectively thereby improving treatment results while reducing corresponding social-health burdens. Informatic healthcare can aid methodical acquisition as well as constituents statistically collecting information about demographic specifications and types of abused drugs alongside consequential impacts or therapy repercussions bettering outcome-records. It all feeds into being run on A digital-like databank approach loop hence offering healthcare executors and scientific scholars valuable insights allowing them to mold strategies regarding both illicit-drug curbing methods and rehabilitation schemes exactly according to population/individual's specific necessities.

Indeed, health informatics can play a pivotal role in the initial recognition and intervention of problems related to substance abuse. Digital patient records or EHRs along with other systems within health informatics have capabilities that include screening abilities for quickly identifying individuals either at risk of falling into addiction issues or those already under its impact (Jha et al. 2009). Speedy identification allows for prompt actions such as therapeutic counseling, directing them toward corrective programs. Tele-healthcare and telemedicine which are crucial elements under the umbrella term "health informatics", present opportunities to expand reachability areas including underserved regions allowing efficient treatment services accessible centered around dependency management on harmful substances. People looking out for help regarding narcotics control could seek therapy-related assistance while receiving support through digital healthcare platforms easily eliminating hurdles like geographical distances and commuting limitations.

Integrating predictive analytics and machine learning into health informatics can play a significant part in tackling substance abuse. These modern techniques assist in identifying addiction patterns, spotting populations at increased risk, and refining treatment protocols. For instance, datasets can be analyzed using machine learning to recognize factors contributing to recurrence hence allowing personalized interventions that are more efficient. This strategy aligns with SDG 3's pledge of advancing mental wellness while minimizing substance misuse impacts. On another note, one major advantage of health informatics is its potential contribution toward better adherence and monitoring during treatments for substance abuse cases. Real-time support systems like mobile apps or wearable devices could merge within these recovery plans providing meaningful reminders as well as tracking advancements (Zheng et al., 2014).

Moreover, the utilization of health informatics plays a vital role in combating global prescription drug misuse. With its ability to track prescribing habits and pinpoint possible indicators of abuse or excessive use. These systems aid healthcare professionals and policy enforcers in averting the illicit redirection of prescribed medicines.

**Universal Health Coverage (Target 3.8)**

Encompassing Universal Health Coverage (UHC) is an essential part of Goal 3 in Sustainable Development—"Good Health and Well-being", pledging to make sure all individuals have availability to quality health services without the fear of facing financial struggles. UHC realization not only stands for social justice but spills over into several other areas, such as decreasing poverty levels, promoting better education systems, gender parity issues, and forming sustainable townships. Amid striving toward this objective, integrating Healthcare IT makes a powerful breakthrough by suggesting innovative methods that help overcome hurdles, encourage accessibility in healthcare facilities while enhancing care standards. Handling healthcare data efficiently proves one impediment when making headway with UHC which can be tackled with health informatics offering potent outcomes through transforming patient records digitally along ensuring their secure retention on digital platforms giving way for steady information circulation across diverse medical environments. With Electronic Medical Recordings(EHRs) coupled with HIS coming onto stages, clinicians find it easier to access individual medical history thereby coordinating effective clinical management leading ultimately lower DX (medical diagnosis code) rates (Jha et al. 2009).

By harnessing the power of health informatics, telemedicine, and telehealth become vital tools to broaden healthcare reach in both underserved areas or locations difficult to access. This digital link between patients and medical professionals bypasses physical location restrictions. Importantly, this plays a major role in realizing Universal Health Care (UHC), offering services even to communities traditionally overlooked due to their remote geographical conditions. Moreover, as a central component for boosting evidence-driven decision-making strategy formation plus resource distribution—two critical aspects indispensable for UHC- health informatics holds significant value alongside technologies such as data analytics or machine learning which permit providers amid the field of medicine together with policy designers to examine large scale volumes. Efficient allocation is made possible through the identification of patterns relating to prevalent ailments & managing sudden disease outbreaks; importantly recognizing the exact placement where providing essential resources has been most required enabling proactive addressing against potential situations that can be prevented ahead therefore helping toward attaining the Targeted objective—Universal Healthcare.

In a UHC setting, health informatics has the potential to mitigate issues of healthcare access and quality disparity. Digital care facilitators like mobile applications and wearable tech devices give individuals governance over their own health status. They can keep an eye on ongoing illnesses, learn from available online health resources, and make informed choices facilitating their wellness journey (Zheng et al., 2014). This augments general wellbeing while engaging all people irrespective of wealth or social standing actively in managing their healthy living criteria. Additionally, adopting such digital strategies into healthcare creates heightened procedural efficiency which becomes significant for realizing UHC goals at large scales. Decisions backed by data processing allow medical practitioners to organize tasks better resulting in decreased administrative hassle with optimal resource use arrangements possible.

**Health Workforce (Target 3.C)**

This goal aims for universal good health and wellbeing. Within this objective, Target 3. c explicitly encourages a substantial augmentation in amassing, developing training programs and retaining healthcare workers especially from low to middle-income nations. In view of these needs creation and inclusion of health informatics allows an array with vast possibilities that can revolutionize how we tackle issues linked to the SDGs as well as those surrounding global professions allied with public wellbeing. Constant learning not only leads to better productive interaction treatments, but also keeps skilled manpower ready to tackle new challenges and providing qualitative patient services in the general parts world aligns well SDG 3's mission to uphold noble cause.

Maintaining a robust staffing of healthcare professionals, particularly in areas with limited resources, presents an ongoing struggle. Elevated rates of staff loss can compromise the consistency and excellence of care provided. The field of health informatics offers solutions to this problem by enhancing working environments and job contentment. For example, digital medical records (EHRs) lessen administrative loads on medical personnel so they can concentrate more deeply on patient wellbeing (Jha et al. 2009). E-services for health not only offer increased flexibility but also reduce professional exhaustion which contributes heavily toward employee attrition rate These workforce retention strategies play crucial roles in ensuring the continuity availability of skilled health practitioners as underscored in SDG 3.c. Additionally another important contribution that Health Informatics brings especially within the context SDG 3c is managing workforces, effectively downsizing workload shifts and evenly distributing all workers. Often rural regions do experience shortages if the availability of workforce is irregular. Digital services like remote consultation narrow the gap and allow specialists to transfer their knowledge and expertise directly to rural distant locations.

Moreover, applying health informatics can boost coordination in medical care across various healthcare professionals. An integrated team effort is a prerequisite to deliver all-embracing patient care. EHRs serve as an example—they permit specialists from numerous fields to view and modify patient's records which advances smooth and coordinated treatment delivery aligning with SDG 3's aim of minimizing fragmentation in healthcare. Data-driven planning also forms an essential aspect of health informatics that aids in effective management of the workforce involved in public service. This allows policymakers along with health institutions by evaluating pertinent data on healthcare, identify regions facing shortage or surplusage practitioners for fitting allocation resources ensuring efficient strategic workforce planning thereby ensuring needed personnel are positioned wherever required fulfilling SDG 3's objective promoting fair provision toward services related to healthcare.

**Research and Development for Vaccines and Medicines (Target 3.D)**

Putting health informatics to use in researching and developing Vaccines and Medicines (Goal 3.d) is a key part of the overarching journey toward attaining Sustainable Development Goals (SDGs). This goal underlines the importance of funneling investments into research as well as innovation for creating vaccines plus medications aimed at illnesses that impact developing regions adversely. In such situations, infusing health informatics turns out crucial for refining R&D processes, enhancing collaborative efforts & data interpreting capacities thereby hastening distribution of essential medical care solutions. Health Informatics involves applying Information Technology besides Data Analytics within healthcare with an array of advantages on offer especially in connection with pharmacological R&D activities. One major benefit would be an ability to collect information and then analyze it further. Electronic Health Records or EHRs integration alongside Clinical Trial Data Management Systems also includes sharing patient info combined together means researchers can have access to abundant clinical records gathered previously hence enabling them to locate possible targets suitable enough for progressing pharma reviews while gaining better understanding about diseases. (Chib et al., 2015).

Health informatics plays a pivotal role in establishing centralized banks and repositories for clinical data, further aiding biomedical research. This amalgamation of information sourced from various areas allows researchers to perform more thorough analyses—an aspect beneficial particularly when dealing with infectious diseases due to the rapid accessibility it provides to epidemiological data, genomic specifics, and medical outcomes which can vastly expedite treatments and developing vaccines. Algorithms powered by AI have proven efficacy in navigating through large amounts of datasets to distinguish potential drug candidates, forecast results associated with clinical trials, while enhancing processes involved in discovering new drugs. Machine learning models are instrumental for scientists trying to decipher genetic factors causing disease along with understanding molecular components critical toward designing targeted therapies or developing vaccinations efficiently. For example, during COVID-19 pandemic breakout, AI enabled expeditious developments related to crafting effective vaccine formulations owing to its ability to predict potent prospective vaccine possibilities thereby catalyzing their trial phases (Reeves et al. 2020).

Effective R&D relies heavily on researchers, agencies, and pharmaceutical corporations working together and sharing data. Platforms for health informatics serve as crucial tools in facilitating this global exchange of secure information across national boundaries. They boost multinational clinical trial's while also serving to disperse research findings or foster joint efforts for vaccine/drug development projects. With digital health platforms and telemedicine at their disposal, researchers can readily connect with participants from various geographical regions including developing countries—granting them capacity to perform large-scale experiments globally. Health informatics has a profound impact on administering clinical trials too—simplifying participant enrollment procedures, monitoring trial progress effectively, optimizing patient's health tracking system making it more efficient than ever before! Through wearable tech devices & telemedicine allow healthcare professionals to maintain real-time updates enabling robust insights generation about patient's wellbeing even when they're situated far off distances especially where healthcare services are scarce. Moreover, Health Informatic systems assist vital functions such as preventing adverse event occurrences during experimental phases of medicine discovery (Zheng et al. 2014).

### *4.4.1 Impact of Health Informatics in Achieving SDGs Goals*

Health informatics, an interdisciplinary field that combines IT and data science with healthcare, carries significant advantages toward progressing the Sustainable Development Goals (SDGs), especially SDG 3—Good Health and Well-being. Its influence over these targets is diverse and can notably amplify health services delivery, reachability, and effectiveness. Health informatics advances both access to medical care along with its quality through improved patient records management systems. The use of Digitized Health Records allows secure housing as well as accessibility of all necessary patient's details by healthcare practitioners. Moreover, organizing digitized disease revelations & epidemic control could be made more feasible via incorporating digital principles comprehensively into public health protocols alongside real-time response mechanisms available due to the collection of data on prevailing disease patterns essential from the perspective of achieving immunization goals set under controlling infectious diseases. (Yogesh et al. 2022).

Health informatics can greatly extend the reach of health services including telemedicine and telehealth, particularly to remote and underserved places. This enables patients to have consultations with healthcare professionals through mediums like mobile apps or videoconferencing thereby reducing geographical distance boundaries while also enhancing access to healthcare. Such features especially benefit expectant mothers in isolated locations by allowing them easy access toward prenatal care along with expert advice corresponding well with SDG 3's aim of cutting down maternal mortality (Target 3.1). Further advancements include using predictive data analytics and machine learning as main parts within health informatics that hold potential for a tremendous transformation in the field of medicine. They help

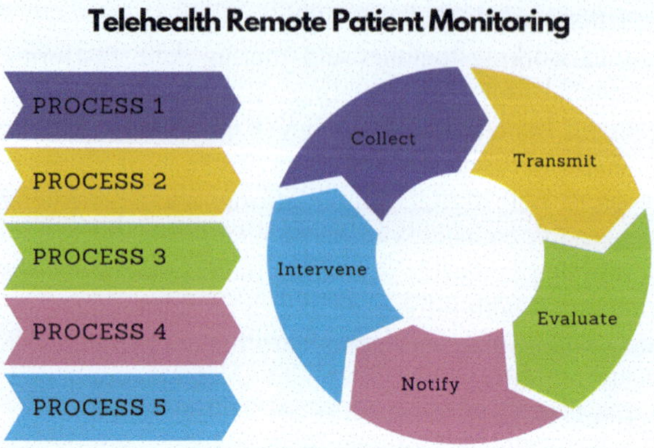


**Fig. 4.3** Processes involved in telehealth remote patient monitoring

foresee disease occurrences, reveal susceptible population groups plus tailor treatments on individual basis suiting accurately toward achieving targets set under SDG goals which include decreasing non-communicable diseases (Target 3.4) as well as promoting mental wellness (Target 3.4). Aiding addiction treatment comes into the picture when tackling substance misuse (Target 3:5) which provides medical practitioners insights upon patient behaviors, and adherence levels relating to therapeutic interventions (Fig. 4.3).

Health informatics serves as a valuable tool in aiding research and development initiatives (Target 3.d) by simplifying the processes involved in gathering data for clinical trials, drug discovery, and genomics studies. This can potentially speed up the creation of medicines and vaccines designed to combat diseases targeted under SDG 3—communicable or non-communicable alike. Fostering patient participation isn't just crucial but is also one of health informatics' key facets perfectly aligned with the principles adopted by SDG 3—no individual should be left behind when it comes to healthcare options available. With innovations like mobile applications and wearable technology devices at their disposal, people are now more capable than ever before (Zheng et al., 2014).

### *4.4.2 Case Study: Rwanda's Health Information System*

The Health Information System of Rwanda (R-HIS) beautifully demonstrates the power and potential of health informatics in reinventing healthcare delivery frameworks, thereby substantially contributing to Sustainable Development Goal 3 "Good Health & Well-being for All" or SDG 3. This forward-thinking project reflects Rwanda's dedication to increasing accessibility in healthcare, bettering health outcomes while lessening disparities by strategically incorporating technology with data-centric decision-making processes. R-HIS is an inclusive digital platform catering toward patient record management, providing a medium for unfettered exchange among medical professionals as well as supporting telemedicine solutions. The integrated utilization of such technological advancements plays a crucial part in combatting some prominent challenges faced by Rwandan Healthcare sectors resonating precisely with goals set forth under SDG 3 (Nishimwe et al. 2016b).

Moreover, the abilities of this system stretch to boosting disease tracking and managing outbreaks—an essential part of SDG 3's pledge against infectious diseases (Target 3.3). With provisions for real-time data gathering, examination, and broadcasting, R-HIS equips public health bodies in controlling infection spread effectively (Eysenbach 2009). Amid COVID-19 outbreak times, it crucially aided coordination efforts while plotting virus impacts; thereby guiding swift governmental measures based on solid data toward lessening severe implications brought by the virus (Reeves et al. 2020). Health services have become more accessible through telemedicine integration into R-HIS especially within remote regions with not enough healthcare facilities due to geographical hurdles—aligning perfectly with universal medical coverage missions under SDG goals (Target 3.8). Now patients interact directly with doctors online or via mobile applications thus bridging distance realities between them ensuring no individual falls behind in maintaining a healthy living standard. The impact leveraged further owing to the inclusion of powerful analytics tools coupled with artificial intelligence. Algorithms were significant contributors enabling us to predict potential disease epidemics, also targeting vulnerable populace alongside framing care programmes (Nishimwe et al. 2016b).

Indeed, the R-HIS illustrates how health informatics can fuel research and progress (Target 3.d). The system simplifies data compilation and investigation in areas such as clinical trials, pharmaceutical discoveries, and gene studies. This not only accelerates vaccine development for SDG 3 specified diseases but also serves broader purposes by setting an example of effective deployment of health informatics by other nations looking to advance their healthcare via this method. Nexus points like patient engagement & empowerment entwined within R-HIS align with SDG's stand on inclusive growth. With embedded digital tools ranging from mobile applications to wearable tech devices in its framework; it offers control back to individuals over personal medical situations empowering them toward individual wellbeing goals (Zheng et al. 2014).

## 4.5 Challenges and Barriers Implementing Health Informatics in the Context of SDGs

Health informatics, lying at the crossroads of medical care and IT, holds a significant position in progressing healthcare services, ameliorating patient results, and facilitating the realization of Sustainable Development Goals (SDGs). These worldwide objectives span numerous facets of sustainable development like SDG 3 which majors on guaranteeing good health & wellbeing for everyone. Despite having substantial potential capabilities—there exist several difficulties in initiatives linked with SDGs' Healthcare. Some of them are:-

### *4.5.1 Data Privacy and Security*

Incorporating health informatics into healthcare systems is crucial for the progress of Sustainable Development Goals (SDGs), especially SDG 3, which endeavors to provide everyone with good health and wellbeing. Health informatics employs data and technology as tools that enhance accessibility, quality, and outcomes in healthcare. However, managing digitized sensitive patient information raises intricate issues regarding its privacy and security. Addressing these difficulties diligently will ensure optimal utilization of health informatics toward accomplishing various assignments related to SDG-oriented medical initiatives. Patient-related data confidentiality particularly takes priority in discussions on challenges faced by health informatics. Electronic Health Records(EHRs) along with other digital platforms contain rich personal medical facts like prior treatments received or diagnosis results which necessitate utmost prudence while handling (Jha et al. 2009). Confidence-building efforts are needed so that patients can be sure that no sensitive data disclosure will happen without their approval. The after-effects following any violation against such private info could be disastrous causing people's faith erosion relating to institutions providing care besides making them hesitant about sharing vital details linked to their wellbeing. With regard to cyber threats, the evolution from paper-based storage models toward more compact, digital ones has increased potential targets for cyber criminals. Owing mostly due to large scale compromising material availability, institutions working around medicinal services are highly prone to being targeted by malicious individuals. Since circumstances accompanying such violations not only facilitate breaching individual rights but also cause detrimental consequences involving entire system operations that facilitate this, tackling this problem should remain our primary concern.

Healthcare data sharing presents sizable hurdles due to its intrinsic complexity. Accomplishing the goals of SDG 3 essentially requires exchanging patient details across medical providers and establishments, thereby promoting harmonized care delivery and guaranteeing universal health benefits. Striking a balance between healthcare professionals accessing this information while maintaining patient confidentiality isn't an easy feat though. To ensure secure transmission of delicate health records only with approved personnel, it's crucial for healthcare bodies to enforce reliable governance structures related to handling data alongside access control checks and encryption safeguards. The issue gets exacerbated when you factor in interoperability challenges linked with privacy concerns over shared health-related information or security risks present within health informatics. Further compounding matters is the complicated nature of the legal framework that directs how electronic medical record-keeping, also known as "E-health" works internationally varying dramatically from nation to nation. Establishing clear-cut regulations is a significant roadblock

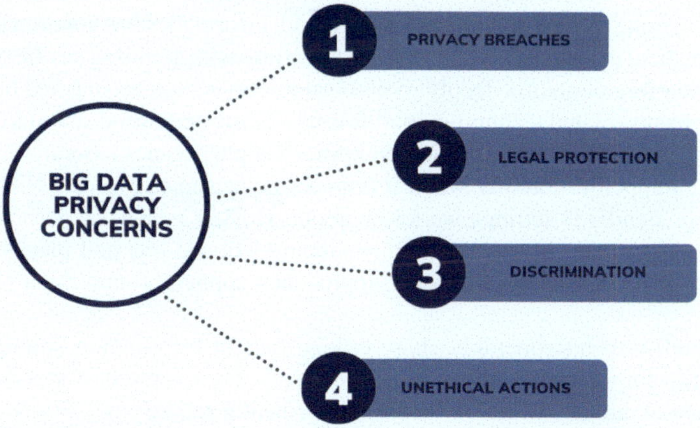


**Fig. 4.4** Big data privacy concerns due to implementation of health informatics

that involves drafting legislative norms centered around topics such as assurance of user-data safety measures maintained throughout the process of deploying telehealth solutions successfully to different regions (Fig. 4.4).

Tackling the issues pertaining to data privacy and security in health informatics demands a comprehensive strategy. Medical institutions have to put cybersecurity safeguards at the forefront, involving encryption, intrusion detection apparatuses, and periodic safety inspections. It's crucial for them also to allot resources toward staff education about best practices concerning data protection as well as risks tied to information leaks. Moreover, healthcare IT systems should adopt privacy-centered design principles that include safeguarding measures throughout all system development phases. Furthermore, adhering strictly to regulatory laws is of prime importance too; medical organizations need continuous awareness of changing legal framework ensuring they align their health-related Information Technology systems along local or global directives. Observance of standards like Health Insurance Portability and Accountability Act (HIPAA) within United States' jurisdiction or European Union's General Data Protection Regulation (GDPR) play an indispensable role both legally & ethically.

### *4.5.2 Interoperability*

The application of health informatics structures within the wider ambit of Sustainable Development Goals (SDGs) indicates a revolutionary chance to enhance healthcare results and push forward the international wellbeing agenda. Nonetheless, meeting interoperability remains as one pressing complication for such schemes in their quest. The term "interoperability" applies to different health systems, technologies, and data collection abilities to exchange information smoothly while putting it into use; this is a crucial footstone toward maximizing what health informatics has got offering SDG-involved medical initiatives. At center stage wrestling with attaining an effective level could be seen fragmentation deeply rooted inside varying interconnected healthcare networks plus its related technology builds up that surround us today—starting from hospitals all way down public protection bodies via pharmacies laboratories clinics, etc., each equipped differently when it comes to electronic records(EHR), diverse devices care standards settings we know about. Having numerous parts unable to communicate properly often results in producing isolated bits and pieces, thereby resulting in scattered patient servicing approaches applied around them (Jha et al. 2009).

The inability to interoperably share crucial patient information among healthcare facilities impedes smooth healthcare delivery. For example, if a person seeks medical care at another clinic or hospital, their health records may not be immediately accessible by the new healthcare provider causing duplicative testing procedures, incomplete clinical histories, and treatment deferrals. These inefficiencies not only risk patients' wellbeing but also escalate expenses in the healthcare sector which pertains directly to Sustainable Development Goal (SDG) 3's commitment to realizing universal availability of health services (3.8). Furthermore, true disease management effectiveness including detection and control activities as detailed under SDG 3's requirement for dealing with contagious diseases, predominantly depends upon systems capabilities of efficient data exchange.

Legal and regulatory frameworks amplify the interoperability puzzle. Data privacy laws, such as America's Health Insurance Portability and Accountability Act (HIPAA) or Europe's General Data Protection Regulation (GDPR), heighten complexity in endeavors of data sharing. While maneuvering through these regulations, healthcare organizations must retain patient-focussed confidentiality whilst ensuring their information security. The challenge of implementing health informatics with seamless connectivity requires a diverse approach. Healthcare stakeholders need to embrace standardizing formats for database systemization that promote uninterrupted knowledge exchange throughout systems. In addition, it is critical to invest in infrastructure allowing safe information transfers among providers which enhances both local efficiency & international compatibility facilitating networked communication effectively without significant issues like safety-privacy, etc.

### *4.5.3 Healthcare Infrastructure and Resources*

Integrating health informatics into healthcare systems is a key move toward reaching the Sustainable Development Goals (SDGs), specifically SDG 3, aimed at promoting global health and wellness. Yet, implementing these technology initiatives correctly will lead to many obstacles due to limitations in available infrastructure and resources for healthcare. These concerns cover different areas such as technical facilities, workforce capacity or financial burdens that can shape the efficacy of proposed measures.

The most pressing issue is the technological infrastructure's quality and accessibility that present significant barriers mainly in rural underdeveloped regions. On one hand, there are robust digital frameworks needed by Health Informatics like constant internet availability, efficient hardware/software solutions, and trustworthy data security services. Where foundational tech amenities fall short creating room for comprehensive core Technologies including telemedicine support/online patient records presents immense challenges. Moreover, the gap between those with access to modern digital tools used in delivering medical care called "The Digital Divide" further complicates solving matters on Healthcare provision/resources. (Yamin et al. 2011).

The challenge involves a crucial aspect related to human resources. The industry of health informatics requires an expertise-laden workforce adept at handling digital healthcare systems, deciphering health data, and applying technology to patient care. An integral part is the training of medical professionals and technicians for the effective deployment of tools provided by health informatics. A probable scarcity might exist in some regions when it comes to IT specialists catering specifically to healthcare needs which would be another stumbling block while establishing or maintaining these sophisticated systems. Financial constraints hover like inhibiting overheads too—setting up, executing, and preserving such a comprehensive system calls for considerable expenditure that consequently compels governments and institutions inclined toward caring for people's healthy lives into allocating hefty portions from their finances toward acquiring technologically advanced equipment required.

The constraints tied to healthcare facilities and resources also involve data safety concerns. Lacking adequate resources, health systems may become susceptible to cyber threats that could compromise patient information security. Amplifying the protection of medical informatics frameworks requires tangible investments. To tackle such problems while endeavoring toward Sustainable Development Goals (SDGs) implementation via health informatics, a multifaceted strategy is essential. Both local governments along with international entities should enhance priority levels for investing in general healthcare infrastructures including reliable power sources and state-of-art technologies. Concurrent endeavors must be made toward creating an experienced IT workforce within sectors dealing with public wellbeing through targeted education sessions. Novel funding methods coupled with private partnerships can alleviate fiscal issues by encouraging Health Informatic initiatives.

### *4.5.4 Cost and Resource Constraints*

In relation to the Sustainable Development Goals (SDGs), applying health informatics is confronted with a major hurdle of financial and resource limitations. Although health informatics has a great capacity for enhancing healthcare availability, standards, and results, these undertakings come attached with significant costs and demands on resources. Establishing a well-functioning health informatics system requires substantial initial capital investment into aspects such as hardware requirements, software components, network foundations alongside qualified personnel's wages— all contributing toward pushing up the total cost involved. These expenses can be notably crippling especially for developing economies where funding options for their healthcare systems may already be limited. (Chib et al. 2015).

Another aspect of the financial issue pertains to workforce elements. The importance of proficient support, inclusive of health IT specialists and data experts, can't be overemphasized for triumphantly ushering in and maintaining health-related information systems. Acquisition and retention strategies involving these professionals carry hefty price tags. Finding top-notch tech talent becomes exponentially difficult within financially limited environments as competitive remuneration packages could present a hurdle for healthcare entities that are economically struggling. Then there's the indispensable role that continuous education plays during system installation procedures dealing with medical informatics: it's compulsory to train relevant staff on effective usage techniques while regularly updating their know-how corresponding to recent developments in the medical informatics field. Healthcare infrastructure needs continual adaptation keeping pace with patient care necessities broadening, leading us into another complication entailing economical considerations regarding scaling up such frameworks.

Limitations on financial means and resources in the enactment of health informatics are capable of causing indirect repercussions. Take for instance, when infirmary expenditures compete with economic requirements from health informatics forcing them to deviate resources from other essential healthcare needs.

### *4.5.5 Sustainability*

The inclusion of health informatics in medical systems offers grand potential for advancing the Sustainable Development Goals (SDGs). Utilizing data, technology, and innovation can heighten access to healthcare services, amplify patient outcomes, and consolidate delivery. However, amidst this hopeful shift, it's valuable to identify and solve sustainability issues that emerge when applying health informatics within SDGs range. The necessity for extended financial commitment is a primary challenge in ensuring longevity while instituting health informatics. Establishing these systems requires substantial fiscal contribution which may offer beneficial returns through improved quality care and efficiency; but at times compete with urgent healthcare demands or priorities. Policymakers alongside other key stakeholders must maneuver around circumspectly balancing between financing initiatives and essential programs related toward overall wellness improvement. There are several constraints regarding resource allocation particularly among low to middle income regions posing severe sustainability obstacles due largely on insufficient tech infrastructure which is required such as lack of reliable internet connectivity.

The flawless transfer of medical data across many systems and providers, identified as interoperability, is yet another critical sustainability issue. It's common for healthcare networks around the globe to lean on an array of tech applications, databases and set standards that can be varied in nature. This unevenness restricts seamless data exchange obstructing efficient care coordination while constraining full utilization capabilities offered by health informatics technologies. An inherent reluctance toward shifting from well-established traditions signifies one major hurdle hampering progress when implementing changes within any system including healthcare. Legal frameworks carry forward further complexities as specific rules & regulations differ between countries.

The perpetuation of health information systems projects has a direct connection with capacity enhancement and personal growth. Talented individuals are required to control, run, and handle concerns involving these systems in healthcare informatics. The creation and retention of such manpower can pose its own set of challenges particularly in areas where access to learning opportunities is scarce. To guarantee the constant availability of competent staff while providing continuous training becomes vital for the sustained achievement they aim at within this field. Moreover, technology becoming quickly out-of-date presents an impending challenge toward sustainability. Due to fast-paced improvements in medical science tech tools, certain solutions or methods will become old-fashioned which causes them not to be effective enough anymore.

## References

Ahmad FN: 2023: Ahmad, T., Gupta, S.K., Singh, S.K., Meraj, G., Kumar, P. and Kanga, S. et al (2023) Unveiling Nature's Resilience: Exploring Vegetation Dynamics during the COVID-19 Era in Jharkhand, India, with the Google Earth Engine. Climate 11(9):187

Borycki EM, Cummings E, Kushniruk AW, Saranto K (2017) Integrating health information technology safety into nursing informatics competencies. Stud Health Technol Inform. 232:222–228 PMID: 28106601

Chib A, van Velthoven MH, Car J (2015) MHealth adoption in low-resource environments: a review of the use of mobile healthcare in developing countries. J Health Commun 20(1):4–34. https://doi.org/10.1080/10810730.2013.864735. Epub 2014 Mar 27 PMID: 24673171

Eysenbach G (2009) Infodemiology and Infoveillance: Framework for an emerging set of public health informatics methods to analyze search, communication and publication behavior on the internet. J Med Internet Res 11(1):e11. https://doi.org/10.2196/jmir.1157

Jha AK, DesRoches CM, Campbell EG, Donelan K, Rao SR, Ferris TG, Shields AE, Rosenbaum S, Blumenthal D (2009) Use of electronic health records in U.S. hospitals. N Engl J Med, 360(16), 1628–1638. https://doi.org/10.1056/nejmsa0900592

Mair FS, May C, O'Donnell C, Finch T, Sullivan F, Murray E (2012) Factors that promote or inhibit the implementation of e-health systems: an explanatory systematic review. Bull World Health Organ 90(5):357–364. https://doi.org/10.2471/BLT.11.099424.PMID:22589569;PMCID:PMC3341692

Meraj FN: 2021: Meraj, G., Kanga, S., Kranjčić, N., Đurin, B. and Singh, S.K. et al (2021) Role of natural capital economics for sustainable management of earth resources. Earth 2(3):622–634

Nishimwe A, Mbarushimana V, Ngenzi JL, Nyssen M (2016) Assessment of health informatics competencies in undergraduate training of healthcare professionals in Rwanda. Rwanda J 3(1):36. https://doi.org/10.4314/rj.v3i1.6f

Reeves JJ, Hollandsworth HM, Torriani FJ, Taplitz R, Abeles SR, Tai-Seale M, Millen M, Clay B, Longhurst C (2020) Rapid response to COVID-19: health informatics support for outbreak management in an academic health system. J Am Med Inform Assoc 27(6):853–859. https://doi.org/10.1093/jamia/ocaa037

Robert KW, Parris TM, Leiserowitz A (2005) What is sustainable development? goals, indicators, values, and practice. Environment 47(3):8–21. https://doi.org/10.1080/00139157.2005.10524444

Sachs JD (2012) From millennium development goals to sustainable development goals. The Lancet, 379(9832), 2206–2211. https://doi.org/10.1016/s0140-6736(12)60685-0

Singh FN: 2021: Singh, S., Singh, H., Sharma, V., Shrivastava, V., Kumar, P., Kanga, S., Sahu, N., Meraj, G., Farooq, M. and Singh, S.K. et al (2021) Impact of forest fires on air quality in Wolgan valley, New South Wales, Australia—A mapping and monitoring study using Google Earth Engine. Forests 13(1):4

Singh SK, Meraj G, Mondal N, Bera A, Verma MK, Tomar JS, Kanga S (2021) Assessment of seasonal vegetation dynamics over parts of thar desert using geospatial techniques. ANGRAU, p.105.

Singh FN, Singh H, Meraj G, Singh S, Shrivastava V, Sharma V, Farooq M, Kanga S, Singh SK, Kumar P et al. (2022) Status of Air Pollution during COVID-19-Induced Lockdown in Delhi. India. Atmosphere 13(12):2090

Yamin CK, Emani S, Williams DH, Lipsitz SR, Karson AS, Wald JS, Bates DW (2011) The digital divide in adoption and use of a personal health record. Arch Intern Med 171(6):568–574. https://doi.org/10.1001/archinternmed.2011.34. PMID: 21444847

Yogesh MJ, Karthikeyan J (2022) Health informatics: engaging modern healthcare units: a brief overview. Front Public Health 29(10):854688. https://doi.org/10.3389/fpubh.2022.854688. PMID:35570921;PMCID:PMC9099090

Zheng Y, Ding X, Poon C, Lo B, Zhang H, Zhou X, Yang G, Zhao N, Zhang Y (2014) Unobtrusive sensing and wearable devices for health informatics. IEEE Trans Biomed Eng 61(5):1538–1554. https://doi.org/10.1109/tbme.2014.2309951

# Chapter 5
# Integrating Ethics and Social Responsibility in Health Informatics

**Chandni Kirpalani and Deepak Kumar**

## 5.1 Introduction

### *5.1.1 Definition of Health Informatics*

Health informatics is an interdisciplinary domain that integrates healthcare, information technology and data science in order to enhance the efficiency and effectiveness of healthcare service provision. The process encompasses the gathering, retention, examination and comprehension of health-related information with the aim of facilitating well-informed decision-making, enhancing the provision of care to patients and ultimately improving overall health outcomes. Health informatics is a field that involves the utilization of electronic health records, medical imaging systems, telemedicine, health monitoring devices, data analytics and artificial intelligence (Ahmad et al. 2023). The primary objective of this initiative is to leverage the capabilities of technology and information systems in order to optimize healthcare procedures, facilitate individualized medical treatments and foster progress in healthcare research and implementation (Sharma et al. 2023).

C. Kirpalani
Department of Chemistry and Environmental Science, Poornima University, Jaipur, India

D. Kumar (✉)
Department of Mechanical Engineering, Swami Keshvanand Institute of Technology, Management and Gramothan, Jaipur, India
e-mail: deepak.kumar@skit.ac.in

© The Author(s), under exclusive license to Springer Nature Singapore Pte Ltd. 2024 
G. Tripathi et al. (eds.), *Sustainability and Health Informatics*, Advances in Geographical and Environmental Sciences,
https://doi.org/10.1007/978-981-97-6706-9_5

### 5.1.2 Significance of Ethics and Social Responsibility in Health Informatics Along with Literature Review

Ethics and social responsibility are integral components within the realm of health informatics, as they serve to uphold moral principles, prioritise the well-being of patients and address the societal requirements associated with technology and data-driven healthcare practices. The importance of ethics and social responsibility in the field of health informatics can be comprehended through the following avenues:

**Patient Welfare**: The ethical considerations and social responsibilities within the field of health informatics place a primary emphasis on safeguarding the welfare and security of patients. The primary objective of these individuals is to establish and execute technology and data management strategies that safeguard patient confidentiality, advance patient self-governance and mitigate potential adverse consequences. Healthcare professionals and informatics specialists can prioritize patient welfare by making decisions and developing systems that adhere to ethical principles such as beneficence and non-maleficence.

**Data Privacy and Security**: Health informatics encompasses the systematic gathering, retention and examination of confidential patient data. The inclusion of ethical considerations is of paramount importance in safeguarding patient privacy and upholding the confidentiality of their health data. In order to reduce the likelihood of unauthorized access, data breaches and inappropriate use of patient information, experts working in health informatics are needed to comply with privacy legislation, develop robust security processes and handle data with the utmost caution. This is done to protect patient information.

**Informed Consent and Autonomy**: Patients have the autonomy and rights to make well-informed decisions regarding their healthcare and the idea of informed consent is an essential ethical rule that supports these rights and respects the autonomy of patients. In the field of health informatics, it is imperative to guarantee that patients are sufficiently educated regarding the utilization, sharing and safeguarding of their health data. Ethical practises encompass the imperative of acquiring informed consent from patients prior to the collection and utilization of their data, ensuring transparency in the handling of such data and affording patients the autonomy to exert control over their personal information.

**Algorithmic Bias and Fairness**: Health informatics is dependent on algorithms and decision-making systems that have the potential to introduce biases and perpetuate discrimination against specific populations. In order to ensure equitable healthcare outcomes, it is imperative to address algorithmic biases, as dictated by ethical considerations. Health informatics professionals should possess knowledge about the potential biases present in data, algorithms and machine learning models. They should actively strive to minimize these biases in order to prevent the reinforcement of existing health disparities and to guarantee equitable access to healthcare for all individuals.

**Governance and Accountability**: The implementation of ethical principles in the field of health informatics necessitates the establishment of governance frameworks and mechanisms that guarantee accountability, transparency and responsible utilization of technology and data. It is imperative for healthcare organizations and informatics professionals to embrace ethical frameworks, codes of conduct and guidelines in order to effectively navigate their decision-making processes and actions. Governance mechanisms serve to facilitate and uphold ethical conduct, facilitate the resolution of ethical dilemmas and ensure the enforcement of accountability for both individuals and organizations.

**Social Justice and Equity**: The concept of social responsibility in the field of health informatics underscores the significance of tackling health disparities, considering social determinants of health and advocating for fair and equal access to healthcare services. Ethical considerations encompass the utilization of health informatics as a means to address the digital divide, extend services to marginalized populations and guarantee equitable access to technological advancements for individuals regardless of their socio-economic status, geographic location, or demographic characteristics.

**Collaboration and Interdisciplinary Approach**: The implementation of ethical principles in the field of health informatics necessitates the cultivation of collaborative efforts and the adoption of an interdisciplinary perspective. Health informatics professionals are required to collaborate with healthcare providers, policymakers, researchers and patients in order to effectively incorporate ethical considerations into the development, execution and assessment of health informatics systems. Collaborative endeavours engender ethical decision-making, facilitate the dissemination of optimal methodologies and advance a comprehensive approach to the provision of healthcare.

**Professional Integrity and Responsibility**: Health informatics professionals bear the responsibility of upholding professional integrity, abiding by ethical guidelines and consistently enhancing their knowledge and skills. Ethical considerations encompass the promotion of ethical conduct, the provision of continuous education and training in ethics and the cultivation of an ethical practise culture within the health informatics community.

Harrison et al. (2007) examine the unintended consequences of information technologies (IT) in the healthcare sector using an interactive sociotechnical analysis. The analysis underscores the dualistic consequences of IT implementation, encompassing enhanced information accessibility and communication as well as augmented work demands and potential safety concerns. The research highlights the significance of taking into account human factors and organizational dynamics in order to maximize the effectiveness of information technology (IT) implementation within the healthcare sector.

Mort et al. (2013) research examines the ethical considerations associated with home telecare for elderly individuals using a participatory approach across multiple sites. Ethical concerns about using telecare technology to provide in-home care for the elderly are explored in this paper, which gives a full theoretical framework to this end. Autonomy, privacy, dignity, trust and the impact of technology on interpersonal dynamics and healthcare delivery are only few of the topics addressed by the framework. This research emphasizes the need for older people, their families and their carers to be involved in decision-making processes to ensure the ethical application of telecare. Overall, the study makes substantial additions to our knowledge of the ethical considerations that must be made before home telecare may be implemented successfully in the field of elder care.

The article authored by Kuziemsky et al. (2020) extensively examines the ethical considerations pertaining to telehealth, with a specific emphasis on the juxtaposition between established guidelines and practical experiences derived from real-world implementation. The present study examines the perspectives of various stakeholders involved in the domain of telehealth, including clinicians, researchers and policymakers. This highlights the importance of aligning guidelines with the practical realities and challenges faced during the implementation of telehealth. The authors suggest the advocacy for the advancement of "learning health systems" as a strategy to prioritize continuous learning and improvement through the ethical examination of telehealth practices. The review underscores the importance of incorporating ethical principles, including but not limited to patient privacy, data security, informed consent and equity, into telehealth initiatives. This study provides key insights into enhancing the ethical implementation of telehealth services, specifically by aligning guidelines with the practical aspects of the telehealth area.

The article by Nittari et al. (2020) offers a thorough examination of the ethical and legal obstacles encountered in the practice of telemedicine. It explains the critically analyse the current status of telemedicine, considering its rapid expansion and widespread adoption, with a particular focus on the impact of the COVID-19 pandemic. This review critically analyses the key features like the important ethical considerations related to patient confidentiality, data protection, informed consent and the development of a strong doctor-patient relationship in virtual healthcare settings. Furthermore, the research also investigates the legal barriers related to telemedicine, encompassing complex matters pertaining to licencing and jurisdiction, the potential liability for medical negligence and the imperative need to adhere to telemedicine regulations. The article also considered the importance of implementing thorough telemedicine protocols and ethical frameworks to ensure the welfare of patients, safeguard confidentiality and foster equitable access to healthcare services. This contribution is valuable for healthcare professionals, policymakers and researchers seeking to expand their knowledge in this area.

Kidholm et al. (2012) present MAST (Model for Assessment of Telemedicine Applications), a comprehensive model specifically developed for the purpose of evaluating telemedicine applications. This research examines the relevance of systematic and structured telemedicine technology evaluation. The MAST framework assesses clinical, user, organizational, economic, technological and ethical dimensions. Analysing each domain helps evaluate telemedicine applications. This study discusses how the MAST framework helps healthcare professionals, policymakers, and researchers make educated telemedicine implementation and adoption decisions. This resource advances and standardizes telemedicine evaluation procedures, improving efficient and evidence-based telehealth practices.

The scholarly article authored by Smith et al. (2019) examines the ramifications of telehealth in the context of worldwide crises, specifically emphasizing the coronavirus disease 2019 (COVID-19) pandemic. This study examines the use of telehealth to treat and control infectious illnesses, especially in emergencies. Telehealth can facilitate remote consultations, triage, and patient monitoring, relieving healthcare facilities and lowering patient and provider exposure. This study addresses the challenges and opportunities of establishing telehealth services during a pandemic, including technological accessibility, privacy issues, and service expansion. This emphasizes the need for strategic planning and legislative support to integrate telehealth into healthcare systems amid global health catastrophes. Telehealth can help fight public health crises like the COVID-19 epidemic, according to the study.

The research paper by Gordon and Hinson (2007) proposes a sustainable computer-based health information system (CHIS) framework for LDCs. The report highlights the difficulties LDCs confront in establishing and maintaining efficient health information systems. Resource and infrastructural constraints cause these issues. The framework emphasizes system simplicity, flexibility, local ownership, and capacity development to meet these difficulties. To maintain its relevance and sustainability, Community Health Information Systems (CHIS) must involve local stakeholders in their design and implementation. It also emphasizes the importance of methodical and progressive system development for expansion and adaptability (Singh et al. 2022). The suggested framework helps policymakers and healthcare practitioners in developing nations construct comprehensive health information systems (CHIS). Thus, this may enhance healthcare and result in resource-constrained contexts (Farooq et al. 2022).

Shiferaw and Zolfo (2012) described a project to provide universal health coverage in Ethiopia using ICT. ICT, particularly telemedicine, can improve healthcare delivery in resource-constrained areas, according to the study. This section discusses the first stages of the telemedicine endeavour and its potential to improve healthcare for geographically isolated populations. This study underlines the necessity of using information and communication technology (ICT) to address geographical and infrastructural disparities, improving healthcare outcomes and achieving universal health coverage in Ethiopia.

Simba and Mwangu (2004) examine how globalization and ICT might improve health information systems in poor nations. This study shows how information and communication technology (ICT) may help developing nations manage health information and improve healthcare delivery. ICT can enhance data collection, retention, and analysis, emphasizing its value. Thus, this technique improves healthcare planning by enabling educated decision-making. Information and communication technology (ICT) may help healthcare stakeholders overcome geographical boundaries and communicate better. It also highlights the necessity of infrastructure, capacity building, and policy support for ICT-based health information systems. The study shows that ICT improves health systems and results in underdeveloped nations.

In conclusion, the importance of ethics and social responsibility within the field of health informatics cannot be overemphasized. Healthcare professionals bear the duty of ensuring the welfare of patients, preserving the confidentiality and security of information, addressing biases, promoting social justice and adhering to professional ethics. By integrating ethical considerations into the implementation of health informatics practises, healthcare professionals and informatics specialists can optimally leverage technology and data to enhance their capacity in advancing the welfare of individuals, communities and society as a whole.

## 5.2 Ethical Principles in Health Informatics

In the field of health informatics, ethics and social responsibility play a key role because they ensure that data-driven healthcare practises and technological advancements comply with moral standards, put patient well-being as a top priority and satisfy societal requirements. One may gain an understanding of the relevance of ethics and social responsibility in health informatics by considering the following:

**Patient Welfare**: The ethical considerations and social responsibilities within the field of health informatics place a primary emphasis on safeguarding the welfare and ensuring the safety of patients. The primary objective of their efforts is to establish and execute technology and data management protocols in a manner that safeguards patient confidentiality, upholds patient self-governance and mitigates potential adverse consequences. Healthcare professionals and informatics specialists can prioritize patient welfare by making decisions and developing systems that adhere to ethical principles such as beneficence and non-maleficence.

**Data Privacy and Security:** Health informatics encompasses the systematic acquisition, retention and examination of confidential patient data. The inclusion of ethical considerations is of paramount importance in safeguarding patient privacy and upholding the confidentiality of their health data. Health informatics professionals are required to comply with privacy regulations, establish strong security protocols and exercise utmost caution in handling data to prevent unauthorized access, data breaches and misuse of patient information.

**Informed Consent and Autonomy**: The principle of informed consent is a crucial ethical tenet that upholds the autonomy and entitlements of patients to make well-informed choices regarding their healthcare. In the field of health informatics, it is imperative to guarantee that patients are sufficiently educated regarding the utilization, dissemination and safeguarding of their health data. Ethical practises encompass the imperative of acquiring informed consent from patients prior to the collection and utilization of their data, ensuring transparency in the usage of such data and affording patients the agency to exercise control over their personal information.

**Algorithmic Bias and Fairness**: The discipline of health informatics employs algorithms and decision-making systems that possess the capacity to introduce biases and perpetuate discriminatory practises targeting particular populations. To ensure the advancement of fair and equitable healthcare outcomes, it is crucial to recognize and confront algorithmic biases in accordance with ethical principles (Debnath et al. 2022; Fayaz et al. 2022). Health informatics professionals are required to possess a thorough comprehension of the inherent biases that may be present in data, algorithms and machine learning models. In order to ensure the avoidance of ongoing health disparities and the provision of equitable healthcare access to all individuals, it is crucial for them to proactively acknowledge and confront these biases.

**Governance and Accountability**: The incorporation of ethical principles within the realm of health informatics requires the creation of governance frameworks and mechanisms that ensure accountability, transparency and responsible utilization of technology and data. Healthcare organizations and informatics professionals must prioritize the adoption of ethical frameworks, codes of conduct and guidelines to ensure the effective management of their decision-making processes and subsequent actions (Sinha et al. 2022). Governance mechanisms play a crucial role in promoting and fostering ethical behaviour, facilitating the resolution of ethical dilemmas and establishing accountability measures for individuals and organizations.

**Social Justice and Equity**: In the realm of health informatics, the idea of social responsibility highlights how important it is to address health inequities, socioeconomic determinants of health and to facilitate equitable access to healthcare services. Ethical considerations encompass the utilization of health informatics to address the digital divide, extend services to marginalized populations and guarantee equitable access to technological advancements, regardless of socio-economic status, geographic location, or demographic characteristics.

**Collaboration and Interdisciplinary Approach**: The implementation of ethical principles in the field of health informatics necessitates the fostering of collaboration and the adoption of an interdisciplinary approach. Health informatics professionals are required to collaborate with healthcare providers, policymakers, researchers and patients in order to incorporate ethical considerations into the development, execution and assessment of health informatics systems. Collaborative endeavours engender ethical decision-making, facilitate the dissemination of optimal methodologies and advance a comprehensive approach to the provision of healthcare.

**Professional Integrity and Responsibility**: Health informatics professionals bear the obligation to uphold professional integrity, abide by ethical principles and consistently enhance their knowledge and competencies. Ethical considerations encompass the promotion of ethical behaviour, the provision of continuous education and training in ethics and the cultivation of an ethical practise culture within the health informatics community (Table 5.1).

In summary, the significance of ethics and social responsibility holds utmost prominence within the realm of health informatics. Healthcare professionals have the responsibility to safeguard the well-being of patients, maintain the confidentiality and security of data, mitigate biases, advocate for social justice and adhere to principles of professional integrity. By incorporating ethical considerations into the implementation of health informatics practises, healthcare professionals and informatics specialists can effectively utilize technology and data to maximize their potential in promoting the well-being of individuals, communities and society at large.

**Table 5.1** Ethical principles in health informatics

| Ethical principles | Description |
|---|---|
| Beneficence | Act in the best interest of patients and promote their well-being |
| Non-maleficence | Do no harm and minimize risks to patients in the use of technology and data |
| Respect for autonomy | Recognize and respect patients' rights to make informed decisions about their healthcare |
| Justice | Ensure equitable access to healthcare services and resources, promoting fairness and social justice |
| Veracity and transparency | Uphold honesty and accuracy in communication, being transparent about potential biases and limitations |
| Privacy and confidentiality | Safeguard patient information and maintain confidentiality to protect patient privacy |

## 5.3 Ethical Challenges in Health Informatics

The ethical dilemmas encountered in the field of health informatics stem from the convergence of technology, data management and the healthcare sector. The primary obstacles encompass ensuring the protection of data privacy and security, acquiring informed consent, mitigating biases in algorithms and decision-making systems, establishing robust governance and accountability mechanisms and upholding professional integrity and responsibility. Health informatics professionals are faced with the task of addressing intricate ethical dilemmas while simultaneously advocating for patient well-being, safeguarding patient autonomy and upholding principles of fairness and equity in healthcare.

### *5.3.1 Data Privacy and Security*

The topic of data privacy and security is of utmost importance in today's digital age. The preservation of data privacy and security poses significant ethical dilemmas within the field of health informatics. The safeguarding of patient information from unauthorized access, data breaches and cyberattacks becomes of utmost importance as healthcare systems increasingly depend on electronic health records and digital data. In order to ensure the confidentiality of sensitive health information, experts working in health informatics are expected to put in place stringent security measures, encryption techniques and access restrictions. The process of protecting an individual's right to privacy in their data entails adhering to legal and ethical standards, obtaining an individual's informed consent before using their data and managing their data with extreme care. The inability to safeguard the privacy of data can result in severe implications, including the breach of patient confidentiality, the loss of public confidence and the potential for injury to people in the event that their sensitive health data is accessed by unauthorized persons.

### 5.3.2 *Informed Consent and Patient Autonomy*

The fundamental ethical principle in the field of healthcare and health informatics revolves around the recognition and regard for patient autonomy, as well as the obtaining of informed consent. The purpose of the concept of informed consent is to ensure that patients are provided with adequate information regarding the objectives, risks, benefits and potential consequences associated with the collection and use of data. Health informatics professionals have a responsibility to effectively communicate information in a manner that is comprehensible and unambiguous, thereby enabling patients to make well-informed choices regarding their involvement in data-sharing or interventions facilitated by technology. The absence of obtaining informed consent can result in a breach of patient trust, infringement upon their rights to exercise authority over their health information and ethical quandaries concerning the appropriate utilization of data.

### 5.3.3 *Bias and Discrimination in Algorithms and Decision-Making Systems*

The presence of algorithmic bias and discrimination poses ethical dilemmas within the field of health informatics, specifically in the context of artificial intelligence (AI) and machine learning (ML) implementations. Unintentional biases may arise from the utilization of biassed training data, resulting in the unequal treatment of patients based on variables such as race, ethnicity, gender, or socioeconomic status. Algorithms possess the potential to perpetuate health disparities, reinforce stereotypes and engender unjust or erroneous healthcare decisions due to inherent biases. In order to uphold fair and ethical healthcare practises, health informatics professionals are required to thoroughly evaluate and address biases present in algorithms and decision-making systems (Table 5.2).

### 5.3.4 *Governance and Accountability*

The intricate characteristics of health informatics require the establishment of well-defined governance frameworks and mechanisms to guarantee accountability and prudent decision-making. The establishment of ethical governance in the field of health informatics entails the formulation of policies, procedures and guidelines that

**Table 5.2** Emerging technologies and ethical implications

| Emerging technologies | Ethical implications |
|---|---|
| Artificial intelligence (AI) | Algorithmic bias, lack of transparency and risks in autonomous decision-making |
| Internet of things (IoT) | Data privacy, security and responsible data use for remote patient monitoring |
| Genomics and personalized medicine | Data privacy and confidentiality, concerns about stigmatization or discrimination based on genetic data |
| Telemedicine and virtual healthcare | Quality of care in remote consultations and patient privacy during virtual visits |

are in accordance with ethical principles and legal obligations. It is imperative for healthcare organizations and informatics professionals to implement protocols for the disclosure of ethical issues, resolution of conflicts of interest and management of breaches in ethical conduct. The absence of effective governance and accountability can potentially result in unethical conduct, improper utilization of data and a dearth of transparency in decision-making procedures.

### *5.3.5 Professional Integrity and Responsibility*

Professional responsibility and honesty are the guiding ethical values in the field of health informatics. Health informatics workers have a duty to follow moral standards, safeguard the confidentiality of patient information, and put patients' needs first. It is essential for people to keep current on the laws, rules, and growing ethical standards that apply to their particular sector. The ethical decision-making process in health informatics entails a careful assessment of the potential risks and benefits of interventions, with an emphasis on prioritizing the welfare of both individuals and society as a whole rather than the interests of individuals or organizations. Negative effects on the reputation of the health informatics field and breaches of confidence can result from failing to sustain professional ethics and accountability (Table 5.3).

In brief, the ethical challenges encountered within the realm of health informatics are intricate and demand ongoing attention and vigilance from professionals operating in this field. It is crucial to engage in a thorough examination of critical domains, including data privacy and security, informed consent, algorithmic bias, governance and professional integrity. Health informatics professionals possess the capacity to offer significant contributions to the conscientious and ethical utilization of technology and data within the healthcare industry. The successful realization of these endeavours necessitates the recognition and resolution of the obstacles inherent in such pursuits, all the while ensuring that their methodologies adhere to ethical standards. By doing this, these professionals may actively promote the wellbeing of those seeking medical treatment, safeguard their privacy, and provide advantages for society as a whole.

**Table 5.3** Ethical guidelines and standards for health informatics professionals

| Ethical guidelines and standards | Description |
|---|---|
| Informed consent and privacy | Obtain informed consent from patients and protect patient privacy throughout data use |
| Transparency and accountability | Be transparent about algorithmic decision-making and be accountable for actions and data use |
| Professional competence | Maintain knowledge and skills to ensure ethical use of technology and data in healthcare |
| Respect for autonomy and justice | Prioritize patient autonomy and consider justice for equitable access to healthcare |
| Collaboration and interdisciplinary approach | Work collaboratively with stakeholders for ethical decision-making and solutions |

## 5.4 Social Responsibility in Health Informatics

In the field of health informatics, the idea of social responsibility draws attention to the moral duty to meet societal needs, advance fair access to healthcare, and use technology and data to enhance health outcomes for all people. This section examines the key tenets of social responsibility in health informatics and how they contribute to patient welfare and the welfare of society at large.

### *5.4.1 Patient Empowerment and Patient-Cantered Care*

An important idea that encompasses the ethical responsibility in the field of health informatics is patient-centred care. The idea entails giving patients' input into healthcare decisions a high priority while taking into account their particular preferences, beliefs, and treatment objectives. The advancement of technology advances and data-centric solutions that enable patients to actively engage in the process of making educated healthcare decisions is made possible in large part by the work of health informatics specialists. The supply of individualized health information, the availability of electronic health records, and the use of technologies that encourage group decision-making may all improve the facilitation of empowerment. Implementing patient-centred care promotes a collaborative relationship between patients and medical professionals, which improves adherence to treatment regimens, patient satisfaction, and health outcomes.

### 5.4.2 Access to Medical Services and Health Information

Maintaining the idea of fair and impartial access to healthcare services and health-related information is important to the development of social responsibility in the field of health informatics. Technology may be a potent tool for addressing the needs of marginalized people and reducing access inequities, especially for those who reside in distant or resource-constrained places. For those who face obstacles linked to location, economy, or mobility, the use of telemedicine, mobile health applications, and remote monitoring equipment shows promise in enhancing healthcare accessibility. Professionals in the field of health informatics must also make sure that health information is delivered in a way that is simple to comprehend and simple to access for a wide range of people, including those with low health literacy or impairments.

### 5.4.3 Increasing Digital Inclusion

The uneven distribution of technology and internet access among distinct demographic groups is referred to as the "digital divide". In order to fulfill its social obligation, the profession of health informatics must work to eliminate this imbalance by ensuring that everyone has fair and equitable access to and use of technological breakthroughs in healthcare. Health informatics specialists must demonstrate a thorough comprehension of the difficulties faced by disadvantaged or marginalized populations and make significant attempts to increase such communities' access to technology and digital health services. This may entail forming partnerships with community-based organizations, decision-makers, and technology companies to create focused projects and interventions aimed at bridging the digital gap and fostering digital inclusion.

### 5.4.4 Addressing Health Disparities and Social Determinants of Health

Health disparities, which are frequently associated with social determinants of health, pose a substantial concern within the healthcare sector. Social responsibility within the field of health informatics encompasses the comprehension and mitigation of these disparities with the aim of advancing health equity. Health informatics professionals have the ability to utilise data analytics in order to identify disparities and their root causes. This empowers policymakers and healthcare providers to create focused interventions and programmes. Health informatics systems can enhance the comprehension of individual and population health by integrating social determinants of health, such as housing status, education, income and access to nutritious food.

### 5.4.5 *Promoting Equity, Diversity and Inclusivity in Health Informatics*

The ethical imperative of promoting equity, diversity and inclusivity holds significant importance in the field of health informatics. Health informatics professionals should prioritize the development of technology and data-driven solutions that demonstrate sensitivity towards cultural variations, linguistic diversity and individual preferences. This entails taking into account the varied needs and viewpoints of patients from diverse backgrounds and ensuring that health informatics systems do not perpetuate biases or discriminatory practises. Through the promotion of equity, diversity and inclusivity, the field of health informatics has the potential to play a significant role in mitigating healthcare disparities and enhancing healthcare accessibility and outcomes for individuals across the board.

In summary, the incorporation of social responsibility within the field of health informatics is a fundamental component in the ethical utilization of technology and data for the betterment of society. Crucial elements of social responsibility encompass patient-centered care and empowerment, ensuring access to healthcare and health information, bridging the digital divide, addressing health disparities and promoting equity, diversity and inclusivity. By incorporating these principles into the practises of health informatics, professionals have the potential to enhance healthcare outcomes, promote equitable access to care and foster improved health for individuals and communities alike. The integration of socially responsible health informatics has the potential to act as a driving force for beneficial transformations in both society and the healthcare sector.

## 5.5 Ethical Decision-Making in Health Informatics

The responsible and ethical utilization of technology, data and information systems in healthcare necessitates the careful consideration and implementation of ethical decision-making principles within the field of health informatics. This section delves into the diverse constituents of ethical decision-making in the field of health informatics. It encompasses frameworks designed for analysis, case studies that exemplify ethical dilemmas and ethical guidelines and standards that govern health informatics professionals.

### 5.5.1 *Frameworks for Ethical Analysis and Decision-Making*

**The Four Principles Approach**: The Four Principles Approach, as posited by Beauchamp and Childress, is a widely employed conceptual framework within the realm of healthcare ethics. The framework comprises four fundamental ethical principles, namely autonomy, beneficence, non-maleficence and justice. Health informatics professionals have the ability to utilize these principles in order to examine ethical predicaments and provide guidance for decision-making processes. When implementing a data analytics system, it is imperative to carefully weigh the potential advantages of enhanced patient outcomes (beneficence) against the potential risks of algorithmic biases (non-maleficence) that could disproportionately impact vulnerable populations (justice).

**The Principlism Approach**: The Principlism Approach, akin to the Four Principles Approach, incorporates a collection of ethical principles; however, it places greater emphasis on a wider array of principles, including veracity (truthfulness) and confidentiality. Health informatics professionals have the ability to employ this methodology in order to analyse ethical quandaries from a multitude of viewpoints, thereby acknowledging the intricate nature and interconnectedness of diverse ethical factors.

**The Ethical Decision-Making Model**: The aforementioned model offers a systematic and methodical framework for the examination and evaluation of ethical predicaments. The process entails the identification of ethical dilemmas, the collection of pertinent data, the examination of alternative courses of action and the assessment of the potential outcomes associated with each choice. This model can be utilized by health informatics professionals in order to methodically assess the ethical ramifications of technological interventions or data utilization, thereby enabling them to make well-informed and ethically appropriate choices.

### *5.5.2 Case Studies Illustrating Ethical Dilemmas in Health Informatics*

Case studies provide valuable insights into ethical dilemmas that are encountered in the field of health informatics. An illustrative instance could involve the examination of the application of artificial intelligence (AI) algorithms in the anticipation of patient outcomes. An ethical quandary emerges when the AI system demonstrates bias, leading to unequal treatment of individuals based on their demographic attributes. In the implementation of these algorithms, health informatics professionals are confronted with the complex issues of fairness, transparency and patient autonomy.

Another potential case study could explore the utilization of patient data for research endeavours without obtaining explicit informed consent. Health informatics professionals are faced with the task of carefully considering the potential advantages of utilizing data to enhance medical understanding, while also upholding the ethical obligation of safeguarding patient autonomy and privacy.

Through the examination of case studies, professionals in the field of health informatics are able to gain a deeper comprehension of the intricate nature of ethical dilemmas and subsequently formulate effective approaches for their resolution in practical settings.

### *5.5.3 Ethical Guidelines and Standards for Health Informatics Professionals*

Ethical guidelines and standards play a crucial role in offering indispensable guidance to health informatics professionals as they navigate through ethical dilemmas. The American Medical Informatics Association (AMIA) and the International Medical Informatics Association (IMIA) have developed ethical codes that delineate ethical principles and optimal approaches for professionals in the field of health informatics.

These guidelines emphasize the importance of:

**Informed Consent and Privacy**: In the realm of health informatics, it is imperative for professionals to acquire informed consent from patients prior to utilizing their health data, while simultaneously ensuring the safeguarding of patient privacy throughout the entirety of the data lifecycle.

**Transparency and Accountability**: It is imperative for health informatics professionals to prioritise and uphold transparency throughout the process of developing and utilizing algorithms, data analytics and decision-making systems. The individuals in question bear responsibility for their actions and are obligated to exhibit transparency regarding potential biases, limitations and uncertainties inherent in the technology they create and employ.

**Professional Competence**: In order to uphold proficiency in the ethical utilization of technology and data within the healthcare domain, it is imperative for health informatics practitioners to consistently uphold and enhance their knowledge and skills.

**Respect for Autonomy and Justice**: In the development and implementation of health informatics systems, it is imperative for health informatics professionals to give utmost importance to patient autonomy and take into account principles of justice. This entails ensuring equitable access to healthcare and its associated benefits.

**Collaboration and Interdisciplinary Approach**: It is imperative for health informatics professionals to engage in collaborative efforts with healthcare providers, researchers, policymakers and patients in order to effectively tackle ethical dilemmas and devise ethically sound resolutions.

It is imperative to uphold these ethical guidelines and standards in order to advance conscientious and ethical health informatics practises that prioritize the well-being of patients, uphold autonomy and address the needs of society.

In summary, the process of ethical decision-making in the field of health informatics entails the utilization of analytical frameworks, the examination of case studies to gain insights into practical dilemmas and adherence to established ethical guidelines and standards. Through the integration of these components, professionals in the field of health informatics are able to make well-informed decisions that give priority to the well-being of patients, uphold ethical principles and guarantee the responsible and ethical utilization of technology and data within the healthcare sector. The responsible and ethical advancement of the field of health informatics is contingent upon the central role of ethical decision-making, which yields benefits for both patients and society at large.

## 5.6 Future Directions and Considerations

The ongoing evolution of health informatics presents a range of novel opportunities and challenges that necessitate meticulous ethical deliberation. This section delves into the prospective avenues and contemplations in the field of health informatics, with a specific emphasis on nascent technologies and their ethical ramifications, ethical deliberations in the realm of research and innovation, the establishment of policies and regulatory frameworks and the significance of education and training in ethics for professionals in health informatics.

### *5.6.1 Emerging Technologies and Ethical Implications*

Artificial Intelligence (AI) and Machine Learning (ML) have emerged as highly promising technologies in the field of healthcare, offering significant potential for enhancing the accuracy of diagnoses, optimizing treatment strategies and ultimately improving patient outcomes. Nevertheless, the utilization of these technologies also gives rise to ethical considerations, including algorithmic bias, insufficiency of transparency in decision-making processes and potential hazards associated with autonomous decision-making systems. Health informatics professionals have the responsibility to guarantee the transparency, fairness and accountability of artificial intelligence (AI) systems, while also ensuring that these systems do not compromise the safety or privacy of patients.

**Internet of Things (IoT) and Wearable Devices**: The practise of remote patient monitoring and data collection holds significant potential with the emergence of the Internet of Things (IoT) and wearable devices. Ethical considerations encompass the protection of data privacy, ensuring security measures and promoting responsible utilization of the data acquired from these devices. Health informatics professionals are required to confront the potential hazards linked to the widespread availability of health data, including unauthorized access, data breaches and concerns regarding data ownership (Kanga et al. 2022).

**Genomics and Personalized Medicine**: The amalgamation of genomics and health informatics facilitates the implementation of personalized medicine that is customized to an individual's specific genetic composition. Ethical dilemmas encompass a variety of concerns, including the safeguarding of data privacy and confidentiality in the context of genomic information, as well as the ramifications associated with the potential for stigmatization or discriminatory practises based on genetic data.

**Telemedicine and Virtual Healthcare**: The implementation of telemedicine holds promise in improving the availability of healthcare services, especially in areas that are geographically remote or underserved. However, ethical considerations encompass the obligation to maintain the quality of care delivered through remote methods, ensure the protection of patient confidentiality during virtual consultations and address the disparities in digital access among various demographic populations.

### *5.6.2 Ethical Considerations in Research and Innovation*

**Informed Consent and Research Ethics**: When carrying out research in the field of health informatics, it is very necessary to obtain informed permission from the people who are participating. Professionals in the field of health informatics have the obligation to see to it that study participants are provided with enough information on the objectives of the investigation, as well as the potential drawbacks and advantages of taking part in it. In addition to this, it is absolutely necessary for these specialists to protect the participants' privacy and confidentiality during the entire process.

**Big Data and Secondary Use**: The use of huge amounts of data in the field of health informatics involves ethical challenges including the ownership of the data, the collection of consent for the use of secondary data and the potential of re-identifying data that has been anonymized (Rather et al. 2022). Establishing and putting into practise all-encompassing data governance systems is an absolute requirement for health informatics experts if they are to guarantee the appropriate and ethical usage of data.

**Bias and Fairness in Research**: In order to ensure results that are both fair and equitable, it is essential for research in the field of health informatics to properly address any biases that may be introduced throughout the process of data collecting and analysis. Health informatics professionals should possess a comprehensive understanding of potential biases and proactively implement measures to mitigate them, thereby preventing the perpetuation of healthcare disparities.

### *5.6.3 Policy and Regulatory Frameworks for Health Informatics*

**Data Privacy and Security Regulations**: It is imperative for governments and healthcare organizations to implement comprehensive data privacy and security regulations in order to safeguard patient information and mitigate the risk of unauthorized access or data breaches. In order to uphold responsible data management practises, health informatics professionals are required to adhere to these regulations.

**Ethical Guidelines and Codes of Conduct**: It is imperative for policy and regulatory frameworks to incorporate ethical guidelines and codes of conduct that are tailored to the field of health informatics professionals. The aforementioned guidelines offer comprehensive direction regarding ethical behaviour, professional obligations and optimal approaches within the field of health informatics.

**Interoperability and Data Sharing**: Policy frameworks ought to incorporate provisions for interoperability standards and data sharing practises in order to promote the responsible and secure exchange of health data among healthcare providers and systems. The establishment of data interoperability is of utmost importance in order to enhance patient care and facilitate advancements in medical research.

### *5.6.4 Education and Training in Ethics for Health Informatics Professionals*

**Integrating Ethics in Health Informatics Curricula**: It is imperative for educational institutions to incorporate ethics into health informatics curricula in order to effectively prepare upcoming professionals with the essential ethical knowledge and skills required to effectively address intricate ethical dilemmas throughout their professional journeys.

**Continuing Education and Training**: It is imperative for health informatics professionals to engage in ongoing education and training in the realm of ethics in order to remain abreast of current advancements, ethical principles and optimal methodologies. The implementation of regular training programmes plays a crucial role in equipping professionals with the necessary knowledge and skills to make well-informed ethical decisions within a rapidly evolving professional domain.

**Ethical Discussions and Case Studies**: The involvement of health informatics professionals in ethical deliberations and the examination of case studies promotes the development of critical thinking skills and ethical decision-making abilities. This aids individuals in comprehending the intricacies of ethical quandaries and cultivating approaches for ethical decision-making.

In summary, the field of health informatics encompasses various emerging technologies that raise ethical concerns. Additionally, it involves ethical considerations in the context of research and innovation, the development of policy and regulatory frameworks and the significance of ethics education and training for professionals in this field. By taking proactive measures to address these aspects, the health informatics community can foster responsible and ethical practises, prioritise patient well-being and make valuable contributions to the advancement of healthcare delivery and research. The practise of ethical decision-making will play a crucial role in effectively addressing the various challenges and opportunities that arise within the dynamic and rapidly changing domain of health informatics.

## 5.7 Conclusion

### *5.7.1 Summary of Key Points*

This scholarly investigation delves into the realm of ethics and social responsibility within the field of health informatics, shedding light on various pivotal aspects. Health informatics, the convergence of healthcare and information technology, presents significant opportunities to transform healthcare delivery, enhance patient outcomes and propel medical research forward. Nevertheless, the presence of these opportunities also brings about ethical challenges that necessitate meticulous deliberation.

The ethical principles of beneficence and non-maleficence in the field of health informatics underscore the responsibility to actively promote the well-being of patients and prevent any potential harm. The principle of autonomy acknowledges and upholds the entitlement of patients to exercise their right in making well-informed choices regarding their healthcare. The idea of justice demonstrates the importance of guaranteeing equitable and impartial availability of healthcare services and resources. The principles of honesty, accuracy and openness in communication are encompassed by the concepts of veracity and transparency. Privacy and confidentiality are fundamental elements that prioritize the safeguarding of patient information and the maintenance of confidentiality.

The importance of ethics and social responsibility in the domain of health informatics cannot be overstated, as they have a pivotal role in fostering responsible and ethical utilization of technology, data and information systems. Healthcare professionals prioritize the welfare of patients, uphold the principle of individual autonomy, advocate for fairness, work towards reducing health disparities and contribute significantly to the overall welfare of society.

The domain of health informatics encompasses a range of ethical dilemmas, primarily centred around the imperative of implementing strong safeguards to protect the privacy and security of confidential patient information. It is crucial to uphold the principles of informed consent and patient autonomy by ensuring the respect for patients' rights to make decisions concerning their health data. Recognizing and addressing inherent biases in algorithms and decision-making systems is crucial in order to prevent discriminatory practises and uphold the principles of fairness. The presence of governance and accountability mechanisms facilitates the exercise of responsible decision-making and promotes transparency in organizational practises. The maintenance of ethical conduct in health informatics practise necessitates the utmost importance of professional integrity and responsibility.

The concept of social responsibility in the field of health informatics revolves around the principles of patient-centered care and empowerment, which place a strong emphasis on addressing the needs and preferences of patients. The promotion of equitable healthcare services is facilitated by the assurance of access to healthcare and health information. The concept of bridging the digital divide pertains to the mitigation of disparities in digital access. The endeavour to mitigate health disparities and address social determinants of health aims to diminish disparities in health outcomes. The promotion of equity, diversity and inclusivity is instrumental in cultivating healthcare practises that are characterized by fairness and justice.

The application of ethical principles in the field of health informatics is a crucial aspect of decision-making. Frameworks that facilitate ethical analysis offer a systematic and organized method to effectively navigate and resolve ethical dilemmas. Case studies serve as valuable tools for professionals, as they provide tangible examples of real-world challenges and offer guidance in making well-informed decisions. Ethical guidelines and standards establish fundamental principles and optimal practises for professionals in the field of health informatics.

### *5.7.2 Call to Action for Ethical Practice and Social Responsibility in Health Informatics*

The ongoing development of health informatics necessitates the active adoption of ethical principles and social responsibility by all individuals and organizations involved in this domain. Health informatics professionals, healthcare organizations, policymakers, educators and researchers collectively contribute significantly to the advancement of ethical conduct and social responsibility within the field of health informatics.

**Health Informatics Professionals**: The primary focus of health informatics professionals should be the prioritization of patient welfare, respect for patient autonomy and the assurance of data privacy and security. It is imperative for individuals to exercise vigilance in identifying potential biases that may exist within algorithms and decision-making systems and to actively strive towards promoting transparency and accountability in these systems. Ongoing education and training in ethics are imperative in order to remain up-to-date with prevailing methodologies and ethical principles.

**Healthcare Organizations**: It is imperative for healthcare organizations to establish ethical guidelines and codes of conduct that are tailored to the unique responsibilities and challenges faced by health informatics professionals. It is imperative to incorporate resilient data privacy and security protocols, while concurrently fostering transparency in the utilization of data. The prioritization of patient-centred care should serve as the fundamental principle guiding healthcare practises, with the aim of ensuring fair and equal access to healthcare services for all individuals.

**Policymakers**: It is imperative for policymakers to develop all-encompassing policy and regulatory frameworks in order to effectively tackle ethical dilemmas in the field of health informatics. The aforementioned frameworks ought to advocate for the responsible management of data, ensure transparency in algorithmic decision-making processes and safeguard the rights of patients. It is imperative for policymakers to allocate resources towards addressing the digital divide in order to facilitate equitable access to technological advancements across all demographic groups.

**Educators and Researchers**: It is imperative for educators to incorporate ethics into health informatics curricula and facilitate avenues for ethical deliberations and analysis of case studies. It is imperative for researchers to undertake studies with meticulous attention to ethical principles, encompassing the acquisition of informed consent from participants and the safeguarding of data privacy.

**Society at Large**: It is imperative for society to advocate for the implementation of responsible and ethical health informatics practises among healthcare providers, technology developers and policymakers. Participating in dialogues regarding the ethical ramifications of health informatics has the potential to cultivate public consciousness and advance societal accountability.

In summary, ethics and social responsibility serve as fundamental principles in the field of health informatics, providing guidance to practitioners in the responsible and ethical utilization of technology and data. Through the adoption of ethical principles and a commitment to social responsibility, the health informatics community has the potential to enhance the provision of healthcare, enhance patient outcomes and foster equal access to healthcare services. Collectively, we possess the capacity to mould a forthcoming era in which health informatics assumes the role of a catalyst for constructive transformation, yielding advantages for individuals, communities and society at large.

## References

Ahmad FN, Ahmad T, Gupta SK, Singh SK, Meraj G, Kumar P, Kanga S et al (2023) Unveiling nature's resilience: exploring vegetation dynamics during the COVID-19 Era in Jharkhand, India, with the google earth engine. Climate 11(9):187

Debnath FN, Debnath J, Meraj G, Das Pan N, Chand K, Debbarma S, Sahariah D, Gualtieri C, Kanga S, Singh SK, Farooq M, Sahu N et al (2022) Integrated remote sensing and field-based approach to assess the temporal evolution and future projection of meanders: a case study on River Manu in North-Eastern India. PLoS ONE 17(7):e0271190

Farooq FN, Farooq M, Mushtaq F, Meraj G, Singh SK, Kanga S, Gupta A, Kumar P, Singh D, Avtar R et al (2022) Strategic slum upgrading and redevelopment action plan for Jammu city. Resources 11(12):120

Fayaz FN, Fayaz M, Meraj G, Khader SA, Farooq M, Kanga S, Singh SK, Kumar P, Sahu N et al (2022) Management of landslides in a rural–urban transition zone using machine learning algorithms—a case study of a national highway (NH-44), India, in the rugged Himalayan Terrains. Land 11(6):884

Gordon AN, Hinson RE (2007) Towards a sustainable framework for computer based health information systems (CHIS) for least developed countries (LDCs). Int J Health Care Qual Assur 20(6):532–544

Harrison MI, Koppel R, Bar-Lev S (2007) Unintended consequences of information technologies in health care—an interactive sociotechnical analysis, JAMIA (J Am Med Inform Assoc) 14(September-October). https://doi.org/10.1197/jamia.M2384

Kanga FN, Kanga S, Meraj G, Farooq M, Singh SK, Nathawat MS (2022) Disasters in the complex Himalayan terrains. In: Disaster management in the complex Himalayan terrains: natural hazard management, methodologies and policy implications. Springer International Publishing, Cham, pp 3–10

Kidholm K, Ekeland AG, Jensen LK, Rasmussen J, Pedersen CD, Bowes A, Flottorp IA, Bech M (2012) A model for assessment of telemedicine applications: MAST. Int J Technol Assess Health Care 28(January). https://doi.org/10.1017/S0266462311000638

Kuziemsky CE, Hunter I, Gogia SB, lyenger S, Kulatunga G, Rajput V, Subbian V, John O, Kleber A, Mandirola HF, Florez-Arango J, Al-Shorbaji N, Meher S, Udayasankaran JG, Basu A (2020) Ethics in telehealth: comparison between guidelines and practice-based experience-the case for learning health systems. Yearb Med Inform 29(2020):044–050. https://doi.org/10.1055/s-0040-1701976

Mort M, Roberts C, Pols J, Domenech M, Moser I (2013) Ethical implications of home telecare for older people: a framework derived from a multisited participative study. Health Expect 18:438–449. https://doi.org/10.1111/hex.12109

Nittari G, Khuman R, Baldoni S, Pallotta G, Battineni G, Sirignano A, Amenta F, Ricci G (2020) Telemedicine practice: review of the current ethical and legal challenges. Telemed & E-Health. https://doi.org/10.1089/tmj.2019.0158

Rather FN, Rather MA, Meraj G, Farooq M, Shiekh BA, Kumar P, Kanga S, Singh SK, Sahu N, Tiwari SP et al (2022) Identifying the potential dam sites to avert the risk of catastrophic floods in the Jhelum Basin, Kashmir, NW Himalaya, India. Remote Sensing 14(7):1538

Sharma FN, Sharma M, Upadhyay RK, Tripathi G, Kishore N, Shakya A, Meraj G, Kanga S, Singh SK, Kumar P, Johnson BA, Thakur SN et al (2023) Assessing landslide susceptibility along India's national highway 58: a comprehensive approach integrating remote sensing, GIS, and logistic regression analysis. Conservation 3(3):444–459

Shiferaw F, Zolfo M (2012) The role of information communication technology (ICT) towards universal health coverage: the first steps of a telemedicine project in Ethiopia. Glob Health Action 5:1–8

Simba DO, Mwangu M (2004) Application of ICT in strengthening health information systems in developing countries in the wake of globalisation. Afr Health Sci 4(3):194–198

Singh FN, Singh H, Meraj G, Singh S, Shrivastava V, Sharma V, Farooq M, Kanga S, Singh SK, Kumar P et al (2022) Status of air pollution during COVID-19-induced lockdown in Delhi, India. Atmosphere 13(12):2090

Sinha FN, Sinha KK, Gupta MK, Banerjee MK, Meraj G, Singh SK, Kanga S, Farooq M, Kumar P, Sahu N et al (2022) Neural network-based modeling of water quality in Jodhpur, India. Hydrol 9(5):92

Smith AC, Thomas E, Snoswell CL, Haydon H, Mehrotra A, Clemensen J, Caffery LJ (2019) Telehealth for global emergencies: implications for coronavirus disease 2019 (COVID-19). J Telemed Telecare 1–5. https://doi.org/10.1177/1357633X20916567

# Chapter 6
# Introduction to Health Informatics (HI): Enhancing Healthcare Through Information Technology

**Ritambhara K. Upadhyay, Poonam Arora, and Sandeep Sundriyal**

## 6.1 Introduction

Healthcare systems worldwide face numerous challenges, such as rising healthcare costs and increased patient volumes. Hence, there is a need for improved healthcare management to provide quality patient care. In this context, HI has emerged as an important field that leverages information technology to meet these challenges and transform healthcare delivery.

To properly handle and utilize healthcare information, the interdisciplinary area of HI combines healthcare, information technology, and data science (Jordan and Mitchell 2015; Data Standards in Healthcare: Codes Documents Exchange Formats 2020).). It encompasses acquiring, storing, retrieving, and using healthcare data, information, and knowledge to support healthcare professionals, patients, and healthcare systems (Ahmad et al. 2023). The significance of HI lies in its potential to enhance healthcare processes, improve patient safety, support evidence-based decision-making, facilitate data exchange, and drive innovation in healthcare. By utilizing information technology and data science principles, HI aims to optimize the management and utilization of healthcare information to improve healthcare outcomes and patient experiences (Future of Health Technology Summits 1996–2000; King et al. 2013).

R. K. Upadhyay
Department of Geology, Panjab University, Chandigarh 160014, India

P. Arora
Chitkara College of Pharmacy, Chitkara University, Rajpura, Patiala, Punjab 140401, India

S. Sundriyal (✉)
Department of Pharmacy, Birla Institute of Technology and Sciences Pilani, Pilani Campus, Rajasthan 333031, India
e-mail: sandeep.sundriyal@pilani.bits-pilani.ac.in

© The Author(s), under exclusive license to Springer Nature Singapore Pte Ltd. 2024

G. Tripathi et al. (eds.), *Sustainability and Health Informatics*, Advances in Geographical and Environmental Sciences,
https://doi.org/10.1007/978-981-97-6706-9_6

The scope of HI is broad and encompasses various aspects of healthcare, including clinical practice, healthcare administration, research, and public health. Health information systems, electronic health records (EHRs), clinical decision support systems, telemedicine platforms, health analytics, and other technologies and tools that enable the efficient use of healthcare data are all included in its design, implementation, and management.

The objectives of this chapter are as follows:

1. To provide an overview of HI: The chapter aims to introduce readers to the field of HI by providing a comprehensive overview of its principles, concepts, and applications. It will explore the foundational components of HI and their relevance to healthcare.
2. To explore the benefits of HI: The chapter will highlight HI's advantages and positive impacts in healthcare. It will discuss how HI enhances patient care, supports clinical decision-making, improves healthcare operations, facilitates data exchange, and contributes to research and evidence-based medicine.
3. To discuss challenges and ethical considerations: The chapter will discuss the difficulties and moral issues surrounding the application of HI. It will look at workforce development, data governance, data privacy and security, and the moral use of health data.
4. To identify future directions and innovations: The chapter will explore the future directions and emerging trends in HI. It will discuss the potential of artificial intelligence (AI), the Internet of Things (IoT), precision medicine, and global health initiatives in shaping the field of HI.

Information technology plays a key role in healthcare by facilitating the gathering, storage, retrieval, and interchange of healthcare data and information. For effective management of healthcare information, software architecture like hospital information systems (HIS), electronic health record (EHR) systems, and picture archiving and communication systems (PACS) are employed. Information technology makes the secure transmission of health information between various healthcare providers possible, providing seamless continuity of care and enhancing care coordination. Technology enables remote consultations, tele-monitoring, and telemedicine services, expanding access to healthcare and improving patient outcomes, particularly in remote or underserved areas. Effective management of health data and information is essential in HI. Various sources, such as patient health records, laboratory test results, medical imaging, and wearable devices, contribute to collecting health data. Standardizing health data formats and terminologies, such as HL7, SNOMED CT, and LOINC, ensures interoperability and facilitates the meaningful exchange and integration of health information. Data quality processes, including data validation, accuracy checks, and data cleaning techniques, are crucial to ensure the reliability and integrity of health data. Policies, regulations, and ethical guidelines are required to govern the collection, use, and disclosure of health data, safeguarding patient privacy, and maintaining data confidentiality (Singh et al. 2022). Electronic health records (EHRs), among other health information systems, are essential elements of HI. They maintain detailed records of patients' medical histories, diagnoses, prescriptions, and test findings (Sharma and Aggarwal 2019). EHRs promote clinical decision-making, eliminate medical mistakes, and increase information accessibility. Clinical Decision Support Systems (CDSS) combine data about the patient with evidence-based suggestions to give healthcare professionals instantaneous clinical advice, enhancing the accuracy of diagnosis and treatment choices. To give complete and coordinated treatment, health information exchange (HIE) enables the secure exchange of patient health records between various healthcare organizations. Standards and interoperability are essential for seamless data exchange and integration in HI. Standardizing health data formats, terminologies, and coding systems ensures consistency and interoperability across healthcare systems and applications. Standards, notably HL7 (Health Level Seven) and DICOM (Digital Imaging and Communications in Medicine), enable the interchange of health data between various healthcare systems, promoting interoperability and data sharing. They store and manage detailed patient data. Frameworks for interoperability, including Fast Healthcare Interoperability Resources (FHIR), offer instructions and resources for developing compatible medical software and systems. Effective data integration and interoperability enable seamless data exchange and sharing between healthcare stakeholders, improving care coordination and supporting data-driven decision-making (Buntin et al. 2011; Menachemi and Brooks 2006).

## 6.2 Applications of HI

Worldwide research activities in HI are being pursued to improve healthcare management and outcomes and gain new insights. As a result, several peer-reviewed journals are published by various publishers reporting diverse applications of HI.

Clinical informatics utilizes HI principles to support clinical practice and enhance patient care. Electronic Health Records (EHRs) facilitate storing, retrieving, and exchanging patient health information, providing healthcare professionals with comprehensive and real-time patient data for clinical decision-making. Clinical Decision Support Systems (CDSS) provide healthcare workers real-time clinical advice to help with diagnosis, therapy planning, and medication management. CDSS does this by using patient data and evidence-based guidelines. Healthcare workers can electronically enter medical orders using computerized physician order entry (CPOE) systems, which reduces prescription errors and boosts workflow efficiency. Telemedicine and telehealth leverage information technology to provide remote healthcare services and improve access to healthcare (Sharma and Aggarwal 2019). Teleconsultation enables healthcare professionals to consult with patients remotely, offer diagnoses, and provide treatment recommendations. This aspect of HI, thus, expands access to specialized care and reduces geographical barriers. Telemonitoring systems collect and transmit patient data remotely, allowing healthcare professionals to monitor patients' vital signs, chronic conditions, and adherence to treatment plans. Mobile Health (mHealth) Applications provide patients with tools for self-management, medication reminders, health tracking, and access to educational resources, promoting proactive healthcare engagement (Gable 1990).

Health analytics involves data analysis and modelling techniques to derive insights and support evidence-based decision-making. Another component of HI is Population Health Management. The latter involves the analysis of health data to assess and manage the health of specific populations, identifying high-risk individuals, and implementing targeted interventions to improve health outcomes (Singh et al. 2021). For instance, machine learning models have been developed to detect and predict frailty in older populations (Leghissa et al. 2023). Predictive analytics modelling employing AI or machine learning (ML) can leverage historical health data to forecast disease patterns. Also, prediction of patient outcomes, resource utilization, proactive interventions, and resource planning can be achieved using these models. Thus, data analytic techniques analyze large healthcare datasets to identify patterns and trends, enabling research discoveries, disease surveillance, and quality improvement initiatives.

## 6.3 Benefits of HI

HI offers several benefits that contribute to improved patient care and safety. HI enables the storage and retrieval of comprehensive patient health records, providing healthcare professionals with access to up-to-date and accurate patient information. This improves clinical decision-making, enhances care coordination, and reduces medical errors. Clinical decision support systems provide healthcare professionals real-time clinical recommendations, alerts, and reminders based on patient data and evidence-based guidelines. This helps identify potential drug interactions, allergies, and adverse events, improving patient safety and treatment outcomes. HI systems facilitate Computerized Physician Order Entry (CPOE) and electronic medication reconciliation, reducing medication errors and improving medication safety by providing accurate medication information to healthcare providers. HI supports healthcare professionals in making informed and evidence-based decisions, leading to improved patient outcomes (Hripcsak et al. 2015). HI systems provide healthcare professionals real-time access to patient data, test results, medical history, and evidence-based guidelines. This enables health professionals to perform swift and more accurate diagnoses, treatment planning, and intervention decisions. Clinical decision support systems analyze patient data and provide healthcare professionals with clinical recommendations, alerts, and reminders. This assists in identifying potential diagnoses, treatment options, and preventive measures, improving clinical decision-making and patient management. HI facilitates integrating clinical research, guidelines, and best practices into clinical workflows. It supports evidence-based medicine, ensuring that patient care aligns with the latest research findings and recommendations. HI optimizes healthcare operations and streamlines workflows, improving efficiency and productivity. Electronic Health Records (EHR) systems enable digitizing patient health records, eliminating paper-based documentation and improving accessibility, retrieval, and sharing of patient information. This streamlines administrative tasks and enhances communication among healthcare providers. HI systems automate routine administrative tasks, appointment scheduling, billing processes, and data entry, reducing manual workload and freeing up time for healthcare professionals to focus on patient care.

HI supports resource allocation and utilization, such as optimizing bed management, surgical schedules, and healthcare staffing based on real-time patient data and demand (Holzinger and Simonic 2011). HI promotes the secure exchange of health information among healthcare providers and stakeholders, leading to improved care coordination and continuity (Sharma and Aggarwal 2019). Health Information Exchange (HIE) systems enable the seamless sharing of patient health records, test results, and treatment plans across different healthcare organizations. This ensures that relevant patient information is readily available to healthcare professionals, supporting coordinated and collaborative care. HI promotes the usage of standardized data formats and terminologies, enabling interoperability between different health information systems. This allows for the exchange and integration of patient data from disparate sources, improving data sharing and continuity of care. By offering a holistic view of patient data across healthcare settings, HI makes care coordination easier. This promotes continuity of treatment for patients, lessens test duplication, and enhances care transitions. HI supports evidence-based medicine and research by facilitating data-driven discoveries and breakthroughs.

HI employs data analytics and modelling techniques to analyze large datasets and identify patterns, correlations, and trends. This supports research initiatives, clinical trials, and population health studies. HI systems capture real-world patient data, enabling the generation of real-world evidence for research and post-market surveillance. This complements traditional clinical trial data and supports evidence-based medicine. HI supports quality improvement initiatives by providing insights into healthcare processes, performance metrics, and outcome measures. This enables healthcare organizations to identify areas for improvement and implement evidence-based practices (Raghupathi and Raghupathi 2014).

Information technology has revolutionized healthcare, as evidenced by the advantages of Health Information Exchange (HI), which include better patient care and safety, more effective clinical decision-making, streamlined healthcare operations, easier exchange of health information, and support for research and evidence-based medicine. Healthcare organizations may improve patient experiences, improve healthcare results, and spur industry innovation by utilizing HI.

## 6.4 Challenges and Ethical Considerations

As HI involves collecting, storing, and exchanging sensitive patient health information, ensuring privacy and security is of utmost importance. Challenges and ethical considerations in this area include data breaches and cybersecurity, patient consent and control, legal and regulatory compliance, data governance and data quality (Fang et al. 2016; King et al. 2013).

The increasing use of digital health systems and interconnected networks exposes health information to potential security breaches and cyberattacks. Protecting health data from unauthorized access, data breaches, and identity theft requires robust security measures, encryption, and ongoing monitoring. HI raises concerns about patient consent and control over their health information. Balancing data sharing and privacy protection requires clear consent processes, transparency in data usage, and mechanisms for patients to control their data-sharing preferences. Lately, the usage of portable electronic devices, such as smartphones, smartwatches, tablets, and wearables, has increased to passively collect and analyze an individual's data. While this data can be used to monitor and improve health-related outcomes, it also provides real-time insight into an individual's social activity and mental state (Maher et al. 2019). Thus, an ethical and legal framework needs to be developed for such data collection methods. HI is subject to a number of legal and regulatory frameworks, including the General Data Protection Regulation (GDPR) in the European Union and the Health Insurance Portability and Accountability Act (HIPAA) in the United States. Patient confidentiality and privacy are protected as long as these rules are followed.

Data governance and quality are critical for the reliable and meaningful use of HI. Challenges and ethical considerations include data integrity and accuracy, ownership and stewardship, sharing and interoperability. HI relies on accurate and reliable data. Ensuring data integrity and accuracy requires appropriate data validation processes, data cleaning techniques, and quality control measures. Determining data ownership and assigning responsibility for data stewardship can be complex, especially when multiple entities are involved in collecting and managing health data. Clear guidelines and agreements are needed to ensure responsible data governance and stewardship. Achieving seamless data sharing and interoperability between various healthcare organizations and systems can be challenging. Standardization of data formats, terminologies, and interoperability frameworks is essential to overcome these challenges and ensure ethical and efficient data exchange.

HI requires a skilled workforce capable of managing and leveraging information technology in healthcare (Herland 2014). Healthcare professionals and IT personnel require specialized education and training in HI to utilize health information systems and technology effectively. Ensuring adequate training opportunities and professional development is crucial to enhance the HI workforce. Promoting diversity and inclusion in the HI workforce is important to ensure that different perspectives and experiences are incorporated into the design and implementation of HI systems (Ravi et al. 2016). This contributes to equitable and culturally sensitive healthcare delivery. Digital Divide and Health Inequities: Health disparities may worsen as a result of the digital divide, which is defined by gaps in access to technology and digital resources. To prevent growing gaps in healthcare outcomes and access, it is imperative to guarantee fair access to HI tools and resources (King et al. 2013).

Ethical considerations must guide the use of health data and technology in HI. Respecting patient autonomy and obtaining informed consent for data collection, sharing, and use are critical ethical considerations (Mandel et al. 2016). Transparency in data practices, clear consent processes, and data anonymization or de-identification are important to protect patient privacy. Health data and technology usage should avoid biases and discrimination based on factors such as race, ethnicity, gender, or socioeconomic status. Ethical considerations involve addressing biases in algorithms, ensuring fairness in data analysis, and promoting health equity. The ethical use of AI in HI involves transparency in AI algorithms, accountability for algorithmic decisions, and continuous monitoring to identify and mitigate biases and unintended consequences.

Addressing these challenges and ethical considerations requires collaboration among healthcare professionals, policymakers, researchers, and technology developers. Ensuring privacy and security, promoting data governance and quality, training a skilled workforce, and upholding ethical standards are essential to maximizing the benefits of HI while safeguarding patient rights and well-being.

## 6.5 A Case Study of Coronavirus Disease in 2019 (COVID-19)

The outbreak of COVID-19 resulted in morbidity and mortality of life on a global scale. Due to the rapid spread of this novel virus, the healthcare system of several countries was overwhelmed. This pandemic situation set the ground for the application of HI. Several researchers reported the development and deployment of HI tools and systems to improve healthcare efficiency, thus benefitting the population at large (Dixon and Holmes 2021).

During the pandemic, swift collection and dissemination of the data were crucial for controlling infection spread and evidence-based decision-making by the policymakers (Massoudi and Sobolevskaia 2021). The researchers from the World Health Organization African Region (WHO AFRO) developed a tool for COVID-19 Data Summarization and Visualization (DSV) to tackle this problem (Ahmed et al. 2020). This tool enables data collection, management, analysis and visualization using simple Microsoft Excel-based spreadsheets. Several WHO AFRO member states effectively used this low-cost, easy-to-use tool.

Clear definitions of disease-related nomenclature and precise vocabulary for data exchange during the COVID-19 pandemic presented another issue. A centralized information management system was described by the Centres for Disease Control and Prevention (CDC) in the US as a "1-stop shop" for COVID information management resources, giving easy shared access to the most recent reference sources for all stakeholders (Garcia et al. 2020). In addition to various other resources, it also has a standardized reporting form for the collection of data on COVID-19-positive individuals. Among various features, the data related to symptoms, exposure, demographics, past medical history, etc., can also be collated and managed using the system.

Restricted movement and isolation of a large population were essential to contain COVID-19 infection and led to the challenges related to ambulatory care. As a result, telehealth systems were upgraded by many health centres and hospitals for triaging patients to minimize the contact between COVID-19-positive and negative populations. For instance, the Comprehensive Cancer Center of the University of California, San Francisco (UCSF) reported increased video consultations from 7–18% to 54–72% between the pre and post-COVID period (Lonergan et al. 2020). Similarly, a chat-assistant-based self-reporting platform was developed for the remote surveillance of marginalized low-income populations in Singapore (Ko et al. 2020). The individuals were trained to use the chat assist and self-report various health parameters such as body temperature, heart rate, and oxygen saturation levels. This study shows the effective management of pandemic situations using telemedicine-based technology, especially for monitoring marginalized populations with distinct cultural and linguistic backgrounds.

## 6.6 Future Directions and Innovations

AI and machine learning ML have immense potential in HI to revolutionize health-care delivery (Kwak and Hui 2019). AI and ML can be used to develop predictive models that identify patients at risk of developing specific diseases or adverse events, enabling proactive interventions and personalized care plans. AI and ML algorithms can analyze medical images, such as radiology scans, and interpret signals from wearable devices to diagnose, detect anomalies, and monitor health conditions (Cohen and Hersh 2005; Weissleder and Pittet 2008). Advanced AI systems can provide real-time clinical decision support by integrating patient data, guidelines, and medical literature, assisting healthcare professionals in making accurate and evidence-based decisions.

The Internet of Things (IoT) and wearable devices offer new avenues for collecting and monitoring health data. These gadgets can continuously monitor patient vitals, activities, and adherence to treatment plans, enabling remote patient monitoring and early detection of health issues. IoT devices can capture real-time data from various sources, including medical devices, environmental sensors, and patient-generated data, providing a holistic view of patient's health and facilitating personalized interventions. Integrating IoT-generated health data with health information systems and analytics platforms enables comprehensive data analysis, improving population health management and personalized care.

Precision medicine aims to tailor medical treatments to individuals based on their genetic makeup, lifestyle, and environmental factors. HI can facilitate the integration of genomic data into patient health records, enabling personalized treatment plans and targeted therapies based on an individual's genetic profile (Weissleder and Pittet 2008). Advanced analytics and machine learning techniques can analyze large-scale genomic and clinical data to generate personalized treatment recommendations, enhancing treatment effectiveness and reducing adverse effects. Predictive risk models that identify people at high risk of acquiring particular diseases can be developed by integrating genomic, lifestyle, and clinical data. This can facilitate early interventions and preventive actions.

HI has the potential to contribute to global health initiatives and address health disparities on a global scale. HI systems can enable real-time disease surveillance, early detection of outbreaks, and rapid response in managing global health crises, such as pandemics (Jamoom et al. 2011). Improving health data sharing and interoperability across different regions and countries enhances collaboration in research, resource planning, and the delivery of healthcare services in underserved areas. Mobile health applications and telemedicine platforms can extend healthcare access to remote and resource-constrained regions, enabling remote consultations, health monitoring, and health education initiatives (Bellazzi and Zupan 2008).

Given these future directions and advancements in HI, the potential to alter healthcare delivery, advance personalized medicine, raise the bar for global health outcomes, and address health inequities is enormous. Continued study, cooperation, and funding in these fields will stimulate innovation and open the door to a more effective, efficient, and fair healthcare system.

## 6.7 Recommendations for Further Research and Implementation

Continued research and development in AI and machine learning applications are needed to improve HI-based clinical decision support, predictive analytics, and image analysis (Holzinger et al. 2017). Exploration of the potential of IoT and wearable devices for remote patient monitoring, real-time data collection, and personalized healthcare. Advancement of precision medicine through integrating genomic data, data-driven treatment recommendations, and predictive risk models. Further research in HI for global health initiatives, focusing on disease surveillance, health data sharing, and mobile health solutions for underserved populations. Investment in workforce education and training to enhance HI capabilities among healthcare professionals and IT personnel. Privacy, data governance, bias, and equity are all addressed in these ethical standards and concerns for the appropriate use of technology and health data. These recommendations should be taken into account by healthcare organizations, governments, and researchers in order to change the way healthcare is delivered, improve patient outcomes, and encourage innovation in the field. A more efficient, patient-centered, and evidence-based healthcare system might be created by HI through the use of information technology, data analytics, and interdisciplinary collaboration.

## 6.8 Conclusion

HI combines healthcare, information technology, and data science to improve the administration and application of healthcare information. This chapter comprehensively introduces HI, exploring its foundations, applications, benefits, challenges, and future directions. Healthcare practitioners, policymakers, and stakeholders can use information technology to change healthcare delivery, enhance patient outcomes, and advance the healthcare field by understanding the principles and promise of HI. Health data and information management, health information systems, standards, and interoperability are all components of the foundations of HI. Applications of HI include clinical informatics, telemedicine, health analytics, and public HI. Benefits of HI include improved patient care and safety, enhanced clinical decision-making, efficient healthcare operations, facilitated health information exchange, and support for research and evidence-based medicine. Multiple reports from scientific literature highlight the benefits and applications of HI during the COVID-19 pandemic. Challenges and ethical considerations in HI encompass privacy and security of health information, data governance, patient consent, workforce quality and training, and ethical use of health data and technology.

## References

Ahmad T, Gupta SK, Singh SK, Meraj G, Kumar P, Kanga S (2023) Unveiling nature's resilience: exploring vegetation dynamics during the COVID-19 Era in Jharkhand, India, with the Google earth engine. Climate 11(9):187

Ahmed K, Bukhari MA, Mlanda T, Kimenyi JP, Wallace P, Okot Lukoya C, Hamblion EL, Impouma B (2020) Novel approach to support rapid data collection, management, and visualization during the COVID-19 outbreak response in the World Health Organization African Region: development of a data summarization and visualization tool. JMIR Public Health Surveill. Oct 14;6(4):e20355. https://doi.org/10.2196/20355.

Bellazzi R, Zupan B (2008) Predictive data mining in clinical medicine: current issues and guidelines. Int J Med Inform 77:81–97. https://doi.org/10.1016/j.ijmedinf.2006.11.006

Buntin M, Burke M, Hoaglin M, Blumenthal D (2011) The benefits of health information technology: a review of the recent literature shows predominantly positive results. Health Aff 30:464–471. https://doi.org/10.1377/hlthaff.2011.0178

Cohen AM, Hersh WR (2005) A survey of current work in biomedical text mining. Brief Bioinform 6:57–71. https://doi.org/10.1093/bib/6.1.57

Data Standards in Healthcare: Codes Documents Exchange Formats (2020) https://www.altexsoft.com/blog/data-standards-healthcare/

Dixon BE, Holmes JH (2021). Section editors for the IMIA Yearbook section on managing pandemics with health informatics. Managing pandemics with health informatics. Yearb Med Inform 30(1):69–74. https://doi.org/10.1055/s-0041-1726504

Fang R, Pouyanfar S, Yang Y, Chen SC, Iyengar S (2016) Computational health informatics in the big data age: a survey. ACM Comput Surv 49:1–36. https://doi.org/10.1145/2932707

Future of Health Technology Summits 1996–2000 (2000) http://www.fhti.org/

Gable CB (1990) A compendium of public health data sources. Am J Epidemiol 131:381–394. https://doi.org/10.1093/oxfordjournals.aje.a115513

Garcia M, Lipskiy N, Tyson J, Watkins R, Esser ES, Kinley T (2020 Sep) Centers for Disease Control and Prevention 2019 novel coronavirus disease (COVID-19) information management: addressing national health-care and public health needs for standardized data definitions and codified vocabulary for data exchange. J Am Med Inform Assoc 1;27(9):1476–1487. https://doi.org/10.1093/jamia/ocaa141

Herland M (2014) A review of data mining using big data in health informatics. J Big Data. 1:1–35. https://doi.org/10.1186/2196-1115-1-2

Holzinger A, Biemann C, Pattichis CS, Kell DB (2017) What do we need to build explainable AI systems for the medical domain? arXiv[Preprint]. arXiv:1712.09923. https://doi.org/10.48550/arXiv.1712.09923

Holzinger A, Simonic KM (eds) (2011) Information quality in e-health, 1st edn. Springer, Graz. https://doi.org/10.1007/978-3-642-25364-5

Hripcsak G, Duke JD, Shah NH et al (2015) Observational Health Data Sciences and Informatics (OHDSI): opportunities for observational researchers. Stud Health Technol Inform 216:574–578

Jamoom E, Beatty P, Bercovitz A, Woodwell D, Palso K, Rechtsteiner E (2011) Physician adoption of electronic health record systems: United States 2011. NCHS Data Brief 98:1–8

Jordan MI, Mitchell TM (2015) Machine learning: trends, perspectives, and prospects. Science 349:255–260. https://doi.org/10.1126/science.aaa8415

King J, Patel V, Jamoom E, Furukawa M (2013) Clinical benefits of electronic health record use: national findings. Health Serv Res 12:49. https://doi.org/10.1111/1475-6773.12135

Ko SQ, Hooi BMY, Koo CY, Chor DWP, Ling ZJ, Chee YL, Jen WY (2020) Remote monitoring of marginalised populations affected by COVID-19: a retrospective review. BMJ Open 31;10(12):e042647. https://doi.org/10.1136/bmjopen-2020-042647

Kwak GH, Hui P (2019) Deep health: review and challenges of artificial intelligence in health informatics. arXiv [Preprint]. https://doi.org/10.48550/arxiv.1909.00384

Leghissa M, Carrera Á, Iglesias CA (2023) Machine learning approaches for frailty detection, prediction and classification in elderly people: a systematic review. Int J Med Inform 8(178):105172. https://doi.org/10.1016/j.ijmedinf.2023.105172

Lonergan PE, Washington Iii SL, Branagan L, Gleason N, Pruthi RS, Carroll PR, Odisho AY (2020) Rapid utilization of telehealth in a comprehensive cancer center as a response to COVID-19: cross-sectional analysis. J Med Internet Res 6;22(7):e19322. https://doi.org/10.2196/19322

Maher NA, Senders JT, Hulsbergen AFC, Lamba N, Parker M, Onnela JP, Bredenoord AL, Smith TR, Broekman MLD (2019) Passive data collection and use in healthcare: A systematic review of ethical issues. Int J Med Inform 129:242–247. https://doi.org/10.1016/j.ijmedinf.2019.06.015

Mandel J, Kreda D, Mandl K, Kohane I, Ramoni R (2016) SMART on FHIR: a standards-based, interoperable apps platform for electronic health records. J Am Med Inform Assoc 23:ocv189. https://doi.org/10.1093/jamia/ocv189

Massoudi BL, Sobolevskaia D (2021) Keep moving forward: health informatics and information management beyond the COVID-19 pandemic. Yearbook Med Inform 30:75–83. https://doi.org/10.1055/s-0041-1726499

Menachemi N, Brooks R (2006) Reviewing the benefits and costs of electronic health records and associated patient safety technologies. J Med Syst 30:159–168. https://doi.org/10.1007/s10916-005-7988-x

Raghupathi W, Raghupathi V (2014) Big data analytics in healthcare: promise and potential. Health Information Sci Syst. 2:3. https://doi.org/10.1186/2047-2501-2-3

Ravi D, Wong C, Deligianni F, Berthelot M, Andreu-Perez J, Lo B et al (2016) Deep learning for health informatics. IEEE J Biomed Health Inform 21:4–21. https://doi.org/10.1109/JBHI.2016.2636665

Sharma M, Aggarwal H (2019) An SDN-based secure mobility model for UAV-ground communications. In: HL-7 based middleware standard for healthcare information system: FHIR: ICCCN. Chandigarh, pp 889–899. https://doi.org/10.1007/978-981-13-1217-5_87

Singh S, Singh H, Sharma V, Shrivastava V, Kumar P, Kanga S, Sahu N, Meraj G, Farooq M, Singh SK (2021) Impact of forest fires on air quality in Wolgan valley, New South Wales, Australia—A mapping and monitoring study using Google Earth Engine. Forests 13(1):4

Singh H, Meraj G, Singh S, Shrivastava V, Sharma V, Farooq M, Kanga S, Singh SK, Kumar P (2022) Status of air pollution during COVID-19-induced Lockdown in Delhi. India. Atmosphere 13(12):2090

Weissleder R, Pittet MJ (2008) Imaging in the era of molecular oncology. Nature 452:580–589. https://doi.org/10.1038/nature06917

# Chapter 7
# Evaluating the Influence of Haryana's Air Quality on Public Health: A Geospatial Approach

**Raghu Sharma, Bhawna Dhiman, Shruti Kanga, Priyanka Roy, Suraj Kumar Singh, Bhartendu Sajan, and Gowhar Meraj**

## 7.1 Introduction

As an important global issue, air pollution has a negative impact on societal well-being, environmental integrity, and human health. To determine the level of pollution and how it affects various societal components, the measurement of air quality is crucial. With the help of information gathered from the Sentinel-5P satellite, this study will examine the air quality in Haryana, India.

Climate change is mostly attributed to air pollution, especially greenhouse gases such as carbon dioxide and methane. Inputs for climate change models and mitigation plans can therefore benefit from quantifying these emissions using Sentinel-5P satellite data (Debnath et al. 2022a; Farooq et al. 2022a). This information can help with the development of focused strategies for reaching carbon neutrality, a crucial component of mitigating climate change, and contribute to a worldwide understanding of emission sources (Debnath et al., 2023c; Farooq et al. 2022b).

R. Sharma · B. Dhiman · P. Roy (✉) · B. Sajan
Centre for Climate Change and Water Research, Suresh Gyan Vihar University, Jaipur-302017, Rajasthan, India
e-mail: priyanka.roy@mygynavihar.com

S. Kanga
Department of Geography, School of Environment and Earth Sciences, Central University of Punjab, VPO-Ghudda, Bathinda-15140, Punjab, India

S. K. Singh
Centre for Sustainable Development, Suresh Gyan Vihar University, Jaipur-302017, Rajasthan, India

G. Meraj
Graduate School of Agricultural and Life Sciences, The University of Tokyo, Tokyo 113-8654, Japan

© The Author(s), under exclusive license to Springer Nature Singapore Pte Ltd. 2024 
G. Tripathi et al. (eds.), *Sustainability and Health Informatics*, Advances in Geographical and Environmental Sciences,
https://doi.org/10.1007/978-981-97-6706-9_7

The data can also have an impact on how cities are planned and how their infrastructure is built. The structure of cities can be optimized for air circulation to lower the concentration of pollutants in populous areas by using knowledge of the polluted landscape. The strategic location of green spaces in metropolitan areas, which serve as the "lungs" of the city by absorbing pollutants and supplying better air, can also be informed by this (Farooq et al. 2022c; Gulati et al. 2023; Kanga et al. 2022a).

When dangerous substances, including chemicals, particulate matter, and biological elements, enter the Earth's atmosphere, air pollution results. Despite the fact that the main constituents of air are nitrogen, oxygen, argon, and trace elements, greenhouse gases such as carbon dioxide and methane can nevertheless influence climate change even in very small amounts. All living things, including people, animals, and plants, depend on the reservation of a balanced air composition for their life and wellbeing (Kanga et al. 2022b; Kotecha et al. 2023; Lahon et al. 2023a). There are many different sources of air pollution, including both gaseous and particle contaminants. Gases including carbon dioxide, methane, and nitrogen dioxide are released as a result of emissions from many different sources, such as fuel-burning companies and automobiles (Debnath et al. 2023b). Both of these options are possible. Particulate pollution, which adversely impacts air quality and includes elements such as soot and black carbon, is visible. (Lahon et al. 2023b; Lavane et al. 2023; Meraj et al. 2022). Although air pollution has always been a problem, this study primarily focuses on pollution that has resulted from different activities since 1900. Wildfires, volcanic eruptions, and dust or sand storms are only a few examples of natural occurrences that have significantly contributed to air pollution. Early human settlements exposed more people to air pollution and its hazards as a result of using biomass and fossil fuels for indoor heating and cooking (Singh et al. 2023a; Sud et al. 2023; Upadhyay et al. 2023). The study of air pollution in Haryana, India, using Sentinel-5P satellite data provides complex insights that go beyond merely identifying pollutants. It enables thorough geographical and temporal analysis, revealing patterns and trends in pollution levels across many locations and periods of time. To strategically deploy intervention measures, it is essential to identify pollution hotspots and times of peak pollution (Meraj et al. 2021b; Nath et al. 2023).

Due to the area's expanding population and rapid industrial development, the assessment of the air quality in Haryana, India, is particularly significant. As a result of advancement in this area, there are now more emissions coming from factories, cars, and other man-made sources, which have raised pollution levels. Sentinel-5P satellite data, which provide an accurate evaluation of air quality, provide key insights into the level of pollution, which are crucial for creating effective mitigation plans (Rafiq et al. 2022; Sajan et al. 2022).

Additionally, Sentinel-5P data can have a large impact on how policies are created. The comprehensive data can be used by decision-makers to create and improve pollution control legislation. This might involve anything from establishing strict emission regulations for businesses and automobiles to offering rewards for adopting greener practices and technologies. Real-world statistics are more likely to inform targeted, effective strategies that maximize benefits to the environment and public health. (Meraj et al. 2021a; Singh et al. 2022b).

The goal of this study is to assess Haryana's air quality using information from the Sentinel-5P satellite. High-resolution data of several atmospheric pollutants are provided by the satellite, allowing for the evaluation of key air quality indicators such as nitrogen dioxide, carbon dioxide, and particle matter. The region's quantitative measurements and pollution maps will be derived through the analysis of these satellite data combined with the use of appropriate models and algorithms. To develop targeted interventions to lessen the negative consequences of pollution on human health, agricultural productivity, and the environment, it is essential to understand the air quality in Haryana. The outcomes of this study will play a crucial role in developing efficient programmes and policies to enhance air quality and promote sustainable development in Haryana, India.

## 7.2 Materials and Methods

### *7.2.1 Study Area*

Haryana, located in northern India, is a landlocked state bordered by Punjab to the northwest, Himachal Pradesh and Uttarakhand to the north and northeast, Uttar Pradesh and Delhi to the east, and Rajasthan to the south and southwest. Its geographical coordinates are approximately between 27.8954° N latitude and 76.3131° E longitude. Haryana is a diverse state in northern India, encompassing a range of geographical features and regions. With Chandigarh as its capital and Faridabad as its largest city, the state was formed in 1966 and covers an area of 44,212 square kilometres. With a population of over 25 million people, Haryana showcases a blend of urban centres, fertile plains, mountains, and semiarid plateaus. Assessing the Air Quality of Haryana, India through Sentinel-5P Satellite Data reveals crucial insights into the air quality conditions in the state. Satellite data from the Sentinel-5P mission have been instrumental in monitoring and analysing air pollution levels in Haryana. Air pollution is a significant environmental concern in Haryana and is primarily attributed to industrial activities, vehicular emissions, agricultural practices, and urbanization. The Sentinel-5P satellite, equipped with a Tropospheric Monitoring Instrument (TROPOMI), provides accurate measurements of key air pollutants, including nitrogen dioxide ($NO_2$), ozone ($O_3$), sulfur dioxide ($SO_2$), carbon monoxide (CO), and aerosols. Figure 7.1 shows the location map of the study area.

### *7.2.2 Data Used*

In this comprehensive study, a diverse range of data sources was employed to conduct a thorough assessment of air quality in Haryana, India, using Sentinel-5P satellite data. The study specifically focused on evaluating the types of pollutants and their

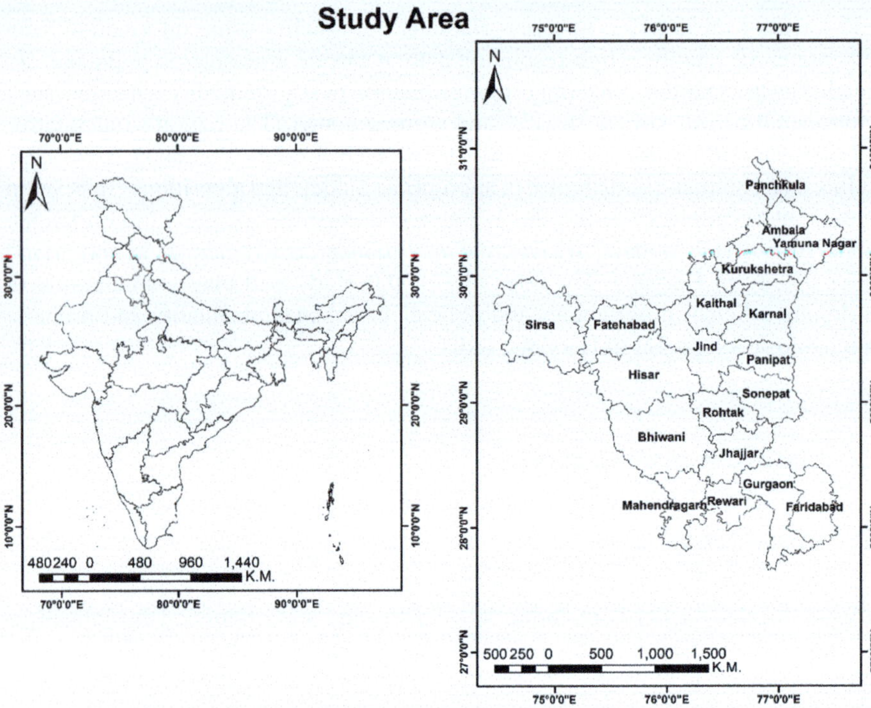


**Fig. 7.1** Location map of the study area

impact on air quality in the region. The primary data sources used in the study were carefully selected, and their descriptions, spatial resolution, and the years of data analysed are summarized in detail in Table 7.1.

The foundation of the methodology employed in this study centres around the utilization of Sentinel 5P satellite imagery for assessing air quality in Haryana, India. The Sentinel 5P satellite, developed by the European Space Agency, is a key tool in detecting and monitoring pollutant levels in the atmosphere, enabling real-time analysis (Singh et al. 2022a).

The imagery captured by Sentinel 5P offers a high level of precision in terms of spatial resolution and coverage, which proves instrumental in accurately identifying

**Table 7.1** Deatails of the dataset

| Data Source | Description | Spatial resolution | Year(s) |
|---|---|---|---|
| Sentinel 5P | High-resolution satellite imagery for detecting atmospheric pollutants from special sensors | 300 m | 2019/ 2022 |
| HPCB | Haryana Pollution Control Board data on air quality parameters for validation | N/A | 2019/ 2022 |

and monitoring pollutants in the air. With a spatial resolution ranging from 300 to 500 m, satellite imagery provides detailed information that aids in effectively characterizing the geographical extent and distribution of pollution areas within Haryana (Singh et al., 2022a, b, c). By relying on Sentinel 5P satellite data, this study gains valuable insight into the air quality conditions specific to Haryana. The satellite's capabilities enable the detection of pollutants, allowing for a comprehensive assessment of their concentration levels and spatial distribution across the region. This information is essential in evaluating the severity and extent of air pollution, identifying pollution hotspots, and facilitating targeted interventions to address and mitigate the environmental and health impacts associated with poor air quality in Haryana.

Accurate air quality data derived from the Sentinel-5P satellite plays a crucial role in assessing the air quality in Haryana. The data help in understanding the impact of various factors, such as biomass burning, road dust emissions, and industrial pollution, on the region's air quality. These activities release harmful gases and pollutants into the atmosphere, leading to a deterioration of air quality. Analysing these data is crucial for comprehending the extent and nature of these environmental concerns in Haryana. By incorporating Sentinel-5P satellite imagery, this study gains valuable insights into the concentration and distribution of air-degrading pollutants in Haryana. The satellite data provide a comprehensive understanding of the sources and intensity of pollution, helping to identify areas with high pollutant levels. This information is crucial for assessing the effectiveness of pollution control measures and formulating targeted interventions to mitigate the adverse effects on air quality (Singh et al. 2023b).

### *7.2.3 Data Processing*

To evaluate the air quality of Haryana, India, using Sentinel-5P satellite data, a thorough data preprocessing methodology was employed. The preprocessing primarily focused on analysing the annual concentration of pollutants in the state, to efficiently manage the extensive volume of Sentinel-5P satellite data, the Google Earth Engine, a cloud-based platform, was employed. This platform offers robust tools for processing and analysing geospatial data on a large scale. By utilizing this platform, yearly composites of the Sentinel-5P imagery for the selected years (2019–2022) were generated. This aggregation process involved consolidating multiple images into composite forms, streamlining subsequent analysis and improving computational efficiency.

Ensuring the reliability and quality of the satellite data was crucial for accurate air quality assessment. A common challenge in satellite imagery analysis is the presence of cloud cover, which can obstruct atmospheric visibility and impact pollutant detection accuracy. To address this challenge, a robust cloud filtering technique was integrated into the data preprocessing workflow. This filtering method aimed to select images with minimal cloud cover, typically below a predefined threshold of 10%. By implementing this quality control measure, the potential influence of cloud cover on the analysis was minimized, enhancing the accuracy and reliability of the air quality assessment, by meticulously conducting data preprocessing, including temporal focus, utilization of the Google Earth Engine platform, and rigorous cloud filtering, the study ensures the integrity and accuracy of the Sentinel-5P satellite data for evaluating the air quality in Haryana, India. These preprocessing steps are vital for obtaining dependable and actionable insights into pollutant levels and their distribution, facilitating informed decision-making and effective implementation of measures to mitigate air pollution in the region.

### 7.2.4 Air Quality Monitoring

The primary objective of this study is to assess the air quality of Haryana, India, using data from the Sentinel-5P satellite. The Sentinel-5P satellite, developed by the European Space Agency (ESA), is widely recognized for its capability to provide valuable information on atmospheric composition and pollution monitoring.

One of the key advantages of the Sentinel-5P satellite is its ability to capture high-resolution data, enabling detailed analysis of air quality at a local scale. With a spatial resolution of up to 7 kms, this satellite offers a comprehensive view of the region, allowing for the identification and quantification of air pollutants with great precision.

In addition to high resolution, the Sentinel-5P satellite collects data using a range of spectrometers that measure sunlight reflected or scattered back to space from the Earth's atmosphere and surface. These multispectral data encompass various spectral bands, enabling the detection and differentiation of different atmospheric constituents and pollutants.

By leveraging the spectral signatures obtained from the satellite's multispectral bands, this study aims to differentiate and analyse key air pollutants in Haryana, such as nitrogen dioxide ($NO_2$), ozone ($O_3$), carbon monoxide (CO), sulfur dioxide ($SO_2$), and formaldehyde ($CH_2O$). These pollutants play a significant role in assessing air quality, as they are associated with human activities, industrial emissions, and vehicular pollution.

Through the utilization of Sentinel-5P satellite data, this study seeks to provide valuable insights into the spatial distribution, temporal trends, and overall air quality status of Haryana. By monitoring and quantifying these pollutants, policymakers, researchers, and relevant stakeholders can gain a comprehensive understanding of the air pollution levels in the region, facilitating informed decision-making and the implementation of effective mitigation strategies.

### 7.2.5 Carbon Monoxide Concentration

The assessment of carbon monoxide (CO) concentration relies on the utilization of the Sentinel-5P satellite dataset. Sentinel-5P, a highly regarded satellite mission developed by the European Space Agency (ESA), provides valuable information on atmospheric composition, including the measurement of carbon monoxide levels.

To determine the concentration of carbon monoxide, specific spectral bands from the Sentinel-5P dataset, sensitive to carbon monoxide absorption, are employed. By applying advanced algorithms and mathematical formulas, the carbon monoxide concentration values are derived. These values represent the amount of carbon monoxide present in the observed atmospheric column. The detection of high carbon monoxide levels is crucial for understanding pollution sources and evaluating air quality. By analysing the carbon monoxide concentration values obtained from the Sentinel-5P dataset, areas with elevated carbon monoxide concentrations can be identified. This information offers insights into the distribution and intensity of carbon monoxide emissions in the study area.

### 7.2.6 Nitrogen Dioxide Concentration

In this study, the evaluation of nitrogen dioxide ($NO_2$) concentration in Haryana was carried out using Sentinel-5P satellite data obtained through the Google Earth Engine platform. The data underwent rigorous quality control measures to ensure their accuracy and reliability, including validation procedures to assess completeness, consistency, and adherence to quality standards. The collected nitrogen dioxide data were then subjected to spatial and temporal analysis to comprehend the distribution and variations in $NO_2$ levels across different regions of Haryana. Statistical techniques were employed to identify trends, patterns, and correlations within the data, involving the calculation of statistical measures such as means, standard deviations, and percentiles. Visual representations, such as maps and graphs, were generated to effectively illustrate the spatial distribution of nitrogen dioxide concentrations. These visualizations provided valuable insights into areas with elevated $NO_2$ levels and offered a clear depiction of pollution patterns throughout the state.

### 7.2.7 Sulfur Dioxide Concentration

In this study, the assessment of sulfur dioxide ($SO_2$) concentration in Haryana was conducted using Sentinel-5P satellite data obtained through the Google Earth Engine platform. The data underwent stringent quality control measures to ensure accuracy and reliability, including validation procedures to assess completeness, consistency, and adherence to quality standards. The collected sulfur dioxide data were then subjected to spatial and temporal analysis to understand the distribution and variations in $SO_2$ concentrations across different regions of Haryana. Statistical analysis techniques were employed to identify trends and patterns within the data, such as calculating statistical measures. To visualize the spatial distribution of sulfur dioxide concentrations, maps and graphical representations were generated. These visualizations provided insights into areas with elevated $SO_2$ levels and depicted the spatial patterns of pollution across the state.

### 7.2.8 Formaldehyde Concentration

In this study, the evaluation of formaldehyde ($CH_2O$) concentration in Haryana was conducted using data derived from the Sentinel-5P satellite and processed through the Google Earth Engine platform. Stringent quality control procedures were implemented to ensure the accuracy and reliability of the data, including validation checks for completeness, consistency, and adherence to quality standards. The collected formaldehyde data were then analysed spatially and temporally to comprehend the distribution and variations in $CH_2O$ concentrations across different regions of Haryana. Statistical analysis techniques were applied to detect trends, patterns, and correlations within the dataset. To visually represent the spatial distribution of formaldehyde concentrations, maps, and graphical visualizations were generated. These visual representations offered valuable insights into regions with elevated $CH_2O$ levels and depicted the spatial patterns of pollution throughout the state.

### 7.2.9 Aerosol Index

In this study, the assessment of the aerosol index in Haryana was conducted using data obtained from the aerosol index measurements derived from the Sentinel-5P satellite and processed through the Google Earth Engine platform. Stringent quality control procedures were implemented to ensure the accuracy and reliability of the data, including validation checks for completeness, consistency, and adherence to quality standards. The collected aerosol index data were then analysed spatially and temporally to understand the distribution and variations in aerosol concentrations across different regions of Haryana. Statistical analysis techniques were applied to identify trends, patterns, and correlations within the dataset, including the calculation of statistical measures such as means and standard deviations. To visualize the spatial distribution of aerosol concentrations, maps and graphical visualizations were created. These visual representations provided valuable insights into regions with elevated aerosol levels and depicted the spatial patterns of aerosol pollution throughout the state.

## 7.3 Results

The assessment of air quality in Haryana, India, involved the utilization of multiple satellite datasets, including nitrogen dioxide ($NO_2$), sulfur dioxide ($SO_2$), carbon monoxide (CO), formaldehyde ($CH_2O$), and aerosol index (AI). These datasets were derived from measurements obtained through the Sentinel-5P satellite and processed using various analytical techniques. The analysis of these datasets provided comprehensive insights into the concentrations and spatial distributions of these air pollutants across different regions of Haryana. The results were summarized in tables and visual representations, allowing for a better understanding of the variations and patterns observed in the air quality parameters. Visualizations such as Figs. 7.4 and 7.5 showcased the spatial distribution of $NO_2$, $SO_2$, CO, $CH_2O$, and AI concentrations, providing graphical illustrations of the observed patterns. These visualizations were instrumental in identifying areas with elevated pollutant levels and gaining insights into the spatial dynamics of air pollution in Haryana.

The utilization of multiple datasets enabled a holistic assessment of air quality, considering the contributions of various pollutants. This approach provided valuable information for policymakers, researchers, and stakeholders involved in addressing air pollution and developing strategies to improve air quality in Haryana. The integration and analysis of these diverse datasets contributed to a comprehensive understanding of the air quality conditions in the region and supported informed decision-making processes to mitigate pollution and promote environmental sustainability.

### *7.3.1 Table Showing Pollution Concentration in Haryana*

Table 7.2 provides an overview of the pollution levels resulting from various sources in Haryana from 2019 to 2022, covering a four-year study period. The analysis reveals significant fluctuations in pollution levels across different years. Sentinel-5P satellite data are used to assess the air quality in Haryana, India. The data in the table show the concentration of pollutants for each year as follows: In 2019, the concentrations were as follows: sulfur dioxide ($SO_2$)—50, nitrogen dioxide ($NO_2$)—43, carbon monoxide (CO)—56, aerosol index (AI)—24, formaldehyde (HCHO)—45. In 2020, the concentrations were $SO_2$—30, $NO_2$—8, CO—50, AI—23, and HCHO—27. For 2021, the concentrations were $SO_2$—46, $NO_2$—40, CO—51, AI—29, and HCHO—30. In 2022, the concentrations were $SO_2$—47, $NO_2$—42, CO—57, AI—30, and HCHO—33.

These findings highlight the annual pattern of pollution in Haryana, with higher levels observed in 2019 and 2021 and lower levels recorded in 2020 and 2022 due to factors such as lockdown measures and government interventions. The data underscore the persistent challenge of rising pollution levels during certain times of the year, emphasizing the need for effective measures to mitigate the environmental and health impacts associated with this issue. The pollutant concentration

**Table 7.2** Pollution concentration (μg/m3) in Haryana (2019–2022)

| Pollutant | Year | Concentration |
|---|---|---|
| $SO_2$, $NO_2$, CO, AI, HCHO | 2019 | 50, 43, 56, 24, 45 |
| $SO_2$, $NO_2$, CO, AI, HCHO | 2020 | 30, 38, 50, 23, 27 |
| $SO_2$, $NO_2$, CO, AI, HCHO | 2021 | 46, 40, 51, 29, 30 |
| $SO_2$, $NO_2$, CO, AI, HCHO | 2022 | 47, 42, 57. 30, 33 |

data obtained from the Sentinel-5P satellite provide valuable insights into the air quality in Haryana during the study period. The measured pollutants include sulfur dioxide ($SO_2$), nitrogen dioxide ($NO_2$), carbon monoxide (CO), aerosol index (AI), and formaldehyde (HCHO). These data help assess air quality and contribute to a better understanding of the environmental effects of these pollutants and other influencing factors. It is essential to consider these fluctuations in pollutant concentrations when devising strategies and policies to improve the region's air quality.

### *7.3.2 Carbon Monoxide Concentration Maps*

The spatial distribution of carbon monoxide (CO) from 2019 to 2022 showed varying patterns in different regions of Haryana in Figs. 7.2 and 7.3a–d shows the graphical representation of carbon concentration from 2019 to 2022. In 2019, the highest concentration of carbon monoxide was observed in the southeast and central parts of Haryana, while the lowest concentrations were recorded in Sirsa, Panchkula, and certain areas of Ambala, Fatehabad, and Bhiwani. In 2020, a significant improvement in air quality was observed across most parts of Haryana due to the coronavirus lockdown, except for Faridabad and some eastern areas of the state. However, in 2021, the air quality declined compared to 2020, with several CO hotspots identified in the central, eastern, and southeastern regions of Haryana. In 2022, the air quality was better than that in the previous year, as many parts of the state experienced cleaner air, except for Faridabad and certain areas of Panipat, Sonipat, and Yamuna Nagar. These observations highlight the temporal and spatial variations in CO concentrations across Haryana during the study period, emphasizing the need for continuous monitoring and targeted interventions to improve air quality in the identified hotspots.

### *7.3.3 Nitrogen Dioxide Concentration Map*

The distribution of nitrogen dioxide ($NO_2$) levels in Haryana varied across different regions from 2019 to 2022, as shown in Figs. 7.4, and 7.5a–d shows the graphical representation of nitrogen dioxide concentrations from 2019 to 2022. In 2019, the highest concentrations of $NO_2$ were observed in Panipat, Gurugram, and Faridabad,

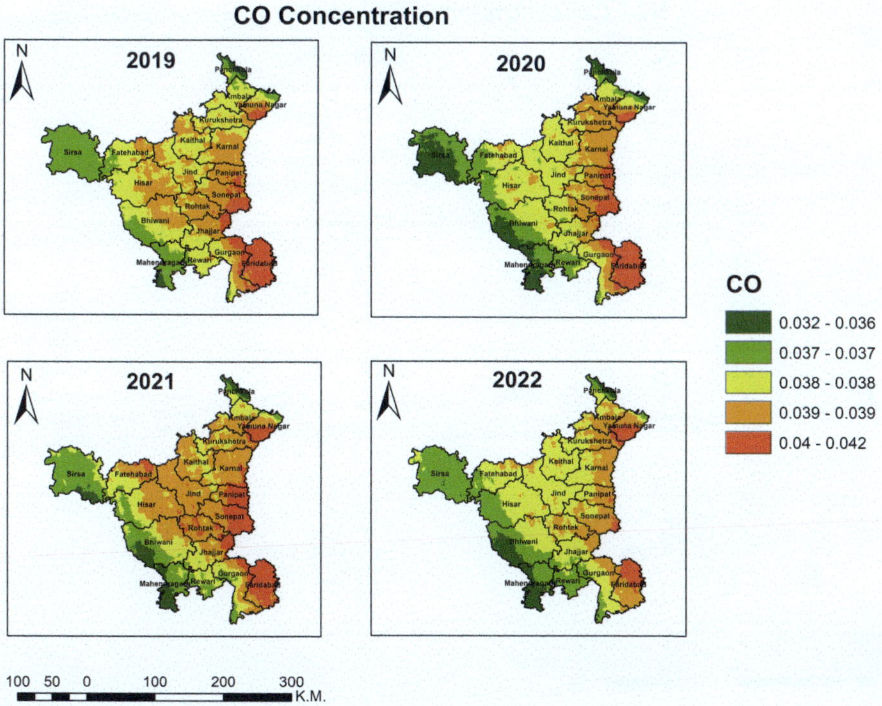


Fig. 7.2 CO concentration

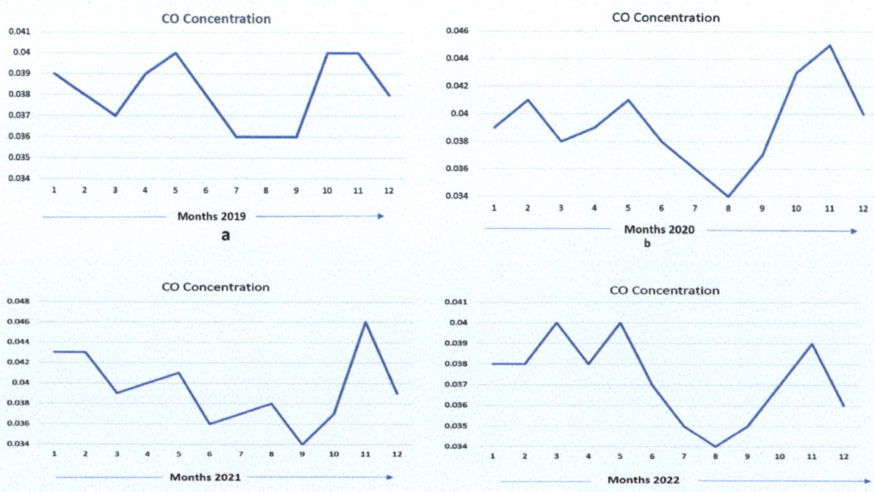


Fig. 7.3 (a–d): CO concentration during 2019, 2020, 2021 and 2022

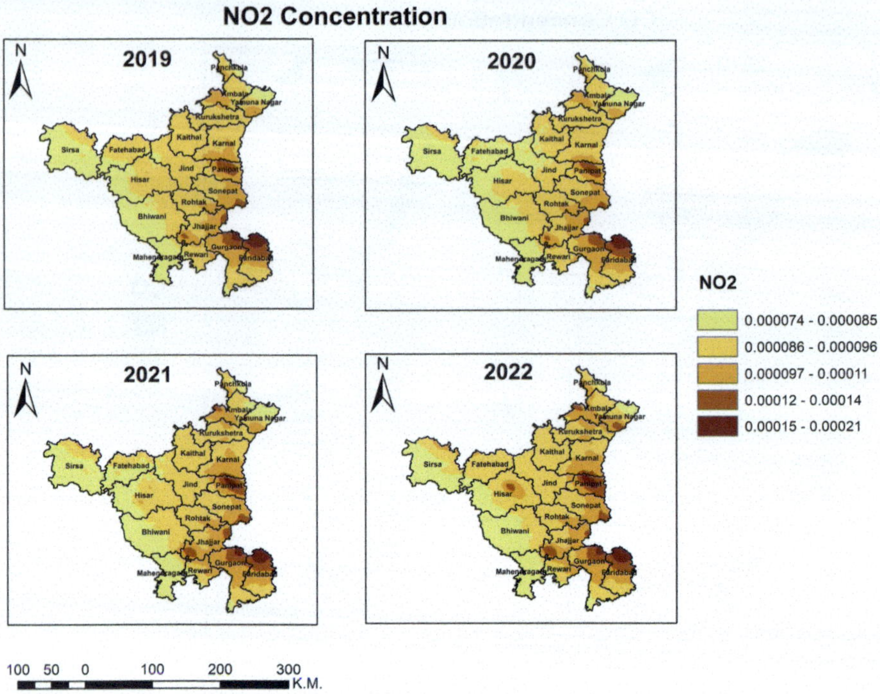


Fig. 7.4 $NO_2$ Concentration

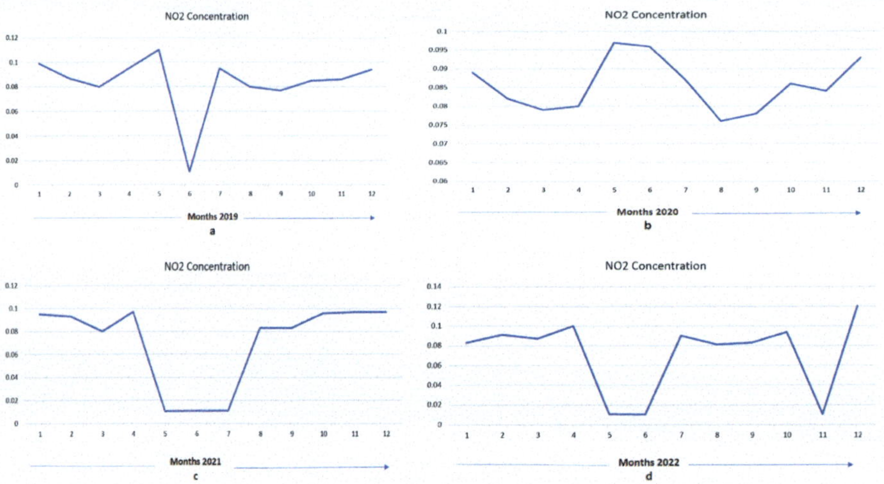


Fig. 7.5 (a–d): $NO_2$ concentration during 2019, 2020, 2021 and 2022

while the rest of Haryana had lower $NO_2$ levels. The year 2020 witnessed a significant improvement in air quality across most parts of Haryana due to the COVID-19 lockdown. However, Faridabad, Gurugram, and Panipat remained areas with high $NO_2$ concentrations, classified as red zones. Unfortunately, in 2021, the air quality worsened compared to 2020, with increased $NO_2$ hotspots identified in the central and southeastern regions of Haryana. Regrettably, in 2022, the air quality deteriorated compared to the previous year, and many parts of the state experienced elevated $NO_2$ levels. Additionally, new $NO_2$ hotspots emerged in central areas of Haryana, particularly in Hisar and Yamuna Nagar. These observations highlight the temporal and spatial variations in $NO_2$ levels throughout the study period in Haryana. They emphasize the need for continuous monitoring and targeted interventions to improve air quality, particularly in identified hotspots.

### *7.3.4 Sulfur Dioxide Concentration Map*

The emission of sulfur dioxide ($SO_2$) in Haryana has shown an upward trend over the years. The potential sources of $SO_2$ emissions include electroplating units and small chimneys in smoke-emitting industries, and the major contributors are vehicles that are more than 10 years old. These older vehicles have the highest impact on $SO_2$ concentration levels. In 2022, the data indicate that nearly all districts in Haryana are classified as danger zones in terms of SO2 levels, except for the western region of the state. Figure 7.6 shows the $SO_2$ concentration from 2019 to 2022, and Fig. 7.7a–d shows the graphical distribution of sulfur dioxide from 2019 to 2022.

### *7.3.5 Formaldehyde Concentration Map*

Formaldehyde plays a significant role in air pollution, as it can undergo conversion into carbon dioxide ($CO_2$) when present in the atmosphere for extended periods. Its emissions not only contribute to pollution but also contribute to the global warming issue, which is a major concern worldwide. In 2019, there was a rising trend of formaldehyde concentration observed in the central parts of Haryana, while the highest levels were found in the northern and southeastern regions. The situation remained relatively unchanged in 2020, with reduced concentrations observed in the northern and southeastern parts of Haryana. However, in 2021, there was a significant increase in formaldehyde concentration levels, and 2022 recorded the highest concentration levels. The northern, eastern, and southeastern parts of Haryana were identified as the major areas with substantial formaldehyde levels. These observations highlight the importance of formaldehyde as a pollutant and its contribution to air pollution and global warming. Continuous monitoring and targeted measures are essential to address the growing concentrations of formaldehyde in specific regions

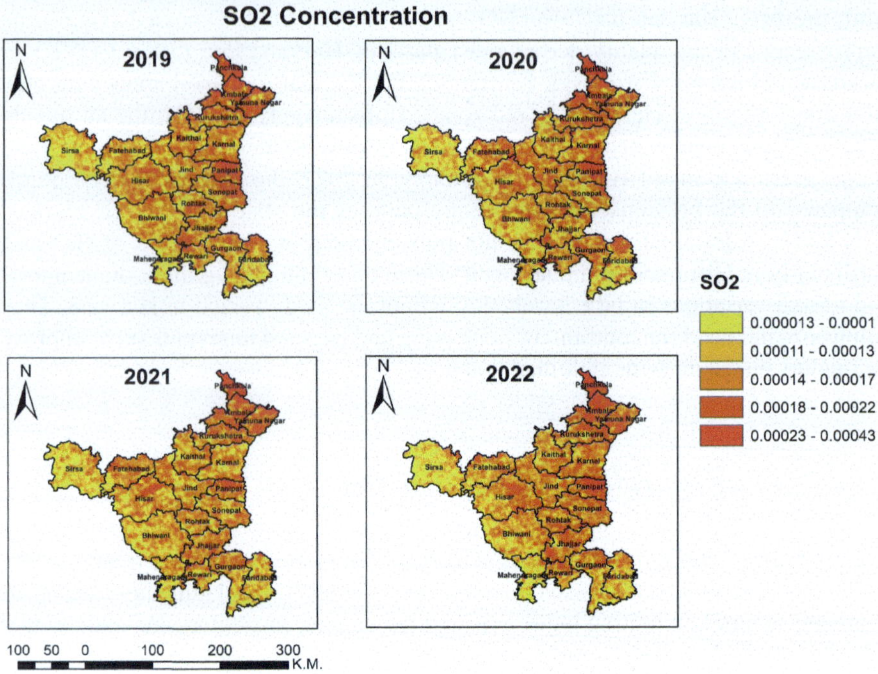


**Fig. 7.6** $SO_2$ concentration

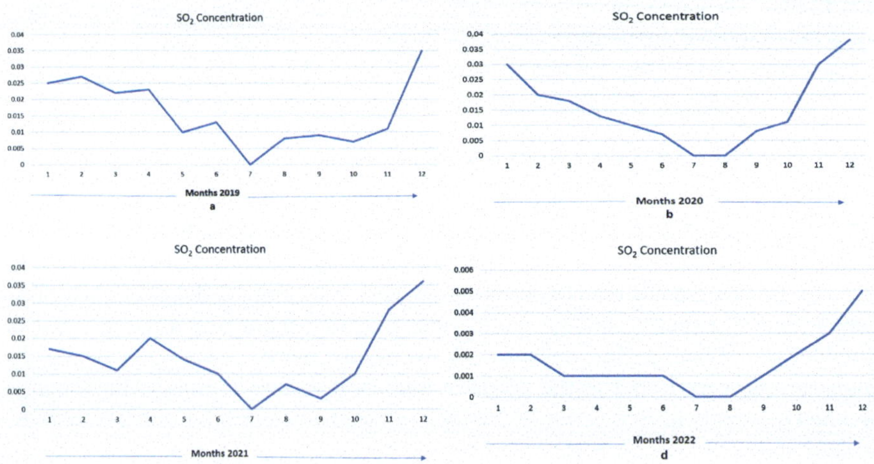


**Fig. 7.7** (**a–d**): $SO_2$ concentration during 2019, 2020, 2021 and 2022

of Haryana, as shown in Figs. 7.8, and 7.9a-d shows the graphical distribution of HCHO from 2019 to 2022.

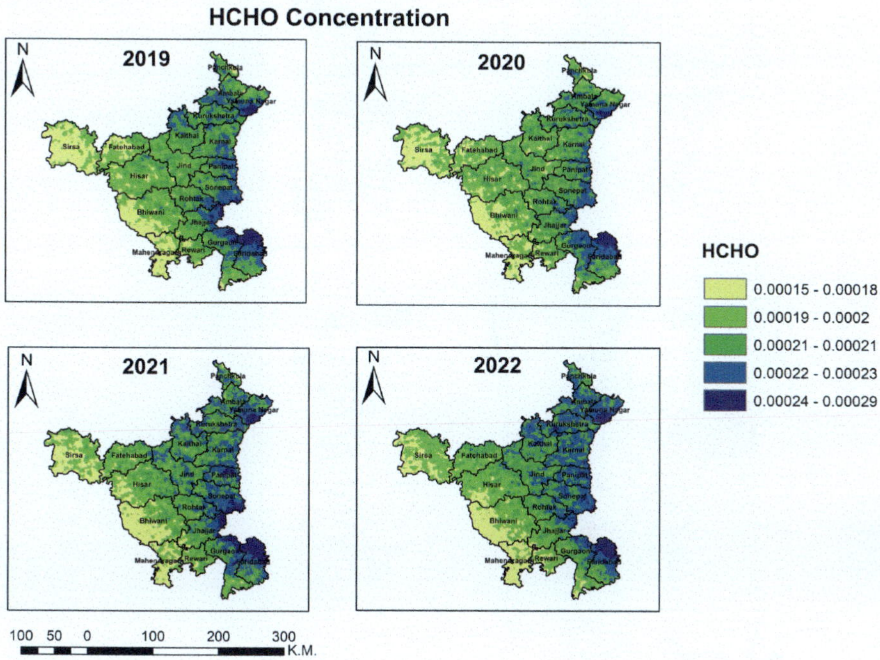


Fig. 7.8 HCHO concentration

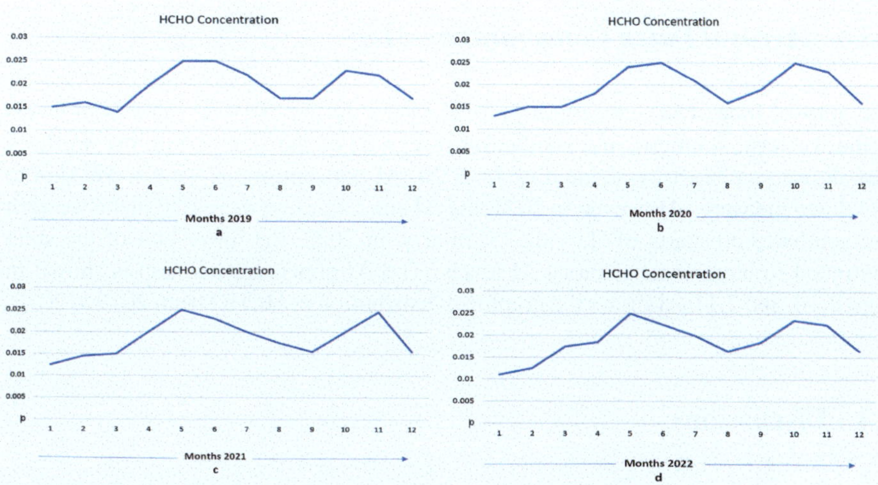


**Fig. 7.9** (**a–d**): HCHO concentration during 2019, 2020, 2021 and 2022

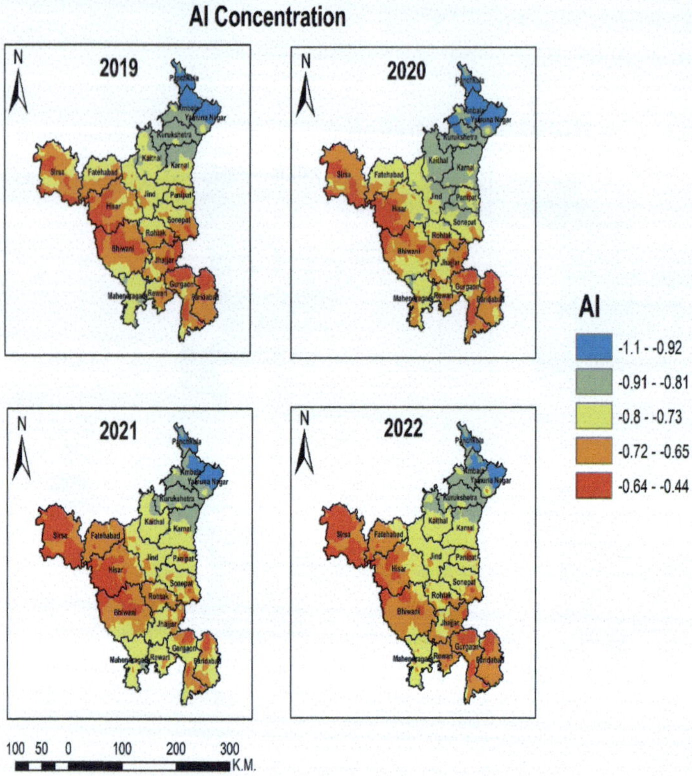


**Fig. 7.10** AI concentration

### *7.3.6 Aerosol Index Concentration Map*

The aerosol index (AI) concentration in Haryana during 2019 revealed high levels in the western, southern, and southeastern regions of the state, while the northern part showed satisfactory air quality. In 2020, AI concentrations were lower due to lockdown measures. However, in 2021, the AI concentration increased in the western and southwestern parts of Haryana. Similarly, in 2022, the lower belt of the state continued to have unsatisfactory AI levels. The AI concentration map is shown in Figs. 7.10, and 7.11a-d shows the graphical distribution of HCHO from 2019 to 2022.

## 7.4 Discussions

The results presented in the research paper provide a comprehensive assessment of air quality in Haryana, India, from 2019 to 2022. The utilization of multiple satellite datasets, including nitrogen dioxide ($NO_2$), sulfur dioxide ($SO_2$), carbon monoxide

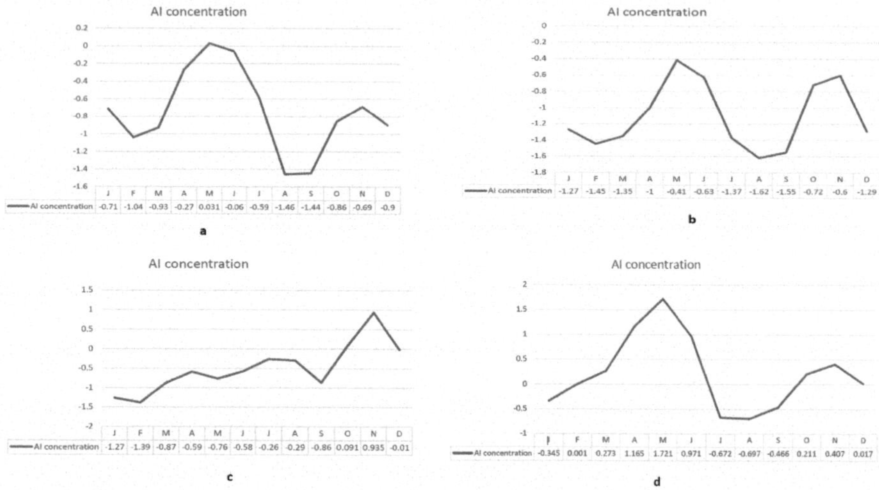


**Fig. 7.11** **(a–d)**: AI Concentration during 2019, 2020, 2021 and 2022

(CO), formaldehyde ($CH_2O$), and aerosol index (AI), allowing for a holistic evaluation of the concentrations and spatial distributions of these pollutants across different regions of Haryana. The analysis revealed significant fluctuations in pollution levels over the four-year study period. In 2019, higher concentrations of pollutants were observed, particularly $SO_2$, $NO_2$, CO, AI, and $CH_2O$. This suggests that pollution levels were relatively high during that year. In 2020, there was a notable improvement in air quality, attributed to the COVID-19 lockdown measures implemented during that period. However, certain areas, such as Faridabad and some eastern regions of Haryana, continued to experience higher pollution levels. The year 2021 saw a decline in air quality compared to 2020, with increased pollutant hotspots identified in the central, eastern, and southeastern regions of Haryana. This indicates the persistence of air pollution challenges in specific areas despite overall improvements. Fortunately, in 2022, air quality improved compared to the previous year, with cleaner air observed in many parts of the state. However, pockets of elevated pollution levels remained in Faridabad and certain areas of Panipat, Sonipat, and Yamuna Nagar. The spatial distribution maps for carbon monoxide (CO), nitrogen dioxide ($NO_2$), sulfur dioxide ($SO_2$), formaldehyde ($CH_2O$), and aerosol index (AI) provided graphical representations of the pollution patterns across Haryana. These visualizations were instrumental in identifying areas with elevated pollutant levels and understanding the spatial dynamics of air pollution. They highlighted the temporal and spatial variations in pollutant concentrations and emphasized the need for continuous monitoring and targeted interventions to improve air quality, particularly in identified hotspots. The findings of this research paper underscore the persistent challenge of rising pollution levels in Haryana and the importance of effective measures to mitigate the environmental and health impacts associated with air pollution. The integration and analysis of diverse datasets from satellite measurements allowed for a comprehensive understanding of air quality conditions in the region. This information is valuable for policymakers, researchers, and stakeholders involved in addressing air pollution and developing strategies to improve air quality in Haryana.

The results also highlighted the impact of external factors, such as lockdown measures, on air quality. The significant improvement in air quality observed during the COVID-19 lockdown in 2020 underscores the potential benefits of reducing human activities and emissions. However, it is essential to recognize that these improvements were temporary and that long-term solutions are needed to sustain clean air. The research paper's findings provide valuable insights into the air quality conditions in Haryana from 2019 to 2022. The analysis of multiple satellite datasets and the visual representations of pollutant concentrations and spatial distributions allowed for a comprehensive assessment of air pollution in the region. The identified temporal and spatial variations in pollutant levels emphasize the need for continuous monitoring, targeted interventions, and policy measures to address the persistent air pollution challenges in Haryana and promote environmental sustainability.

## 7.5 Conclusions

The assessment of air quality in Haryana, India, utilizing Sentinel-5P satellite data has yielded significant findings. Through this literature review, it has been demonstrated that the Tropospheric Monitoring Instrument (TROPOMI) on Sentinel-5P provides accurate measurements of air pollutants, including nitrogen dioxide ($NO_2$), carbon monoxide (CO), and aerosols. These measurements have enabled the identification of temporal and spatial variations in air quality across different regions of Haryana. Furthermore, strong correlations have been observed between satellite-based measurements and ground-based monitoring stations, validating the reliability and accuracy of Sentinel-5P data for air quality assessment in Haryana.

This literature review has contributed to the existing knowledge on assessing air quality in Haryana using Sentinel-5P satellite data. Synthesizing previous research has highlighted the capabilities and limitations of satellite-based remote sensing for monitoring air pollutants in the region. The review has provided insights into the heterogeneity of pollution sources in Haryana, such as industrial emissions, vehicular traffic, and agricultural activities, and has demonstrated the value of Sentinel-5P data in capturing and analysing these sources. These findings have enhanced our understanding of the air quality situation in Haryana and its implications for public health and the environment. The practical implications of utilizing Sentinel-5P satellite data for assessing air quality in Haryana are significant. Accurate and timely measurements of air pollutants facilitate evidence-based decision-making by policymakers and environmental agencies. By identifying pollution hotspots and understanding temporal patterns, targeted strategies can be developed to mitigate air pollution and protect public health effectively. Integrating Sentinel-5P data with ground-based monitoring networks enhances monitoring capabilities, providing a comprehensive understanding of air quality dynamics in Haryana.

Based on the findings of this literature review, several recommendations for future action can be proposed. Continual monitoring of air quality in Haryana using Sentinel-5P satellite data is essential to assess long-term trends and evaluate the effectiveness of pollution control measures. Improving the spatial resolution of satellite data and refining algorithms for retrieving air pollutant concentrations would enable more precise identification and characterization of pollution sources, facilitating targeted interventions. Collaboration between satellite data providers, researchers, and local environmental agencies should be encouraged to enhance the utilization of satellite data for air quality management in Haryana. Additionally, further research on the specific health effects of air pollution in Haryana and its contribution to climate change is warranted, by implementing these recommendations, stakeholders can maximize the potential of Sentinel-5P satellite data to address air quality challenges in Haryana. This will lead to improved environmental management and the protection of public health in the region.

## References


Debnath J, Sahariah D, Lahon D, Nath N, Chand K, Meraj G, Kumar P, Singh SK, Kanga S, Singh SK (2022a) Geospatial modelling to assess the past and future land use-land cover changes in the Brahmaputra Valley, NE India, for sustainable land resource management. Environ Sci Pollut Res, 1–24

Debnath J, Sahariah D, Lahon D, Nath N, Chand K, Meraj G, Kumar P, Singh SK, Kanga S, Farooq M (2023a) Assessing the impacts of current and future changes of the planforms of River Brahmaputra on its land use-land cover. Geosci Front. https://doi.org/10.1016/j.gsf.2023.101557

Debnath J, Sahariah D, Saikia A, Meraj G, Nath N, Lahon D, Annayat W, Kumar P, Chand K, Singh SK, Kanga S (2023b) Shifting sands: assessing bankline shift using an automated approach in the Jia Bharali River. India. Land 12(3):703

Farooq M, Meraj G, Rishabh K, S., Nathawat, R., Singh, S. K., & Ranga, V. (2022a) Slum Categorization for Efficient Development Plan—A Case Study of Udhampur City, Jammu and Kashmir Using Remote Sensing and GIS. Geospatial Technology for Landscape and Environmental Management: Sustainable Assessment and Planning. Springer International Publishing, Cham, pp 283–299

Farooq M, Mushtaq F, Meraj G, Singh SK, Kanga S, Gupta A, Kumar P, Singh D, Avtar R (2022b) Strategic slum upgrading and redevelopment action plan for Jammu City. Resources 11(12):120

Farooq M, Rashid H, Meraj G, Kanga S, Singh SK (2022c) Assessing the microclimatic environmental indicators of climate change of a temperate valley in the Western Himalayan region. In A. R. Siddiqui and Avijit Sahay (Eds.), Climate Change, Disaster and Adaptations—Contextualizing human responses to ecological change (pp. 47–61). https://link.springer.com/chapter/https://doi.org/10.1007/978-3-030-91010-5_5

Gulati B, Sharma R, Kanga S, Singh SK, Sajan B, Meraj G, Kumar P, Ramanathan AL (2023) Unravelling the relationship between stubble burning and air quality degradation in punjab: a temporal and spatial analysis (2019–2022). J Clim Chang 9(2):43–53

Kanga S, Meraj G, Farooq M, Singh SK, Nathawat MS (eds) (2022a) Disaster management in the complex Himalayan Terrains. Geogr Phys Environ. Springer, Cham

Kanga S, Singh SK, Meraj G, Kumar A, Parveen R, Kranjˇ N, Đurin B (2022b) Assessment of the impact of urbanization on geoenvironmental settings using geospatial techniques: a study of Panchkula District, Haryana. Geographies, 2(1), 1–10. ISSN No. 2673–7086

Kotecha MJ, Kanga S, Pankhaniya SK, Agrawal S, Meraj G, Singh SK (2023) Urban Heat Island (UHI) Resilience Plan in Varying Climatic Conditions Using Geospatial Approach: A Case Study of Rajkot City. In: Change C (ed) Agriculture and Society: Approaches Toward Sustainability. Springer Int Publ, Cham, pp 307–338

Lahon D, Sahariah D, Debnath J, Nath N, Meraj G, Farooq M, Kanga S, Singh SK, Chand K (2023a) Growth of water hyacinth biomass and its impact on the floristic composition of aquatic plants in a wetland ecosystem of the Brahmaputra floodplain of Assam. India. Peerj 1–32. https://doi.org/10.7717/peerj.14811

Lahon D, Sahariah D, Debnath J, Nath N, Meraj G, Kumar P, Hashimoto S, Farooq M (2023b) Assessment of Ecosystem Service Value in Response to LULC Changes Using Geospatial Techniques: A Case Study in the Merbil Wetland of the Brahmaputra Valley, Assam. India. ISPRS Int J Geo-Inf 12(4):165

Lavane K, Kumar P, Meraj G, Han TG, Ngan LHB, Lien BTB, Ty TV, Thanh NT, Downes NK, Nam NDG, Minh HVT, Singh SK, Kanga S (2023) Assessing the effects of drought on rice yields in the mekong delta. Climate 11(1):13

Meraj G, Farooq M, Singh SK, Islam MN, Kanga S (2022) Modelling the sediment retention and ecosystem provisioning services in the Kashmir valley, India, Western Himalayas. Model Earth Syst Environ 8(3):3859–3884

Meraj G, Kanga S, Kranjčić N, Đurin B, Singh SK (2021a) Role of natural capital economics for sustainable management of earth resources. Earth 2(3):622–634

Meraj G, Singh SK, Kanga S, Islam MN, Singh SK (2021b) Modelling on comparison of ecosystem services concepts, tools, methods and their ecological-economic implications: A review. Model Earth Syst Environ, 1–20

Nath N, Sahariah D, Meraj G, Debnath J, Kumar P, Lahon D, Chand K, Farooq M, Chandan P, Singh SK, Kanga S (2023) Land use and land cover change monitoring and prediction of a UNESCO world Heritage site: Kaziranga Eco-Sensitive zone using cellular Automata-Markov model. Land 12(1):151

Rafiq M, Meraj G, Kesarkar AP, Farooq M, Singh SK, Kanga S, Bhardwaj V (2022) Hazard mitigation and climate change in the himalayas–policy and decision making. In Kanga S, Meraj G, Farooq M, Singh SK, Nathawat MS (Eds.), Disaster Management in the Complex Himalayan Terrains. Geogr Phys Environ. Springer, Cham. https://doi.org/10.1007/978-3-030-89308-8_12

Sajan B, Mishra VN, Kanga S, Meraj G, Singh SK, Kumar P (2022) Cellular automata-based artificial neural network model for spatial-temporal dynamics and land use/land cover change prediction. Agronomy 12:2772

Singh A, Singh SK, Meraj G, Kanga S, Farooq M, Kranjcic N, Đurin B, Sudhanshu. (2022a) Designing geographic information system based property tax assessment in India. Smart Cities 5:364–381

Singh H, Meraj G, Singh S, Shrivastava V, Sharma V, Farooq M, Kanga S, Kumar P (2022b) Status of air pollution during COVID-19-Induced lockdown in Delhi. India. Atmosphere 13(12):2090

Singh SK, Gupta SK, Meraj G, Kumar P, Kanga S (2023a) Uncovering the hydrometeorological drivers responsible for forest fires utilizing geospatial techniques. Theoret Appl Climatol. https://doi.org/10.1007/s00704-023-04497-y

Singh SK, Meraj G, Mondal N, Bera A, Verma MK, Tomar JS, Kanga, S (2022) Assessment of seasonal vegetation dynamics over parts of thar desert using geospatial techniques. ANGRAU, 105

Singh S, Singh SK, Prajapat DK, Pandey V, Kanga S, Kumar P, Meraj G (2023b) Assessing the impact of the 2004 Indian ocean tsunami on south andaman's coastal shoreline: a geospatial analysis of erosion and accretion patterns. J Mar Sci Eng 11(6):1134

Sud A, Kanga R, Singh SK, Meraj G, Kanga S, Kumar P, Ramanathan AL, Bhardwaj V (2023) Simulating groundwater potential zones in mountainous Indian Himalayas—A Case Study of Himachal Pradesh. J Geophys Res Earth Surf 10(3):65

Upadhyay RK, Tripathi G, Đurin B, Šamanović S, Cetl V, Kishore N, Sharma M, Singh SK, Kanga S, Wasim M, Rai PK, Bhardwaj V (2023) Groundwater potential zone mapping in the Ghaggar River Basin, North-West India, using integrated remote sensing and GIS techniques. Water 15(5):961

# Chapter 8
# Food Security Through Sustainable Agriculture: A Prospective Study in the Indian Context

**Ankit Singh and Anurag Gupta**

## 8.1 Introduction

Food security has evolved in response to shifts in mainstream policy thinking over the last few decades. The global food crisis of 1972–1974 is largely responsible for the current concern over food security. During the World Food Conference in 1974 that took place in Rome, it marked the first formal recognition of this problem affecting all individuals. "Every man, woman, and child has the inalienable right to be free from hunger and malnutrition to develop fully and maintain their physical and mental faculties. As a result, eradicating hunger is the shared goal of all countries in the world community, particularly developed countries, and others in a position to assist."

The 1983 FAO analysis focused on the concept of food security, which sought to strike a balance between supply and demand. The term "food security" encompasses both physical and economic access to food. According to the World Food Summit of 1996, there is food security "when all people at all times have access to sufficient, safe, nutritious food to maintain a healthy and active life." Food security, in accordance with the WHO, is made up of three elements: "food availability," "food access," and "food use." The fourth factor was also introduced by FAO, which it describes as "the stability of the first three dimensions of food security over time" (Fig. 8.1).

Ensuring food security requires a comprehensive approach that addresses four key elements:

- Food Availability: Sufficient quantities of appropriate food must be physically accessible, whether through domestic production or imports. This aspect focuses on the "supply side" of food security.

A. Singh (✉) · A. Gupta
Department of Geography, Deen Dayal Upadhyaya Gorakhpur University, Gorakhpur, India
e-mail: ankit.geo@ddugu.ac.in

© The Author(s), under exclusive license to Springer Nature Singapore Pte Ltd. 2024

G. Tripathi et al. (eds.), *Sustainability and Health Informatics*, Advances in Geographical and Environmental Sciences,
https://doi.org/10.1007/978-981-97-6706-9_8

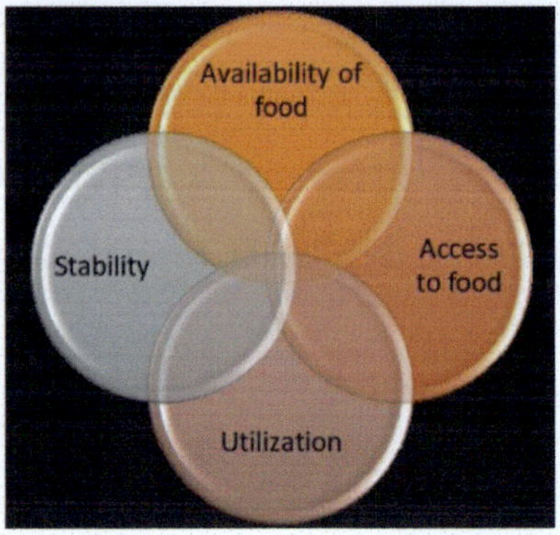


**Fig. 8.1** Dimensions of food security

- Food Accessibility: Individuals must have the means and resources (entitlements) to acquire the necessary food for a nutritious and balanced diet. This element emphasizes the ability to obtain food.
- Food Utilization: Utilization encompasses the body's ability to utilize the various nutrients found in food. This requires a balanced diet, access to clean water, proper hygiene practices, sanitation, and healthcare.
- Food Stability: Food security must be sustained over time, ensuring that individuals, households, and communities have consistent access to adequate food. This element emphasizes the resilience of food systems.

Achieving the Sustainable Development Goals and other ambitious targets is inextricably linked to ensuring adequate food and nutrition. While only SDG 2 ("End hunger, achieve food security and enhanced nutrition, and promote sustainable agriculture"), explicitly mentions "nutrition," its influence extends far beyond this single goal. Nutrition plays a crucial role in advancing progress toward all SDGs, underscoring its cross-cutting significance (Ipe et al. 2022). The 2030 Sustainable Development Agenda includes a transformational vision that anticipates the changes our world is undergoing and the new problems it will face. Achieving a world free from food insecurity, malnutrition, and hunger necessitates addressing significant challenges, including the immense strain placed on global food supplies and natural resource availability due to rapid population growth (Kotecha et al. 2024). In the previous two decades, economies have become more integrated and globalized, which has resulted in record harvests, new technical advancements, income enhancement, and food diversification (IPCC 2019). Progress towards achieving Zero Hunger, the second Sustainable Development Goal, is alarmingly slow. With its burgeoning population and staggering hunger statistics, India faces a daunting food security challenge. The Food and Agriculture Organization estimates that over 190 million Indians face hunger daily, accounting for 15% of the world's undernourished population. India's Global Hunger Index (GHI) ranking of 107th out of 121 countries and its Global Food Security (GFS) Index score of 58.9, placing it 68th out of 113 nations, underscore the urgency of addressing food insecurity. Therefore, several global organizations are asking for immediate action to transform the food systems to tackle hunger, alleviate food insecurity, and increase accessibility, not just in India but globally (Rai et al. 2021).

Studies have long focused on agriculture because of its importance to food security and its ability to employ a sizable portion of the working-age population. Agriculture holds immense significance in driving economic growth, particularly for poverty reduction, as it serves as the primary livelihood for a substantial portion of the rural population, where poverty is more prevalent. Indian agriculture, traditionally a subsistence activity, has played a pivotal role in safeguarding the nation's food security (Ahmad et al. 2023; Debnath et al. 2024b). The Green Revolution, marked by the adoption of fertilizers, high-yielding variety seeds, and irrigation techniques, has enabled many states to achieve surplus food production, outpacing population growth. The so-called "Green Revolution," which commenced in late 1967–1968, was a major turning point that led to a tremendous growth of crop production, particularly cereals, and transformed India's condition regarding food security. The Revolution had the greatest influence on the wheat crop among the food grains; it tripled the production of wheat in the decades that followed, and cereal production doubled, resulting in a reduction of almost 50% in both poverty and food insecurity in the nation. However, the First Green Revolution's weariness is now obvious, and practically all agricultural yields have plateaued in terms of both production and yield. Declining crop yields due to unsustainable agricultural techniques over the years, along with a rapid rate of population growth, have put a tremendous strain on the country's food supply situation. Therefore, a second Green Revolution is required in India to increase food production and guarantee food security (Swaminathan 2008).

By managing natural resources in a way that preserves ecological harmony while providing for present and future human needs, sustainable agriculture may play a significant part in helping to solve this problem by ensuring that there is always an ample supply of nutritious food that is accessible to everyone.

## 8.2 Objectives

- To analyze India's food production trend.
- To highlight the food demand in the future for India.
- To analyze the factors entailing the need for sustainable agriculture in India.
- To delineate the sustainable agricultural practices that will promote and safeguard food availability in India.

## 8.3 Methodology

The data owing to the study were obtained from secondary sources. Various government publications and international publications, viz. Agricultural Statistics at a glance (GOI 2022), a report by NITI Ayog (2018), Report by National Council on

Integrated Water Resources Development (NCIWRD 1999), Reports published by FAO (1991, 2010, 2021), etc., research papers, and articles have been cited for this study. Tabulation and Graphical techniques have been used for trend analysis of various factors.

## 8.4 Trends of Food Production in India

Agriculture in India is mostly concerned with food grain production and livestock development to boost economic growth. In light of the problem of food security that most emerging nations face, the production of food grains is crucial. Given the population size, rate of population growth, and shift in dietary patterns, increased production of foodgrains and livestock products becomes critical to ensuring food security in the Indian context.

### *8.4.1 Foodgrain Production in India*

India had to import a large amount of foodgrains from the First World's "developed" nations, with the United States serving as the main provider. While India's efforts to achieve food grain self-sufficiency began in the early 1970s, it was the groundbreaking contributions of the Indian Agricultural Research Institute (IARI) that ultimately enabled the nation to attain and sustain this critical milestone. Table 8.1 highlights the decadal variations in Foodgrain Production and Yield in India from 1951 to 2021 to comprehend the Foodgrain Production in India (comprising wheat, rice, coarse cereals, and pulses) (Fig. 8.2).

**Table 8.1** Production and yield of Foodgrains in India (1951–2021)

| Year | Production (in million tonnes) | Yield (kg/ha) | Average per annum growth rate in yield (%) |
|---|---|---|---|
| 1951 | 50.82 | 522 | – |
| 1961 | 82.02 | 710 | 3.60 |
| 1971 | 108.42 | 872 | 2.28 |
| 1981 | 129.59 | 1023 | 1.73 |
| 1991 | 176.39 | 1380 | 3.48 |
| 2001 | 196.81 | 1626 | 1.78 |
| 2011 | 244.49 | 1930 | 2.05 |
| 2021 | 310.74 | 2394 | 2.40 |

*Source* GOI (2022). *Agricultural Statistics at a Glance*. Ministry of Agriculture & Farmers Welfare, Department of Agriculture & Farmers Welfare, Economics & Statistics Division

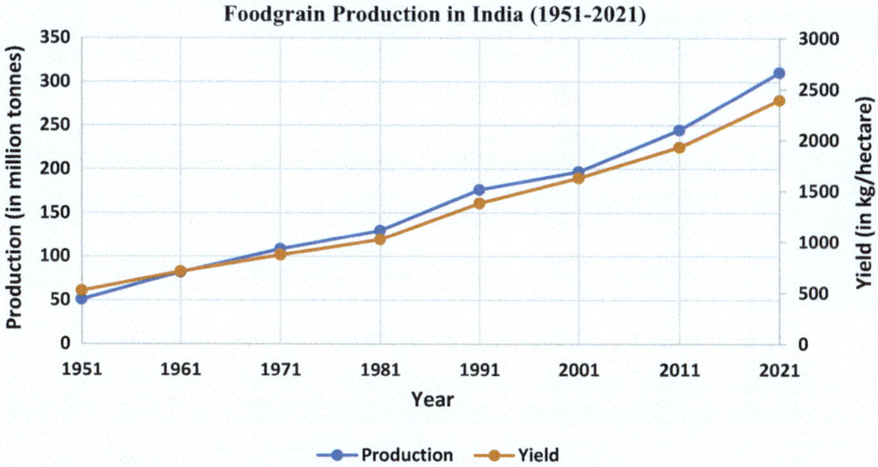


**Fig. 8.2** Trends of Foodgrain production in India (1951–2021)

The data presented above illustrates a remarkable transformation in India's foodgrain production. From a stagnant state in the 1950s, foodgrain output surged six-fold following the advent of the Green Revolution, leaping from 50.82 million tonnes in 1951 to an impressive 310.74 million tonnes in 2021. There has been a steady rise in foodgrain productivity, which has more than quadrupled over the past seven decades, from 522 kg/ha in 1951 to a substantial 2394 kg/ha in 2021. This substantial boost in yield can be attributed to the widespread adoption of high-yielding variety (HYV) seeds, the implementation of advanced irrigation techniques, and the utilization of a variety of other inputs.

### *8.4.2 Per Capita Availability of Foodgrains*

Table 8.2 illustrates the net availability of food grains per capita in India from 1951 to 2022. The table shows that there has been a modest rise in foodgrain availability per person over the past few decades. Foodgrain availability per person was 171.1 kg/year in 1961, but it was unchanged in 1971. In 1981, this decreased to 166 kg. In the following ten years, net availability grew by more than 12%, reaching 186.2 kg per person per year in 1991. This gain was transient, though, as it dropped to 151.1 kg/year in 2001. We were unable to reach the level reached in 1991 till 2020, but only in 2021 and 2022(P) did India reach 187.8 kg/year.

While food security in India has historically been linked to foodgrain security, it is crucial to recognize that the availability of high-value horticulture and animal products is gradually replacing staple food grains as the primary source of nutrition for individuals (Fig. 8.3).

**Table 8.2** Net availability of foodgrains per capita (kg/year)

| Year | Rice | Wheat | Other cereals | Cereals | Pulses | Food grains |
|---|---|---|---|---|---|---|
| 1951 | 58 | 24 | 40 | 122 | 22.1 | 144.1 |
| 1961 | 73.4 | 28.9 | 43.6 | 145.9 | 25.2 | 171.1 |
| 1971 | 70.3 | 37.8 | 44.3 | 152.4 | 18.7 | 171.1 |
| 1981 | 72.2 | 47.3 | 32.8 | 152.3 | 13.7 | 166 |
| 1991 | 80.9 | 60 | 29.2 | 171 | 15.2 | 186.2 |
| 2001 | 69.5 | 49.6 | 20.5 | 141 | 10.9 | 151.9 |
| 2011 | 66.3 | 59.7 | 23.9 | 149.9 | 15.7 | 170.9 |
| 2021 | 73.3 | 67 | 30.2 | 170.5 | 16.3 | 186.8 |
| 2022 | 69.6 | 68.8 | 29.8 | 168.2 | 19.6 | 187.8 |

*Source* GOI (2022). *Agricultural Statistics at a Glance.* Ministry of Agriculture & Farmers Welfare, Department of Agriculture & Farmers Welfare, Economics & Statistics Division

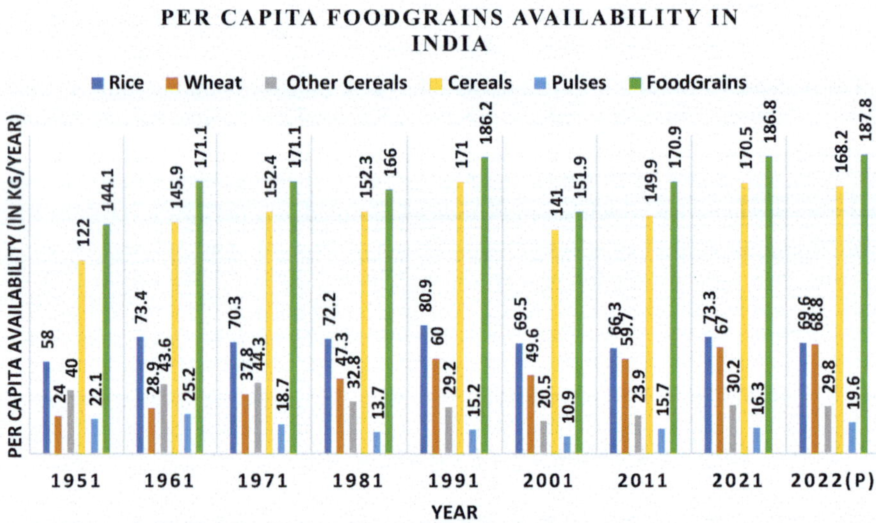


**Fig. 8.3** Net availability of foodgrains per capita in India

## *8.4.3 Livestock and Poultry Production*

Cattle and poultry are the mainstays of Indian agriculture, which is essentially an economic symbiosis of crop and livestock production. About 70% of the rural populace depends on the livestock sector in a variety of ways for their livelihood. In addition to ensuring the farmers' financial viability, raising livestock and poultry is crucial for India's nutritional security (Singh and Singh 2022). In 1927, the Royal Commission on Agriculture established increased production per animal and decreased animal numbers as goals for livestock development in India. However, there wasn't much improvement in livestock development until the late 1980s. However, there was minimal development in cattle up until the late 1980s. Only in recent years has the livestock sector developed as an important element of the Indian economy's developing and diverse agriculture sector.

India's demand for livestock products has witnessed a significant surge in recent years, driven by factors such as population growth, rising per capita income, and urbanization. Gandhi and Mani (1995) analyzed the changing demand patterns for animal products between 1970–1971 and 1989–1990. Their study revealed a notable increase in the demand for animal products (particularly milk, egg, and meat) throughout the nation. Notably, the demand for these products outpaced that of vegetables, edible oil, and pulses. The researchers also observed that the demand for these products exhibited a strong positive correlation with income levels, surpassing the growth of other food categories. Milk emerged as the most sought-after product, followed by eggs, fish, and meat (Tisdell and Gali 1999).

## Livestock and Poultry Population in India

Table 8.3 shows the livestock population in India has increased steadily over the past few decades, except for chickens, which experienced a dramatic surge in the late 1980s and early 1990s. The number of cattle, buffaloes, and goats increased by 24.57, 153.11, and 215.4%, respectively, between 1951 and 2019. Buffaloes are preferred for milk production due to their higher milk yield, longer lactation periods, better milk quality, and adaptability to various climates. The poultry sector in India is also experiencing rapid growth. The poultry population was estimated at 347 million birds in 1997, accounting for 8.7% of Asia's and 0.1% of the world's poultry stocks. The chicken population has grown at an annual rate of approximately 4.3% since 1980, reaching an estimated 852 million birds in 2019. This remarkable growth is attributed to advancements in breeding technology and favorable government policies (Tisdell and Gali 1999) (Fig. 8.4).

## Production of Livestock Products in India (Including Milk, Eggs, and Meat)

As indicated by Figs. 8.5 and 8.6, despite the relatively slow growth in livestock numbers (except for chickens and pigs), livestock production in India has witnessed a remarkable surge, outpacing the growth in livestock populations. This indicates a substantial increase in the productivity of individual animals. Milk and milk products are essential components of the human diet, providing vital nutrients for a healthy life. In India, milk production has witnessed an impressive growth trajectory, with output increasing from a mere 17.0 million tonnes in 1950–1951 to a staggering 209.9 million tonnes in 2021, as depicted in Table 8.4. This remarkable growth can be attributed to two primary factors: a gradual increase in the number of milking

**Table 8.3** Livestock and poultry population in India (in million numbers)

| Species\year | 1951 | 1956 | 1961 | 1966 | 1972 | 1977 | 1982 | 1987 | 1992 | 1997 | 2003 | 2007 | 2012 | 2019 |
|---|---|---|---|---|---|---|---|---|---|---|---|---|---|---|
| Cattle | 155.3 | 158.7 | 175.6 | 176.2 | 178.3 | 180 | 192.45 | 199.69 | 204.58 | 198.88 | 185.18 | 199.08 | 190.9 | 193.46 |
| Buffalo | 43.4 | 44.9 | 51.2 | 53 | 57.4 | 62 | 69.78 | 75.97 | 84.21 | 89.92 | 97.92 | 105.34 | 108.7 | 109.85 |
| Sheep | 39.1 | 39.3 | 40.2 | 42.4 | 40 | 41 | 48.76 | 45.7 | 50.78 | 57.49 | 61.47 | 71.56 | 65.07 | 74.26 |
| Goats | 47.2 | 55.4 | 60.9 | 64.6 | 67.5 | 75.6 | 95.25 | 110.21 | 115.28 | 122.72 | 124.36 | 140.54 | 135.17 | 148.89 |
| Pigs | 4.4 | 4.9 | 5.2 | 5 | 6.9 | 7.6 | 10.07 | 10.63 | 12.79 | 13.29 | 13.52 | 11.13 | 10.29 | 9.06 |
| Poultry | 73.5 | 94.8 | 114.2 | 115.4 | 138.5 | 159.2 | 207.74 | 275.32 | 307.07 | 347.61 | 489.1 | 648.88 | 729.21 | 851.81 |

*Source* GOI (2015). Basic Animal Husbandry & Fisheries Statistics. Ministry of Agriculture & Farmers Welfare, Department of Animal Husbandry, Dairying and Fisheries

GOI (2022). *Agricultural Statistics at a Glance*. Ministry of Agriculture & Farmers Welfare, Department of Agriculture & Farmers Welfare, Economics & Statistics Division

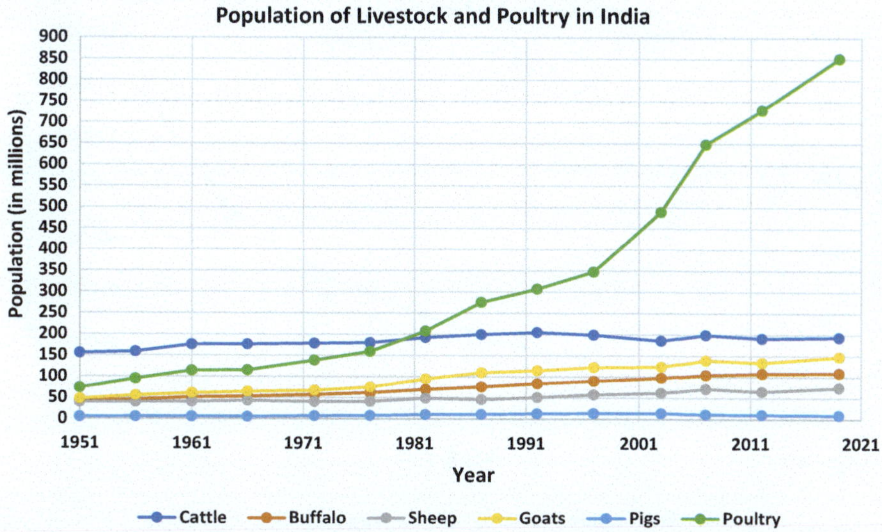


**Fig. 8.4** Population of livestock and Poultry in India

animals and significant technological advancements that have enhanced individual animal productivity. Since the 1960s, India has spearheaded ambitious initiatives like "Operation Flood" and established the National Dairy Development Board at Anand, playing a pivotal role in transforming the country into a global dairy powerhouse. These efforts have enabled India to meet the growing demand for milk in its rapidly expanding urban centers, ensuring that millions of consumers have access to this essential food source. Poultry development is an essential activity in the livestock industry. In 1950–1951, the nation produced 1832 million eggs in total, and since that time, as shown in Fig. 8.5, egg production has been steadily increasing. Egg production in India surged significantly following the implementation of the Silver Revolution (1969–1978) and several government incentives, reaching a peak of over 122,048 million eggs in 2021.

India's meat production has witnessed a steady upward trend, with an output of 1.9 million tonnes recorded in 2000 and projections indicating a rise to 8.6 million tonnes by 2020. Despite this growth, several factors have hindered the pace of production. These include a scarcity of high-yielding livestock breeds, inadequate availability of nutritious feed in remote regions, poor animal health, particularly among young animals, and insufficient processing and marketing infrastructure in many areas.

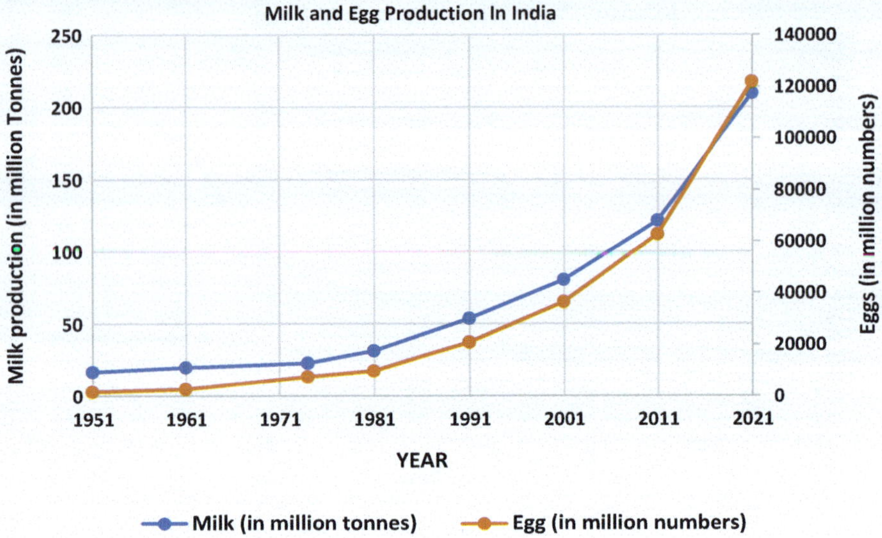


**Fig. 8.5** Milk and egg production in India

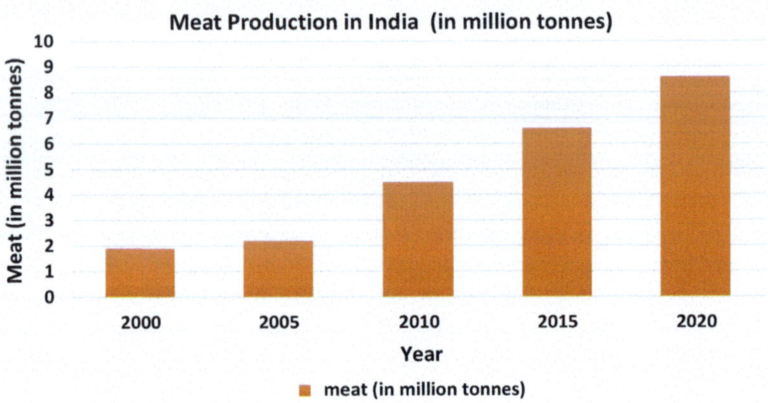


**Fig. 8.6** Meat production in India

**Table 8.4** Livestock products in India

| Animal products\year | 1951 | 1961 | 1974 | 1981 | 1991 | 2001 | 2011 | 2021 |
|---|---|---|---|---|---|---|---|---|
| Milk (in million tonnes) | 17 | 20 | 23.2 | 31.6 | 53.9 | 80.6 | 121.8 | 209.9 |
| Egg (in million numbers) | 1832 | 2881 | 7755 | 10,060 | 21,101 | 36,632 | 63,024 | 122,048 |

*Source* GOI (2015). Basic Animal Husbandry & Fisheries Statistics. Ministry of Agriculture & Farmers Welfare, Department of Animal Husbandry, Dairying and Fisheries

## 8.5 Future Food Demand in India

Since India gained independence, its population has grown quickly, and with that growth has come a multiple-fold rise in food demand. The population of India and its decadal growth rate from 1951 are shown in Table 8.5. According to the table, Over the seven decades from 1951 to 2021, India's population experienced a remarkable fourfold increase, soaring from 361.1 to 1393.4 million people. The decadal growth rate also showed an upward trend up to 1971, but after that it showed a downward trend, falling from 24.8% in 1971 to 15.07% in 2021, pointing to a future of population stabilization. However, according to UN estimates, India's population will surpass 1668 million by 2050 and 1718 million by 2060 (Fig. 8.7).

**Table 8.5** Indian population and decadal growth rate (1951–2021)

| Year | Population (in million) | Decadal growth rate (%) |
|---|---|---|
| 1951 | 361.1 | – |
| 1961 | 439.2 | 21.6 |
| 1971 | 548.2 | 24.8 |
| 1981 | 683.3 | 24.7 |
| 1991 | 846.4 | 23.9 |
| 2001 | 1028.7 | 21.5 |
| 2011 | 1210.9 | 17.7 |
| 2021(P) | 1393.4 | 15.07 |

*Source* Census of India
(P)—Projected by U.N. (2022)

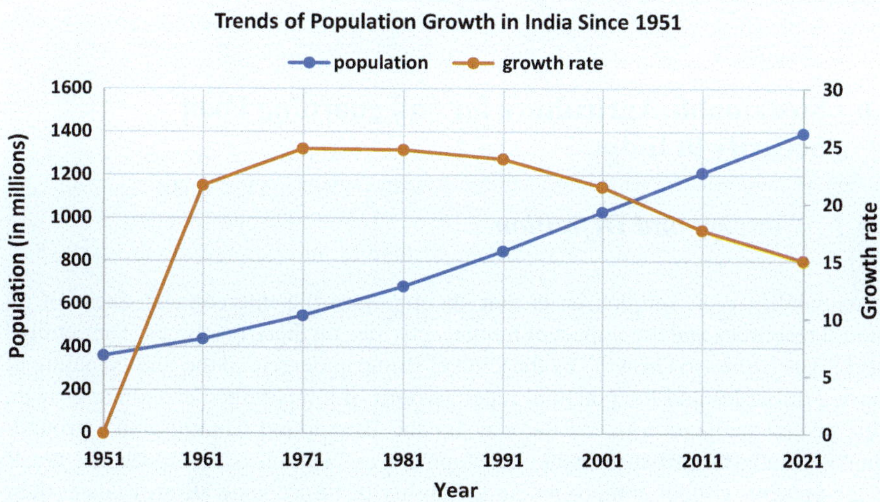


**Fig. 8.7** Trends of Indian population and decadal growth rate since 1951

**Table 8.6** Demand of Foodgrains in India

| Demand (in million tonnes)\year | 2033[a] | 2050[b] | 2065[b] |
|---|---|---|---|
| Foodgrains | 337.01 | 481 | 515 |

*Source* [a]NITI Aayog (2018). Demand & Supply Projections Towards 2033; Crops, Livestock, Fisheries and Agricultural Inputs; The Working Group Report
[b]Estimated by NCIWRD (National Council for Integrated Water Resources Development 1999)

**Table 8.7** Demand for milk, egg, meat, and fish in India

| Commodity demand (in million tonnes)\year | 2029 | 2030 | 2033 |
|---|---|---|---|
| Milk and milk products | 245.61 | 256.43 | 292.15 |
| Eggs, meat and fish | 27.73 | 29.45 | 35.52 |

*Source* NITI Aayog (2018). Demand & Supply Projections Towards 2033; Crops, Livestock, Fisheries, and Agricultural Inputs; The Working Group Report

In 2021, India's foodgrain demand is estimated to be around 261.6 million tonnes (mt). The demand is anticipated to increase to 337.01, 481, and 515 mt by 2033, 2050, and 2065, respectively, due to the population's rapid growth (Table 8.6). Additionally, according to Table 8.7 by 2033, the demand for dairy products will be around 292.15 million tonnes, and that for eggs, meat, and fish will be around 35.52 million tonnes due to rising per capita income, urbanization, changing dietary trends, and rising consumer spending on livestock products (Table 8.7).

India's escalating food demand necessitates the adoption of sustainable agricultural practices to safeguard food and nutritional security while navigating the challenges of limited land and water resources.

## 8.6 Sustainable Agriculture for Safeguarding Food Security in India

### *8.6.1 Concept and Definition*

Sustainability is a complex issue that encompasses the interconnectedness of all natural resources and the impact of human activities on these resources. The seminal work "The Limits to Growth" by the Club of Rome underscored the limited nature of Earth's resources and the potential repercussions of unrestrained economic growth. This pivotal realization paved the way for the Brundtland Commission's groundbreaking definition of sustainable development as "a process that meets the needs of the present without compromising the ability of future generations to meet their own needs." This definition emphasizes the importance of balancing resource use, economic development, and social equity to ensure a sustainable future for all.

Sustainable agriculture has taken the forefront since WCED popularized the idea of sustainable development as an organizing principle for equitable relations between generations.

The 1991 Conference on Agriculture and the Environment in the Netherlands, organized by the Food and Agriculture Organization (FAO), marked the genesis of the concept of sustainable agriculture and rural development. At its core, this approach emphasizes the preservation of fertile agricultural lands, ensuring food security, improving current living standards, safeguarding the well-being of future generations, and establishing harmonious mechanisms that promote a thriving society through balanced agricultural and economic development (Schaller 1993). Sustainable agriculture "over the long term, enhances environmental quality and the resource base on which agriculture depends; provides for the basic needs of human food and fiber; is economically viable; and improves the quality of life for farmers and society as a whole" (Crews et al. 1991).

Embracing sustainability demands a comprehensive approach that acknowledges the intricate interplay between environmental, social, and economic considerations. The Triple Bottom Line (TBL) framework provides a valuable lens for evaluating and optimizing sustainable practices (Abubakar and Attanda 2013). The 1991 FAO definition of sustainable development highlights the crucial role of managing and conserving natural resources to meet the present and future needs of humanity. This definition emphasizes the need for environmentally sound, technologically feasible, economically viable, and socially acceptable approaches to agricultural development to safeguard precious natural resources like land, water, plants, and genetic diversity.

- **Ecological Sustainability**: Sustainable agriculture practices emphasize the conservation and improvement of land, air, water, and biodiversity to protect and maintain the ecological systems that support agricultural production.
- **Economic viability**: If agriculture is to be sustainable, the farmer must be profitable. Although economic viability and productivity are closely related, neither of these factors is static or stable because consumer demand for agricultural products changes over time in response to the changing trends and individual demands as well as government policies.
- **Social Acceptability**: It includes a wide range of both definite and ambiguous criteria, including impoverishment, cultural aspects, education, social capital, equality and justice, value system, and government initiatives.

Sustainable agriculture, as defined by Mac Rae et al. (1990), encompasses both a philosophical approach and a set of agricultural practices. It is rooted in a deep understanding of social and ecological interconnectedness, coupled with a commitment to act responsibly in accordance with this understanding. To conserve all resources, cut down on waste, and safeguard the environment while maintaining and enhancing agricultural productivity, it places a strong emphasis on design and management strategies that support organic processes.

### 8.6.2 *Why Do We Need Sustainable Agriculture in India?*

The production of enough food for India's expanding population has been the country's most significant problem in agriculture, especially after independence. The Green Revolution propelled India towards food self-sufficiency, not only ensuring the sustenance of its growing population but also creating an abundance of food grains for export to other nations. Even though India currently produces more food than it needs, future food and nutritional security in India will be a major concern due to Green Revolution shortcomings, the country's rapidly growing population, a lack of arable land and water, climate change, the emergence of epidemics, and war.

#### The Green Revolution's Shortcomings

- There was a notable disparity in the development of food production among various goods and ecologies. Rice and wheat production received all of the attention during the Green Revolution, while other crops received little attention, which led to a decrease in crop diversification in India.
- While food grain production has witnessed a remarkable six-fold increase over the past seven decades, this surge has been accompanied by a dramatic rise in fertilizer usage. Despite this significant boost in grain output, certain crops have exhibited yield stagnation attributed to diminishing soil organic matter and uncontrolled fertilizer application.
- India's current fertilizer application pattern, with an NPK ratio of 10:2.7:1, is significantly skewed compared to the recommended ratio of 4:2:1. This imbalanced and excessive fertilizer usage is raising serious concerns among experts, as it is rapidly deteriorating soil health.
- Another flaw with the Green Revolution focused on cereal was its lack of attention to environmental concerns. The high-external-input agriculture that served as the foundation for the green revolution's technologies began to show its detrimental ecological effects in the 1980s. Integrating high-yielding production technologies has led to concerns regarding the long-term sustainability of farming, environmental contamination, toxic effects of pesticides, and the health of soil (Yadav et al. 2013).

#### Climate Change

- India's increased dependence upon agriculture, disproportionate usage of natural resources, and insufficient strategies for coping with climate change have made it more susceptible. Another reason to consider and adapt to certain alternative approaches in the agriculture sector is the severe climate variability, such as frequent occurrences of drought, abrupt temperature rises, and erratic patterns of rainfall.

- Rising temperatures and shifting weather patterns pose significant threats to India's agricultural sector. A 1 °C increase in average temperature is projected to impact wheat yields, particularly in areas already experiencing water shortages and reduced rainfall. Numerous researches have been done under the National Innovations in Climate Resilient Agriculture (NICRA) initiative indicate that rainfed rice yields may decline by 2.5% by 2050, while irrigated rice yields could drop by 7% in 2050 and 10% by 2080 (PIB 2021).
- Moreover, the demand for irrigation is expected to increase by 10% for every 10 °C rise in temperature, particularly in dry and semi-arid regions (FAO 2010). Coastal communities and fishermen are also vulnerable to the impacts of rising sea levels. If no adaptation measures are implemented, the long-term consequences of climate change could be severe, potentially jeopardizing food security and livelihoods to a greater extent than other economic sectors (NMSA draft 2010).

### Limited Availability of Net Cultivable Area

India's burgeoning population will not only lead to a rise in the future demand for food but will also escalate the conversion of agricultural land into industrial and residential lands, resulting in the decline of Per capita availability of Net Cultivable Area from 1990 to 2020. Table 8.8 shows the trend of cultivable land and per capita cultivable land. The table reveals the declining trend of Net Cultivable Area as it decreased from 156.70 million ha in 1990 to 153.67 million ha in 2020. The picture gets even more clear with the inclusion of the population factor as the Per capita Availability of Net Cultivable Area declined from 0.19 to 0.11 ha (Fig. 8.8).

As per the Projection made by NCIWRD, the Net cultivable area in 2050 will be around 145.0 million ha and by that time the projected Indian population will be around 1668 million which eventually will further decrease the Per Capita Net Cultivable Area to around 0.08 ha.

**Table 8.8** Trends of cultivable land and per capita cultivable land in India

| Classification | 1991 | 2001 | 2011 | 2020 (P) | 2050 (Pr) |
|---|---|---|---|---|---|
| Net cultivable area[a] (in million hectares) | 156.7 | 156.12 | 155.75 | 153.67 | 145 |
| Per capita Net Cultivable Area (in hectares) | 0.19 | 0.15 | 0.13 | 0.11 | 0.08 |

*Source* GOI (2022). *Agricultural Statistics at a Glance*. Ministry of Agriculture & Farmers Welfare, Department of Agriculture & Farmers Welfare, Economics & Statistics Division
*Note* [a]Net Cultivable Area = Net sown Area + Current Fallow land
(P)—Provisional
(Pr)—Projections made by NCIWRD

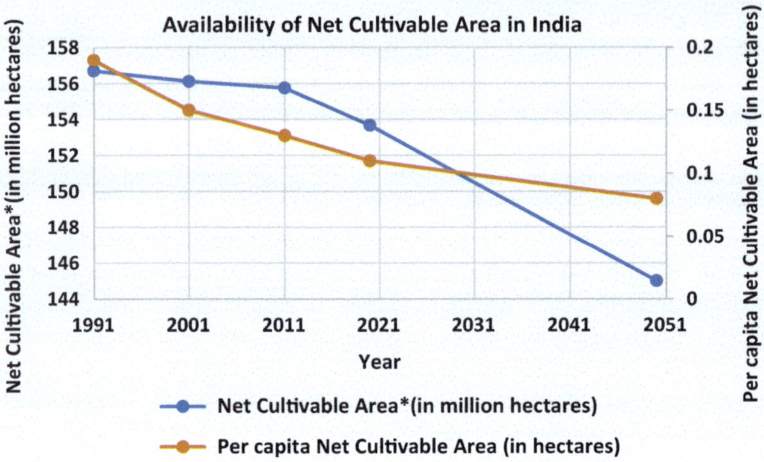


**Fig. 8.8** Trends of cultivable land and per capita cultivable land in India

## Limited Availability of Water for Agricultural Purposes

Water and energy security are crucial issues for India and the rest of the world, and food and nutrition are of utmost importance when combined with these difficulties. There is a moderate to severe lack of water, with the majority of river basins in India and elsewhere drying up or closing. Despite receiving an average annual rainfall of 1170 mm, India faces a growing water scarcity due to escalating demand from various sectors, including agriculture, industry, and urban development. While agriculture currently consumes approximately 83% of total water resources, this proportion is expected to decrease to 68% by 2050 due to the increasing water demands from other sectors. As a result, India's water demand is outpacing its available water resources. It is predicted that by 2050, India's per-person water availability will continue to drop, hovering around the level of scarcity at 1140 $m^3$/person/year. Additionally, it is anticipated that by 2050, the country's irrigation water consumption will rise to approximately 1745 billion L per day (BLD), up from 1658 BLD in 2000, to feed an increased population. According to projections, India's water demand will rise by over 50% during the next 12–15 years, but the supply will only rise by 5–10%. The majority of people, especially those who depend on agriculture, would suffer as a result of the water shortage crisis (IARI 2013).

## Covid Outbreak and War

The pandemic caused a catastrophe in the health, economy, and food industries. The poorest and most vulnerable nations, as well as communities, are most severely impacted. The availability, cost, and quality of food have all been impacted by the significant interruptions to food supply networks that have resulted from lockdown measures.

The ongoing turmoil brought on by the ongoing conflict between Russia and Ukraine, along with the possible implications it may have on human security, follows previous difficulties that have already impacted supply chains and prices, including the pandemic of COVID-19, a supply shortage of energy, shipping limitations, and the recent severe weather events brought on by climate change. A significant portion of the world's fertilizer supply has been cut off by Russia and Belarus, and the conflict has prevented these countries from exporting wheat, corn, and barley. The FAO estimates that if this issue is not addressed quickly, global food and feed prices will rise by 8–22% beyond their high baseline levels (Ipe et al. 2022). The convergence of various factors, including limited land and water resources, second-generation problems arising from the Green Revolution, and an expanding population, poses a significant threat to India's food and nutritional security in the coming years. To address this challenge, India must focus on enhancing agricultural productivity while maintaining ecological sustainability. Sustainable agricultural practices are crucial to achieving this goal, ensuring India's perpetual food security.

### *8.6.3 Adoption of Sustainable Agricultural Practices in India*

Four categories of determinants, viz. Physical determinants (climate, water availability, soil characteristics), biological determinants (pests, weeds, diversity of microbes), socio-economic determinants (population density, government schemes, technological availability), and cultural determinants (ethics, beliefs, value systems, knowledge, and work culture); were identified by Raman (2006) as crucial to the organization and management of agricultural and farming systems.

The aforementioned determinants of the agricultural system need to be intervened with agricultural techniques that are an amalgamation of both traditional and modern knowledge; thus, producing an agricultural system that is within the pragmatic pursuit of ecological sustainability, economic viability, and social acceptability vis-à-vis capability of meeting the future food requirements. Farming and natural resource management, improved crop seeds and livestock, improved and diversified farm practices, pest management, nutrient management, and the PEST (Political, Economic, Social, and Technological) factors are some of the particular approaches required for achieving the basic objectives of sustainable agriculture.

#### **Farming and Natural Resources Management**

Optimizing agricultural productivity within a specific ecological region hinges on the effective management of soil and water resources, the cornerstones of plant and animal growth.

**Soil**:

- Unsustainable agricultural techniques, excessive irrigation, and illegitimate use of fertilizers have caused numerous problems like soil erosion, water logging, soil salinization and alkalization, and loss of fertility.
- Various soil conservation techniques exist, including minimizing tillage, optimizing irrigation to reduce water loss and runoff, installing subsurface drainage to prevent waterlogging, and maintaining soil cover with vegetation or mulch. Reducing or eliminating tillage practices can enhance soil conservation without compromising productivity levels (Fabrizzi et al. 2005).

**Water**:

- Achieving sustainable food security requires addressing three critical issues: ensuring adequate irrigation water supplies to support agricultural growth, safeguarding water access for marginalized farmers, and providing sufficient financial incentives to encourage farmers to maximize productivity while minimizing environmental impact (Kumar 2003).
- Enhancing water management practices can be achieved through several strategies, including cultivating drought-resistant crops and adopting efficient techniques of irrigation such as micro-irrigation, sprinkler irrigation, watershed management, and water harvesting and recycling initiatives (Raman 2006).
- Biotechnology has a significant potential to increase yields and provide stability in challenging conditions. In India, technical advancements to increase crop yields and increase water productivity appear to have greater short- and long-term prospects. In addition, despite potential losses due to evaporation and evapotranspiration, the productivity of irrigation water can be increased by skilfully supplementing it with rainwater, especially in locations with heavy rainfall (Rao 2002)
- Farmers should be encouraged to share knowledge and integrate it with indigenous knowledge systems to improve water resource sustainability over time. This will help farmers use water resources sustainably and preserve them from degradation. Effective implementation of government initiatives like the Pradhan Mantri Krishi Sinchai Yojana (PMKSY), which aims to expand irrigation coverage under the "Har Khet Ko Paani" initiative and enhance water use efficiency through the "More crop per drop" approach, is crucial. Promotion of Dryland agriculture and millet production in water-stressed regions will ensure food and nutritional security. A subsurface drainage system needs to be developed for flood-prone and over-irrigated land thus, leading to the reduction of excessive salinity and alkalinity in soil (Debnath et al. 2024a).

**Energy**:

- Agriculture In the contemporary context is significantly reliant on energy that is not renewable, mainly petroleum, which is unable to last forever because it is apparent that the supply is finite. As biomass is a carbon–neutral resource, sustainable farming practices ought to seek out and utilize the various potentials for sustainable energy production and consumption that can help mitigate climate change. The inherent accessibility, dependability, and environmental friendliness of unconventional and renewable energy sources like wind energy, solar energy, and bioenergy make them particularly advantageous for rural areas.

### Improved Crop Seeds and Livestock

- Biotechnology serves as a powerful instrument for engineering genetic resources with improved resilience to evolving environmental conditions. It harbors significant potential to address vulnerabilities in fisheries, livestock, and crop production. Genetic engineering of enzymes like Rubisco and the incorporation of higher carbon (C4) pathways into low carbon (C3) plants will augment $CO_2$ utilization efficiency and contribute to greenhouse gas (GHG) emission mitigation (NAPCC 2008).
- India's abundant indigenous genetic resources, encompassing resistance to adverse climatic conditions, pests, and pathogens in crops and livestock, warrant preservation, cataloging, and profitable utilization while safeguarding the nation's agricultural heritage. Leveraging appropriate technologies will expedite the development of superior crop, livestock, and fish varieties. Fostering public–private partnerships in the creation, management, and distribution of improved varieties is crucial.

## Improved and Diversified Agricultural Practices

- **Crop diversification**

A variety of strategies are used in sustainable farming techniques. Crop diversification is the most crucial strategy for the development of sustainable agriculture. Diversification is the process of switching from a less profitable agricultural system to a more remunerative cropping system. Agriculture diversification, in general, is the process of producing a variety of crops to meet the rising need for cereals, pulses, oilseeds, fiber, fuel, and feed by moving away from the geographical or temporal dominance of a single crop.

Crop diversification disrupts the cycle of diseases, inhibits the growth of weeds, and encourages the interaction of good soil microbes. Crop diversification improves the physical and chemical characteristics of soil, which in turn increases crop productivity and the efficiency of land use. Crop diversification provides an excellent means of addressing issues such as the comeback of weeds and pest insects, soil degradation, pollution of the environment, salinity of the soil, declining farm profits, and climate change. Examples of agricultural diversification strategies include increased agricultural diversity, more diverse crop rotations, mixed cropping, livestock-integrated agricultural practices, the cultivation of grain legumes in typically cereal-dominated systems, perennial leys or grassland, and regionally adapted varieties or variety combinations.

- Crop rotation, an age-old agricultural practice, involves cultivating two distinct crop types sequentially, such as alternating between grains and legumes (Patel et al. 2020). This approach effectively suppresses weeds, pests, and diseases. Crop rotation promotes the restoration of soil nutrients, diversifies the agricultural landscape, and contributes to improved overall yield (Gaudin et al. 2015). Crop rotation can be implemented repeatedly and sustainably to promote soil health, mitigate soil erosion, and boost agricultural productivity.
- Preserving crop diversity enhances soil nutrient balance, bolsters natural pest control mechanisms, and improves resilience against adverse weather conditions. Intercropping, the practice of cultivating two or more crop types simultaneously in the same field, has gained prominence in modern agriculture as a means of boosting productivity under low input and adverse weather conditions. Intercropping outperforms monocultures due to its unique adaptability to climate change. The interplay between diverse crop species fosters a varied agricultural landscape characterized by high yields and enhanced soil fertility.
- Crop-livestock systems are cyclical processes in which animals and crops are integrated to enhance overall productivity and sustainability. Crop residues provide a supplementary food source for livestock, while animal manure serves as a valuable organic fertilizer for crops. This integrated approach not only boosts yields but also offers financial security to farmers by diversifying their income sources. Additionally, raising livestock at the farm level reduces reliance on synthetic fertilizers, thereby lowering greenhouse gas emissions (Abhilash 2015).

- **Conservation Agriculture**

Managing agricultural ecosystems through conservation agriculture aims to raise and sustain production, boost earnings, and guarantee adequate nutrition while protecting and strengthening the natural environment and resource base. The following three related principles serve as its distinguishing features:

1. Consistently minimal mechanical disturbance of the soil
2. A persistent organic soil layer
3. Crop species diversification grown sequentially or in associations

Biodiversity and above- and below-ground ecological processes are enhanced by conservation agriculture. Organic farming practices strive to enhance the natural ability of the agricultural system by avoiding the use of agrochemicals, synthetic fertilizers, or excessive mechanical soil tillage (Hegde and Sudhakar Babu 2016). The adoption of sustainable agricultural practices opens up opportunities for integrating diverse production sectors, such as incorporating trees and fields into agricultural landscapes and combining crop and livestock production. Implementing conservation agriculture practices and employing optimal crop management strategies can significantly improve input utilization efficiency and enhance the sustainability of agricultural production systems (Gathala et al. 2013).

### Nutrient Management

- The vital nutrients needed to support crops are present in the soil. A major technological obstacle to guaranteeing adequate food supply and promoting rural growth is nutrient management for plants, which aims to increase soil nutrients and hence raise crop output. To maintain and increase crop output, satisfy the requirement for raw materials and food, and protect the integrity of the water and land resources, plant nutrition management is crucial. It has been discovered that using integrated nutrient management approaches increases the effectiveness of nutrient usage.
- The greatest way to prevent the improper utilization of external inputs is to deliver inputs based on the need to limit their wastage. Instead of making general suggestions, it is advisable to modify fertilizer recommendations for site-specific nutrient management. Inputs like nutrients and water must be used more effectively using established techniques to achieve sustainable crop production.
- Biofertilizers, comprising microorganisms like bacteria, fungi, and algae, hold promise as alternatives to chemical fertilizers in large-scale agricultural practices. These eco-friendly and cost-effective solutions promote plant growth, enhance nutrient availability, and suppress soil-borne diseases, all while preserving soil structure and biodiversity. Nitrogen-fixing bacteria, such as rhizobia, are particularly beneficial for legumes and other crops. Additionally, Azolla and blue-green algae contribute to the nitrogen budget in agricultural systems. Biofertilizers also offer additional benefits, including the production of compounds that promote plant growth, disease resistance, and drought tolerance (Giri et al. 2019).

- Organic composting provides an eco-friendly and sustainable solution for waste management. It effectively handles the substantial amount of agricultural waste generated after harvesting (Singh and Singh 2017).
- Properly applied nano-fertilizers readily permeate seeds, enhancing nutrient uptake and accessibility for developing seedlings, resulting in extended root and shoot growth (Pandey 2018). Nano-fertilizers augment nutrient availability and distribution to plant branches and leaves, promoting enhanced vegetative growth (Adhikari et al. 2010).
- Furthermore, farmer awareness of soil health and the aforementioned procedures is a prerequisite for proper and sustainable soil nutrient management, which can be achieved by proper implementation of government schemes such as the Soil Health Card Scheme, PRANAM (Promotion of Alternate Nutrients for Agricultural Management) scheme, Organic manure scheme.

**Pest Management**

Pesticide use has increased over time in India, and its improper application has led to issues like the proliferation of pest species, exposure to the farmers, elimination of parasites and other beneficial microbes, and contamination of pesticides in the food chain and surroundings. Animal health is affected by the presence of residual pesticide compounds in the feed and water they consume as a result of both direct and indirect exposure during pest control operations.

IPM, or integrated pest management, can be used to manage pests sustainably. IPM principles include:

- Prevention—IPM focuses on preventing pest problems before they occur. It is possible to ensure optimal plant nutrition and health by putting methods like crop rotation and choosing pest-resistant cultivars into practice. The need for chemical interventions is decreased by creating an environment that is less pest-friendly.
- Monitoring—In IPM, it's crucial to regularly monitor insect numbers and crop health. This enables prompt action and aids in the identification of pests and their severity for farmers. Using monitoring methods like trap crops, pheromone traps, and visual examination, farmers may make well-informed pest control decisions.
- Biological Control—IPM encourages the use of natural enemies such as beneficial insects, parasites, and predators to control pests. Farmers can efficiently control insect populations without the use of synthetic pesticides thanks to the existence of these natural enemies. Encourage pest-eating insects that are not hazardous to crops.
- Cultural Techniques—IPM promotes the use of cultural techniques that improve crop health and lessen insect pressure. Pest breeding grounds are eliminated by techniques like effective irrigation management, weed control, and sanitation. Pests have less of an impact on crop growth when the environment is averse to them. The environment of the pest can be destroyed through soil cultivation. Pest

damage can be avoided by growing strong, healthy plants that receive the right amount of water and fertilizer.
- Chemical Control—Chemical pesticides are employed sparingly and only as a last option. IPM focuses on using less hazardous pesticides in focused applications. Materials for pest management should be utilized and applied in a way that reduces hazards to human health and is advantageous to the application's purpose. Non-targeted pests and the environment should not be affected.

**Animal Husbandry**

To expand production and meet future food demand, smallholder farmers, particularly those in developing nations, will need to make better and effective utilization of the genetic potential of plants and animals they already have. The demand for livestock has surged due to changing diets, rising incomes, and population expansion, making it a rapidly expanding sector of agriculture in India. So, in the years to come, sustainable development of livestock and poultry will be crucial for India's ability to ensure its own food security.

However, the lack of quality germplasm, inadequate vaccination and disease management among livestock and poultry, insufficient availability of finance to sustain livestock, farmers' poor perception of raising livestock as an additional source of income, and the recurrence of floods are some of the constraints for livestock production (Singh and Singh 2022).

However, India has enormous potential to increase the number of livestock and increase productivity, which can be fully realized by implementing specific measures.

- Better breeds of cattle and goats with improved germplasm will lead to increased milk and meat production.
- Improving the management of common grazing areas and pastures, training farmers on the growing of forage crops, strengthening the supply chain for fodder seeds, and making certified fodder seeds accessible would all help to increase the output and productivity of feed and fodder. (ICAR-IGFRI 2021).
- Enhancing vaccine efficacy requires a multi-pronged approach that encompasses disease control strategies for livestock, public education initiatives, and research into the genetic underpinnings of innate disease resistance. Implementing field-based diagnostic kits, adopting precise vaccination schedules, and investigating the biological mechanisms and genes responsible for innate immunity can significantly improve disease control in livestock populations. Simultaneously, implementing effective educational programs can raise public awareness of critical preventive measures, further reducing the spread of diseases.
- The availability of insurance for livestock and poultry to farmers, as well as an increase in the flow of short-term finance for the purchase of livestock all of which promote risk management.
- Supporting skill-based training and applied research in the chicken, sheep, goat, feed, and fodder priority sectors.

## The PEST (Political, Economic, Social and Technological) Factors

A variety of control variables, generally referred to as "PEST" factors, are present in the socioeconomic setting created by humans on the farm.

- Political factors (government priorities and policies regarding both agricultural and non-agricultural matters, such as free markets and imports).
- Economic factors (costs, subsidies, prices charged for acquisition, credit, etc.).
- Social factors (geographic preferences, labor limits, cultivators' skill, farmers' awareness, and so on).
- Technological factors (level of technological advancement, as well as ancillary services including prompt availability of inputs, infrastructure, and markets).

The term "food system" refers to a diverse group of stakeholders, including producers, researchers, suppliers, laborers on farms, unions, advisers for farmers, processing units, traders, consumers as a whole, and policy framers, all of whom have various and frequently competing agendas. As new technologies bring about changes in the economy, society, and politics, relationships between various actors evolve throughout time.

A model for sustainable agricultural development is presented in Fig. 8.9 with the intervention of all the above-mentioned sustainable agricultural practices in the present-day agricultural system.

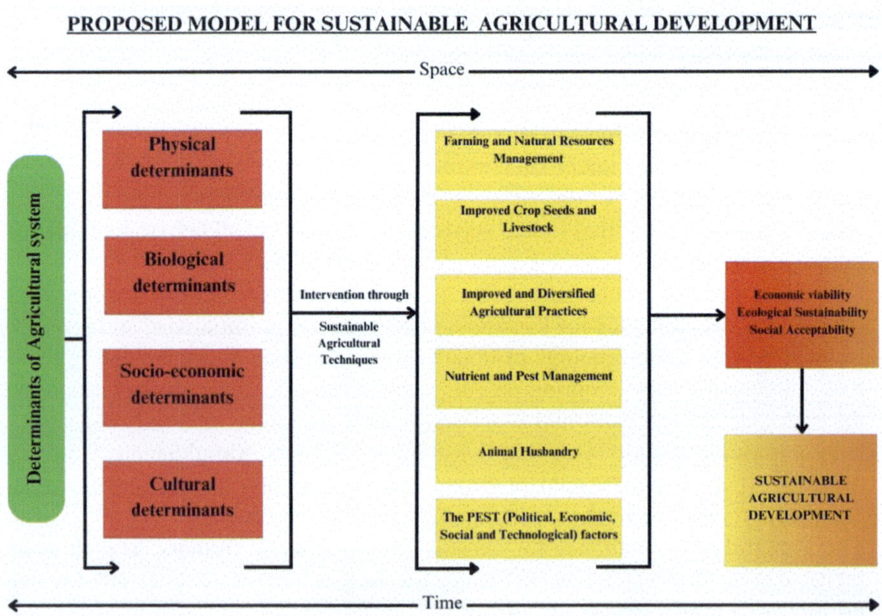


**Fig. 8.9** Model for sustainable agricultural development

## 8.7 Conclusion

It has been noted that the practice of sustainable agriculture is essential for a developing nation like India because it increases productivity, efficiency, employment, and ecological sustainability, and will therefore be crucial in safeguarding India's food and nutritional security moving ahead. However, several issues impede India's path to sustainable agriculture and food security, including farmers' slow adoption of modern technologies, unorganized and unscientific farming practices, lack of implementation of government policies, food wastage during procurement, lack of storage facilities, lack of credit and incentives for farmers, land fragmentation, and lack of knowledge. Certainly, the Indian Government has a distinct array of strategies to address practically all of the needs and concerns associated with the promotion of sustainable agriculture. The seamless execution of these programs, however, is the key to the solution and success. To facilitate the adoption of the initiatives, a practical operational and institutional framework must be put in place. Agriculture, a dynamic socioeconomic sector, has seen several continuing initiatives to support growth and boost productivity. Building a food system that is more environmentally sustainable requires the use of a variety of methods and strategies. These will range from targeted, concentrated efforts to support agricultural research, capacity building, and infrastructure development. Food and agricultural policies, food waste reduction, land, labor, rural community development, development of adequate food storage facilities, consumers and the food chain, and advances in technology like application of biotechnology, nanotechnology, and remote sensing and satellite imagery in agriculture are a few areas where change is most urgently required. The sustained implementation of these initiatives will safeguard India's food security for the foreseeable future, ensuring resilience against unforeseen disruptions such as wars and epidemics.

The interplay between food security, sustainability, and health informatics forms a dynamic and mutually reinforcing relationship within the framework of sustainable agriculture. Sustainable agricultural practices, which prioritize soil health, biodiversity, and efficient resource utilization, directly contribute to food security by ensuring a continuous supply of food. Health informatics, on the other hand, plays a critical role in guaranteeing the safety and quality of the produced food. Food security, sustainability, and health informatics are fundamental pillars of a future in agriculture and global food provision that is more secure, sustainable, and health-conscious. Sustainable agriculture not only ensures food availability but also safeguards the environment and promotes economic and social well-being. Health informatics, utilizing data-driven approaches, enhances food safety and transparency, thereby protecting public health. The harmonious interplay of these elements is indispensable to navigating the intricate challenges plaguing our food systems and forging a more prosperous future for all.

## References

Abhilash PC (2015) Towards the designing of low carbon societies for sustainable landscapes. J Clean Prod 87:992–993. https://doi.org/10.1016/j.jclepro.2014.09.057

Abubakar MS, Attanda ML (2013) The concept sustainable agriculture: challenges and prospects. IOP conference series: materials science and engineering, vol 53. https://doi.org/10.1088/1757-899X/53/1/012001

Adhikari T, Biswas AK, Kundu S (2010) Nano-fertilizer—a new dimension in agriculture. Indian J Fertilisers 6(8):22–24

Ahmad T, Gupta SK, Singh SK, Meraj G, Kumar P, Kanga S (2023) Unveiling nature's resilience: exploring vegetation dynamics during the COVID-19 era in Jharkhand, India, with the Google earth engine. Climate 11(9):187

Crews T, Mohler C, Power A (1991) Energetics and ecosystem integrity: the defining principles of sustainable agriculture. Am J Altern Agric 6(3):146–149. https://doi.org/10.1017/S0889189300004124

Debnath J, Sahariah D, Nath N, Saikia A, Lahon D, Islam MN, Hashimoto S, Meraj G, Kumar P, Singh SK, Kanga S (2024a) Modelling on assessment of flood risk susceptibility at the Jia Bharali River basin in Eastern Himalayas by integrating multicollinearity tests and geospatial techniques. Model Earth Syst Environ 1–27

Debnath J, Debbarma J, Debnath A, Meraj G, Chand K, Singh SK, Kanga S, Kumar P, Sahariah D, Saikia A (2024b) Flood susceptibility assessment of the Agartala Urban Watershed, India, using machine learning algorithm. Environ Monitor Assess 196(2):110

Fabrizzi KP, Garcia FO, Costa JL, Picore LI (2005) Soil water dynamics, physical properties and corn and wheat responses to minimum and no tillage systems in the southern Pampas of Argentina. Soil Tillage Res 81:57–69. https://doi.org/10.1016/j.still.2004.05.001

FAO (1991) The den Bosch declaration and agenda for action on sustainable agriculture and rural development. Food and Agriculture Organisation, Rome

FAO (2010) Globally Important Agricultural Heritage Systems (GIAHS). Food and Agriculture Organisation, Rome. http://www.fao.org/nr/giahs/giahs-home/home-more/en/

FAO (2021) FAO in India: India at a glance. Food and Agriculture Organisation, Rome. http://www.fao.org/india/fao-in-india/india-at-aglance/en/

Gandhi VP, Mani G (1995) Are livestock products rising in importance? A study of the growth and behaviour of their consumption in India. Indian J Agricult Econ 50(3):283–293

Gathala MK, Kumar V, Sharma PC, Saharawat YS, Jat HS, Singh M, Kumar A, Jat ML, Humphreys E, Sharma DK, Sharma S, Ladha JK (2013) Optimizing intensive cereal-based cropping systems addressing current and future divers of agricultural changes in the north western Indo-Gangetic Plains of India. Agr Ecosyst Environ 177(1):85–97. https://doi.org/10.1016/j.agee.2013.06.002

Gaudin AC, Tolhurst TN, Ker AP, Janovicek K, Tortora C, Martin RC, Deen W (2015) Increasing crop diversity mitigates weather variations and improves yield stability. PLoS One 10(2):e0113261. https://doi.org/10.1371/journal.pone.0113261

Giri B, Prasad R, Wu Q, Varma A (2019) Biofertilizers for sustainable agriculture and environment. Soil Biol. https://doi.org/10.1007/978-3-030-18933-4

GOI (2015) Basic animal husbandry & fisheries statistics. Ministry of Agriculture & Farmers Welfare, Department of Animal Husbandry, Dairying and Fisheries. Government of India. https://dahd.nic.in/sites/default/filess/BAH_%26_FS_Book.pdf

GOI (2022) Agricultural statistics at a glance. Ministry of Agriculture & Farmers Welfare, Department of Agriculture & Farmers Welfare, Economics & Statistics Division. Government of India. https://desagri.gov.in/wp-content/uploads/2023/05/Agricultural-Statistics-at-a-Glance-2022.pdf

Hegde DM, Sudhakara Babu SN (2016) Sustainable agriculture. In: Rana DS, Ghosh PK, Shivay YS, Singh G (eds) Modern concepts of agronomy. Published by The Indian Society of Agronomy, Division of Agronomy, IARI, New Delhi, pp 1–36

IARI (2013) Vision 2050, Indian Institute of Agricultural Research, New Delhi, India. https://krishi.icar.gov.in/jspui/bitstream/123456789/1827/1/DRWA%20Vision%202050.pdf

ICAR-IGFRI (2021) Fodder resources development plan for Uttar Pradesh. ICAR-Indian Grassland and Fodder Research Institute Jhansi. https://igfri.icar.gov.in/wp-content/uploads/2021/09/UP_Plan.pdf

IPCC (2019) Climate change and land: an IPCC special report on climate change, desertification, land degradation, sustainable land management, food security, and greenhouse gas fluxes in terrestrial ecosystems [Shukla PR, Skea J, Calvo Buendia E, Masson-Delmotte V, Pörtner H-O, Roberts DC, Zhai P, Slade R, Connors S, van Diemen R, Ferrat M, Haughey E, Luz S, Neogi S, Pathak M, Petzold J, Portugal Pereira J, Vyas P, Huntley E, Kissick K, Belkacemi M, Malley J (eds)] (In Press)

Ipe BT, Shubham S, Satyasai KJS (2022) Food and nutritional security in India: charting the way to a robust agri-food system. Department of Economic Analysis and Research, NABARD. https://www.nabard.org/auth/writereaddata/tender/2501230131nrs-35-food-and-nutritional-security-in-india.pdf

Kotecha MJ, Bakori D, Agarwal S, Meraj G, Kanga S, Singh SK, Farooq M (2024) Geoinformation for integrated urban water resource management. In: Earth observation in urban monitoring. Elsevier, pp 93–111

Kumar MD (2003) Food security and sustainable agriculture in India: the water management challenge. International Water Management Institute (Working paper 60), Colombo, Sri Lanka

Mac Rae R, Hill S, Henning J, Bentley A (1990) Policies, programs, and regulations to support the transition to sustainable agriculture in Canada. Am J Altern Agric 5(2):76–92. https://doi.org/10.1017/S0889189300003325

Meadows DH, Meadows DC, Randers J, Behrens WW (1972) The limits to growth. New York Universe

National Action Plan on Climate Change (2008) Prime Minister's Council on Climate Change, Government of India, New Delhi

National Mission for Sustainable Agriculture (2010) Strategies for meeting the challenges of climate change. Department of Agriculture and Cooperation, Ministry of Agriculture, New Delhi

NCIWRD (1999) Report of the national commission for integrated water resources development. Ministry of Water Resources, GOI, New Delhi. https://cwc.gov.in/sites/default/files/nciwrd-hashim-report-vol-i.pdf

NITI Aayog (2018) Demand & supply projections towards 2033; Crops, livestock, fisheries and agricultural inputs; The Working Group Report. https://www.niti.gov.in/sites/default/files/2023-02/WG-Report-issued-for-printing.pdf

Pandey G (2018) Challenges and future prospects of agri-nanotechnology for sustainable agriculture in India. Environ Technol Innov 11:299–307. https://doi.org/10.1016/j.eti.2018.06.012

Patel SK, Sharma A, Singh GS (2020) Traditional agricultural practices in India: an approach for environmental sustainability and food security. Energy, Ecol Environ 5(4):253–271. https://doi.org/10.1007/s40974-020-00158-2

PIB (2021) Effect of climate change on agriculture (n.d.). https://pib.gov.in/PressReleasePage.aspx?PRID=1696468

Rai SK, Jaiswal A, Sarvade S, Shrivastava A (2021) Sustainable agriculture for food security. In: Emerging tools for sustainable agriculture and food security. Deepika Book Agency, pp 221–235

Raman S (2006) Agricultural sustainability: principles, processes and prospects. Food Products Press, New York, USA, pp 474

Rao CHH (2002) Sustainable use of water for irrigation in Indian agriculture. Econ Pol Wkly 37(18):1742–1475. https://doi.org/10.2307/4412072

Schaller N (1993) The concept of agricultural sustainability. Agr Ecosyst Environ 46(1–4):89–97. https://doi.org/10.1016/0167-8809(93)90016-I

Singh R, Singh GS (2017) Traditional agriculture: a climate-smart approach for sustainable food production. Energ Ecol Environ 2:296–316. https://doi.org/10.1007/s40974-017-0074-7

Singh A, Singh R (2022) Spatio–temporal analysis of livestock & poultry: a case study of eastern uttar pradesh. J Asiatic Soc Mumbai 96(10). ISSN: 0972-0766

The Economist Impact Unit (2022) Global Food Security Index, Strengthening food systems and the environment through innovation and investment. https://impact.economist.com/sustainability/project/food-security-index

Tisdell C, Gali J (1999) Trends and developments in India's livestock industry, economics, ecology and the environment. Working Paper No. 31, The University of Queensland, pp 1–12

U.N. (2022) World population prospects: the 2022 revision, methodology of the united nations population estimates and projections. United Nations, Department of Economic and Social Affairs, Population Division, New York

World Development Report (2008) Agriculture for development. Washington, DC. https://openknowledge.worldbank.org/handle/10986/5990

Yadav SK, Babu S, Yadav MK, Singh K, Yadav GS, Pal S (2013) A review of organic farming for sustainable agriculture in Northern India. Int J Agron 2013:1–8. https://doi.org/10.1155/2013/718145

# Chapter 9
# Environmental Sustainability and Its Impact on Public Health in Ladakh-India

**Aijaz Ahmad Khanday, G. M. Rather, Mushtaq Ahmad Kumar, and Adnan Hussain Lone**

## 9.1 Introduction

Sustainability means "Meeting the needs of the present without compromising the ability of future generations to meet their own needs" (Brundtland Commission, 1987). It is a condition of balance, resilience, and interconnectedness that allows human societies to satisfy their needs while neither exceeding the capacity of its supporting ecosystems to continue to regenerate the services necessary to meet those needs nor our actions diminishing biological diversity (Morelli, 2011). It requires information on sustainability that is supported by science, providing rational, cogent, and transparent decision support for sustainable consumption and production patterns. In areas where there is limited agreement, human behaviour, and societal context are more important for social and economic sustainability (Jeelani et al., 2023; Ahmad et al. 2023). The environmental pillar discusses ecosystems and how they support human life. Environmental science has a better degree of predictability and scientific agreement in this area, allowing assessments to be made on it (Dong & Hauschild, 2017). The term "sustainability" has a variety of bases, including ecological carrying capacity, resource reserves, and technological criticism. Each of these research disciplines has its own history and distinct goals, such as minimizing the effects of technological advancement and remaining within ecological carrying capacities (Kidd, 1992). Environment Sustainability is defined as "the capacity to preserve things or qualities that are valued in the natural environment". Whenever there is a threat of difficult or irreparable loss of the things or environmental features that people value, sustainability challenges become evident. And if such threats exist, there is some sense of urgency to respond (Sutton, 2004).

A. A. Khanday (✉) · G. M. Rather · M. A. Kumar · A. H. Lone
Department of Geography and Disaster Management, School of Earth and Environmental Sciences, University of Kashmir, Srinagar 190006, India
e-mail: kaaijaz121@gmail.com

© The Author(s), under exclusive license to Springer Nature Singapore Pte Ltd. 2024

G. Tripathi et al. (eds.), *Sustainability and Health Informatics*, Advances in Geographical and Environmental Sciences,
https://doi.org/10.1007/978-981-97-6706-9_9

Ecosystem services, which include both renewable and non-renewable resources as well as the ability to absorb trash, are the foundation of the concept of environmental sustainability. These services assist humans and hence raise their standard of living. Humanity must learn to adapt to the restrictions of the biophysical environment to benefit from and utilize the services throughout the years (Gulati et al. 2023). The three pillars of sustainability (socio-economic and environmental) were born out of the concept of sustainable development, in 1980 (Moldan, 2012). To make development sustainable, sustainability development should include socio-economic and environmental sustainability. "Improving the quality of human life while remaining within the ecological systems' carrying capacities" (Goodland, 1995).

Due to resource depletion, climate change, and biodiversity loss, People's impact on ecosystems is now so severe that it threatens nature's ability to regenerate itself, endangering humankind's ability to improve the quality of life. Health is determined by social dynamics, lifestyles, eating habits, and the bio-physical environment in addition to the absence of disease. The interaction of social and economic systems, the environment, and personal characteristics and behaviours is intricate (Singh et al. 2021). The goal of health promotion is to emphasize strength, resilience, and health assets while moving away from a bio-medical approach and towards a socio-ecological one. The significant negative consequences of climate change on world health have been underlined by the UCL Lancet Commission. They contend that the biggest health risk of the twenty-first century is climate change. The Commission names several challenges, such as population increase, migration, extreme weather, changing disease patterns, water, and food scarcity, and fragile settlements and housing. These effects could worsen existing health disparities. The Commission's report emphasizes the necessity of taking climate change's broader health implications into account (Kjærgård et al., 2014). Globally, pollution is a major contributor to morbidity and mortality. Air pollution, which causes 1 in 8 fatalities worldwide, is currently connected to the majority of these environmental-mediated deaths. An estimated 150,000 people die each year owing to health risks brought on by climate change, including extreme weather, deteriorating air quality, food- and water-borne illnesses, vector-borne diseases, food shortages, and social instability. Climate change is predicted to cause an additional 250,000 fatalities annually between 2030 and 2050 (Sherman et al., 2020). Global population health is being affected by radical changes in the biosphere and human contact with the environment more and more. Through the use of globalized transportation systems, diseases are fast spreading beyond the species barrier (Meraj et al., 2021). Public health professionals must recognize the association between the sustainability of the environment and the sustainability of the human species due to the threat of global pandemics. Since environmental preservation necessitates the development of preventive healthcare measures, healthcare institutions should provide a framework for corporate social accountability and promote environmental sustainability (Carnero, 2015).

### *9.1.1 Environmental Sustainability and Its Impact on Public Health in Ladakh*

The region of Ladakh, nestled in the northern part of India, is known for its unique geographical features and rich cultural heritage. However, the fragile ecosystem of Ladakh faces numerous sustainability issues as a result of societal, geopolitical, climatic, and economic changes that have an overall impact on the environment, water resources, population trends, and health in specific. This chapter explores the concept of environmental sustainability and its profound impact on the well-being of the local population in Ladakh.

## 9.2 Environmental Sustainability in Ladakh

### *9.2.1 Overview of Ladakh's Environmental Landscape*

#### Temperature, Precipitation, and Vegetation

Ladakh is bounded by the Great Himalayan Range on the south and the Karakoram Range on the north. The former range chunks most of the monsoon-bearing clouds, making the region dry. Winter precipitation is mostly in the form of snow, which is associated with the extra-tropical disturbances of mid-latitudes known as Western Disturbances. The southwestern part of the region gets a little more precipitation compared to the east and the northeast and thus is greener. The vegetation of eastern Ladakh is largely dominated by hemicryptophytes, followed by Therophytes and Chamaephytes (Farooq et al. 2022). The Ladakh's vegetation is a 'dry alpine scrub' that is characterized by the complete absence of forest cover, except relict patches of juniper *Juniperus polycarpous* stands in some parts. Other prominent trees are willow *Salix* spp. and poplar *Populus* spp. which are restricted mainly to cultivated areas along rivers (Namgail et al., 2012). Ladakh's average temperature has risen by 1.7 °C over the past century, and the warming trend increased in the 1990s, causing glaciers to retreat quickly. Along with a temperature rise, Ladakh also saw an increase in precipitation, which was accompanied by more frequent intense snowfall and storm events that brought on devastating floods. The mean winter, spring, and fall temperatures in Ladakh exhibit positive trends over the long term (1901–2014) of meteorological records, with noticeable warming since the mid-1990s. Since the mid-1990s, statistics show a significant rise in both the overall amount of precipitation and summer temperature. The overall trend has been confirmed by in-situ climatic recordings from the years 2008 to 2014. For June, there was an increase in mean daily maximum Temperature of 2.43 °C per year, for July, it was 14.42 °C per year, for August, it was 0.95 °C per year, and for September, it was 1.02 °C each year. The summer months, except August, saw daily mean Temperature data show similar trends. The average summer temperature, measured from June to August, rose 0.59 °C annually. Additionally, the diurnal temperature range increased. The only significant change in the winter month temperature was cooling in February (0.38 °C per year) (Dolezal et al., 2016). The northwest monsoon does not bring rain to Ladakh because it is located in the Western Himalayas' rain shadow. Extremely low temperatures, scant precipitation, a diurnal pattern of temperature changes, and a short growing season with negligible growing season precipitation are the region's climatic characteristics (Husaini et al., 2013). The ecology, flora, fauna, hydrology, cryosphere, and even human society will all be significantly impacted by global warming and changes in precipitation patterns, even though Ladakh is a cold desert region.

## Biodiversity and Ecosystem

Biodiversity, the variety of life on Earth, is distributed unevenly. It is a remarkable and diverse aspect of our planet, sparking scientific curiosity (Behera et al., 2014). Due to the harsh climate and short growing season, the Trans-Himalayan mountains only have 1 20% coverage of vegetation, although they are recognized to be home to a rare combination of flora and fauna (Joshi et al., 2006). The unique and highly endemism floral and faunal elements that characterize this region are directly related to maintaining both the ecology and human life. This cold desert requires specialized research and a coordinated effort in terms of resource management because it is a unique biome, particularly in light of its vulnerable ecosystems and severely depleted natural resource status (Meraj et al. 2022). Ladakh's environment and biodiversity are severely stressed as a result of intense demands. Both the human and cattle populations have grown significantly. Additional factors that have contributed to the loss of wildlife in the area include the increased agricultural and development operations. Since nearly every large organism in this area is listed as endangered, the situation is dire (Sidhu et al., 2012). In terms of ecology, biodiversity, and local socio-economic factors, the alpine pastures form a significant component of India's cold deserts. The distribution of local and regional vegetation and biodiversity is subsequently determined by topographic and climatic factors that have an impact on community structure and ecosystem activities (Behera et al., 2022). The biosphere's ecosystems are dynamic, self-sustaining systems that support life. They can provide a variety of main and important amenities necessarily meant for maintaining human health, welfare, livelihoods, and subsistence. The inhabited area is an improved version of the mountain environment that has remained more enhanced to support agricultural practices and cater to the needs of the locals daily. Numerous essential Ecosystems from the surrounding mountains are utilized by the inhabitants of the several settlements of Ladakh (Hussain et al., 2022).

## Water Resources and Glaciers

Traditional societies, through their knowledge and practises have significantly contributed to the management and development of their local surroundings. In the past, people's interaction with their immediate surroundings influenced the landscapes and their way of life. Sustainability study has a lot to gain from this reciprocal link between people and their natural settings, both historically and currently. The Ladakh (Himalayan region) stores a vast quantity of frozen water in the form of ice, glaciers, permafrost, and snow. Glacial meltwater is essential to the survival of the population in Ladakh because the livelihood of the population depends on it. In Ladakh, subsistence agriculture has been practised for many years. The main focus of conventional agricultural methods is on controlling glacial meltwater for irrigation. Due to the dearth of natural resources, the availability and management of water in Ladakh affects both village settlement patterns and agricultural practices. The Indus basin which is locally known as Sindhu is the main drainage basin. Over the past few decades conventional forms of development have brought about changes in land use and demographic changes in terms of population size, livelihood and urbanization, coupled with the changing climatic conditions, threaten the integrity of the society and ecology. Water resources have been critically affected by the above-mentioned factors and the increasing tourism in the region has led to a high demand for water as the region is already water scarce. More environmental changes affect the availability of water and, consequently, the possibility for irrigation, which in turn will affect the socioeconomic situation.

The high-altitude Himalayan region experiences changes in the glaciers that are typically brought on by variations in temperature, direction, debris, and black carbon, among other factors. These changes are also linked to a decrease in winter precipitation (Soheb et al., 2020). Although it is situated in a higher altitude zone between the Karakoram and western Himalayan regions, the glacierized region is losing mass at a faster rate. Due to their quick response times, the Ladakh glaciers are most likely more susceptible to regional climate changes, including rising temperatures and decreases in solid precipitation (Schmidt & Nüsser, 2017). Due to the region's greater dependence on glacial meltwater streams, the decline of the glaciers in this setting poses a major threat to the livelihood, ecology, and socioeconomic aspects of the area. Due to a reduction in water resources, the locals have seen a downturn in agricultural production, the region's primary source of economic growth (Barrett & Bosak, 2018). The recent growth in tourism activity and the lack of water have caused a shift away from agricultural to other business activities (Muller et al., 2020). In addition to agricultural activity in response to water constraints, some Ladakhis have created novel water management strategies known as "ice reservoirs" (Nusser et al., 2019). These are also called artificial glaciers that conserve the wasted water in winter and store it in huge mountains of ice then in spring these glaciers melt and feed the farmers when they need the most. It boosts the groundwater recharge, revitalizing the spring and supplying water for irrigation.

## 9.3 Environmental Challenges in Ladakh

### *9.3.1 Climate Change and Its Effects*

One of the most contentious issues in recent decades is climate change. The reasons for global warming trends include land use and land cover changing patterns, as well as a rise in greenhouse gases and aerosols (Bhutiyani et al., 2010). The Himalayan region has experienced fluctuations in temperature and precipitation patterns, which have had an impact on the availability of water, glaciers, ecology, agriculture, and other factors (Dimri & Dash 2012; Shekhar et al., 2010). As water reservoirs that replenish mountain rivers, the glaciers of the Himalayas have a significant influence on the hydrology, ecology, and climatology of the Indian subcontinent in general and Ladakh in particular. A rapid rise in temperature and a variety of precipitation patterns in current decades suggests a further changing climate with a higher probability of unforeseen events over the region in the coming years, according to amplified variations in temperature and precipitation patterns that are noted in such regions. This shift in the climate pattern could have disastrous effects on the area and be irreversible. Any location may adapt to its consistent climate over a long period of time, but a changing climate is troublesome because it disturbs the delicate balance between the region's ecosystem and historical climate (Chevuturi et al., 2018). Agriculture, livelihoods and water resources are extremely exposed to hazards and risks brought on by climate change. Agriculture continues to be the primary source of livelihood and food security, despite the growing significance of non-agricultural economic activity. Traditional irrigation systems driven by glacier and snowmelt water are the chief source of agriculture in the region. Agriculture practices, productivity, and sustainability are being impacted by climate-related changes in seasonal timings, temperature increases, variations in precipitation patterns, increasing frequency of pest attacks, and water-related risks (Tuladhar et al., 2023). Self-sufficiency, links to place, social cohesiveness, and a sense of belonging are thus important elements in creating societal frameworks for responding to climate change and ought to be incorporated into deeper understandings of vulnerability and associated adaptation activities (Barrett & Bosak, 2018).

### *9.3.2 Waste Management and Pollution*

The management of solid waste is becoming a significant global environmental problem. This is directly related to both production and consumption. In the city and tourist areas where it is managed using antiquated and traditional methods, the issue is more severe (Kumar et al., 2023). As tourist destinations, the Ladakh towns of Leh and Kargil have experienced a massive influx of visitors that occasionally outnumber the native populace. As a result, a significant amount of waste is produced quickly, escalating environmental degradation and attracting stray dogs, mice, and flies that detract from the aesthetic appeal of the locations. Despite being a popular tourist attraction in India, the Trans Himalayan region of Ladakh is also ecologically vulnerable. As Ladakh's high consumption patterns contribute to a dirty environment and deter future tourism growth, improper waste disposal presents a huge environmental concern for the development of the tourism industry in general and health in particular.

Due to a large number of tourists, solid waste management becomes problematic during the summer. Large volumes of trash in the form of plastic cans, bottles, papers, and wrappers are produced as a result of the shipment of packaged food and drinks to meet the demands of tourists, residents, and army troops (Gondhalekara et al., 2013). Due to the quantity and variety of the wastes, the socioeconomic and environmental issues resulting from inaccurate solid waste disposal sites are complicated. Therefore, it is important to manage solid waste using innovative methods to preserve the environment for the long term. Solid garbage is carelessly dumped in unsuitable places in hilly areas, endangering the ecology and the local economy. However, choosing an appropriate disposal site for solid waste can be quite difficult and expensive, especially in hilly areas. Therefore, owing to their capacity to manage significant capacities of geographical and non-spatial data from many sources, GIS and remote sensing techniques are providing to be particularly effective tools for these types of investigations (Dar et al., 2019), as it is the most menacing and disenable pollution in the trans-Himalayan region (Alaxander, 2005).

### 9.3.3 *Water Scarcity and Quality*

In rapidly expanding economies like India, access to clean drinking water is already a critical problem for big urban populations. Climate change is also having an impact on this, which is why there are more diseases related to the water. An essential need exists for integrated urban planning, particularly of water resources, in cooperation with other sectors like energy and considering health. Due to tourist and economic growth, Ladakh, which is located in an ecologically delicate semi-arid region of the Himalayas, is changing drastically. The already critical environmental problems are being exacerbated by a significant increase in water consumption, an inadequate water supply, and poor wastewater treatment. Modern agriculture and water management methods are progressively giving way to new ways of living and alternate sources of income, especially as the tourism sector expands quickly. To increase their attractiveness and, consequently, their revenue from tourism, tourist establishments are progressively constructing en suite bathrooms with flush toilets. Hotels and guesthouses in Ladakh generally, and in its towns in particular, do not have a sewage facility and they typically dispose of their waste in poorly maintained septic tanks and soak pits. Because the aquifer beneath Ladakh is utilized for drinking water and is nourished by glacier and snow melt water, the massive increase in tourist accommodations and the resulting increase in wastewater may pose a risk to human health (Gondhalekar et al., 2015).

### 9.3.4 Tourism Industry and Its Impacts

Ladakh has experienced three distinct tourist influxes over the past 30 years: Phase I (1974 to 1989) with a gradual increase peaking at 25,000 visitors, Phase II (1989 to 2002) Phase II (1989–2002) with significant variation and Phase III (2002 to present) with exponential growth peaking at 50,000 visitors in 2006. These phases were influenced by policies, accessibility, and political events. Ladakh's economy has benefited from mountain tourism, which also poses a threat to the environment due to ecosystem vulnerability and poor infrastructure and policies (Geneletti & Dawa, 2009). The rapid growth of Ladakh's tourism since its opening in 1974 has had considerable impacts on the ecosystems and ecology of the region. The increase in tourist numbers, from about 20,000 in 2001 to about 150,000 in 2011, has led to an increase in infrastructure development and the consumption of resources beyond the capacity of local ecosystems, resulting in overexploitation and degradation of the environment. One of the major issues is improper sewage disposal, which has caused pollution and environmental degradation. The influx of tourists has also caused changes in land use and an increase in livestock density, further exacerbating the environmental challenges. Additionally, the region is facing water and solid waste management problems due to the rapid development associated with tourism (Rajashekariah & Chandan, 2013).

However, it is important to identify the role of tourists in Ladakh's development and their potential implications on public health and the economy. As the tourism industry in Ladakh continues to grow, it brings both opportunities and challenges. On one hand, tourism contributes to the local economy, generating income and employment opportunities for the people. It promotes cultural exchange, fosters understanding, and raises awareness about the exclusive heritage and natural exquisiteness of Ladakh. On the other hand, unregulated tourism can put a strain on the ecology and public health. Increased tourist activities can lead to overcrowding, waste generation, water scarcity, and pollution if not managed adequately. These factors can have detrimental effects on the local ecosystem, public health, and social cohesion.

To address these problems, there is a necessity for sustainable tourism practices that prioritize the preservation of Ladakh's ecosystems and promote community involvement. Eco-tourism and community tourism initiatives have been recognized as potential solutions to mitigate the adverse effects of tourism and ensure the sustainable development of the region. These approaches focus on responsible tourism practices, local participation, and the conservation of natural and cultural heritage. Efforts are being made by different organizations and stakeholders to promote responsible tourism and address the environmental issues in Ladakh. These include initiatives to enhance waste management, promote sustainable practices, and raise awareness among tourists and local communities about the importance of preserving the region's ecosystems. By adopting sustainable tourism practices, Ladakh can ensure that the tourism industry contributes positively to the local economy while preserving the unique environmental and cultural legacy of the region. It is crucial to strike an equilibrium among tourism development, environmental conservation, and public health considerations to create a sustainable and harmonious future for Ladakh and its inhabitants.

## 9.4 Public Health Implications

Ladakh is a cold desert with very low precipitation and devoid of vegetation except in river valleys. The region has a subsistence of agriculture and frequent dust storms. It is a water-scarce region with irregular water supplies and a harsh and dry climate. The socio-economic conditions are low, houses are poorly ventilated, and health infrastructure is inadequate. The terrain is inaccessible, and roads are blocked during winter. Due to these characteristics, some notable diseases are most prevalent in the region which include;

1. Acute respiratory Infection
2. Acute diarrheal disease

Acute Respiratory Infection (ARI)

Acute respiratory infection (ARI), which has a high morbidity and mortality rate, is a significant global health issue. According to estimates from the World Health Organisation (WHO), ARI accounted for three million fatalities worldwide approximately in 2016, with 40 deaths per 100,000 (Li et al., 2021). In India, ARI-related diseases take the lives of over 400,000 children under five every year. This number represents 13–16% of all paediatric hospital admissions involving child fatalities. ARI is a substantial public health concern in India as it accounts for about one-fourth of all infant deaths under five years old worldwide each year (Hasan et al., 2022).

Acute Respiratory Infection (ARI) is the most prevalent disease in Ladakh, with a significant number of cases reported each year. In 2020, Ladakh recorded a total of 30,554 ARI cases indicating a prevalence rate of 103.74. However, the incidence of ARI varied across different medical blocks of Ladakh. The district of Leh reported a higher prevalence of ARI, (22,392 cases), with a prevalence rate of 180.34 per 1000 population. While as, the district Kargil reported 8,182 cases, with a lower prevalence rate of 48.03 per 1000 population. In comparison, the total ARI cases in 2019 were 46,318, with a prevalence rate of 157.26. The district of Leh had a notably high prevalence rate of 301.88, significantly higher than the district of Kargil, which had a prevalence rate of 51.8 per 1000 population (CMO Leh & Kargil).

Several factors could contribute to the variation in ARI cases between different years and districts in Ladakh. In 2020, the lesser number of cases related to 2019 could be ascribed to the COVID-19 lockdown, which restricted people's access to healthcare facilities and potentially reduced the overall prevalence rate. The higher prevalence of ARI in the district of Leh might be associated with the significant influx of tourists, which directly or indirectly increases the prevalence of the disease. Additionally, the distribution of the population in Leh is more scattered compared to the district of Kargil, which could also contribute to the variation in disease prevalence between the two districts. Ladakh is a cold desert region with harsh winters and frequent dust storms in the summers. To cope with the cold, people in Ladakh minimize ventilation to keep rooms warmer, using the same fireplace for cooking and heating. This practice directly impacts respiratory health, leading to an increased risk of infections. During summers, the combination of frequent dust storms and high traffic movements exposes people to sand particles containing silicosis, further contributing to respiratory diseases.

*Acute Diarrhoeal Disease (ADD)*

The 2017 Sustainable Development Goal set a target date of 2030 for everyone to have access to sanitation and clean water. However, over 2.5 billion people in developing countries do not have access to appropriate sanitary conditions, and yet there are 780 million people without access to safe water globally. People are at risk for health problems because of contaminated water and bad sanitation practices (Kumar et al., 2022). Diarrhoeal disease, which is characterized by a decrease in the consistency of stool, loose stools, and/or passing three or more stools in a day, continues to be the second leading cause of paediatric death worldwide, particularly in underdeveloped nations. According to estimates from the World Health Organisation (WHO) and United Nations Children's Fund (UNICEF), there are roughly 2 billion diarrheal episodes annually and 1.9 million deaths of children under the age of five worldwide in 2013. These estimates increased to 525,000 deaths in 2017 and 370,000 deaths in 2019 for children under the age of five (Charoenwat et al., 2022). 29 per cent of all child fatalities worldwide (and 50 per cent of all child fatalities in India) are caused by pneumonia and diarrhoea. Diarrheal disease is strongly correlated with a variety of sociodemographic characteristics, including caste, religion, socioeconomic status, and women's educational attainment. The disease affects most children of economically disadvantaged families. Environmental aspects of the home, such as drinking water sources, toilet amenities, how to handle child stools, and the materials used for the roof and floors of the home are additional risk factors for diarrhoeal illness (Saha et al., 2022).

Acute diarrhoeal disease is the second leading disease in Ladakh. During 2020 the total cases of ADD in Ladakh were 10,917, signifying a prevalence rate of 37.07 with the highest incidence noticed in district Leh compared to district Kargil. The prevalence rate of ADD in District Leh was 50.58 per 1,000 population while Kargil with a low prevalence rate of 27.22 (CMO Leh & Kargil). In Ladakh except in the main towns, ADD is related to unhygienic conditions which are due to improper drainage systems, and irregular water supply, especially during winters when water gets frozen. In addition to this, the main towns in Ladakh, particularly Leh, have undergone significant urbanization, which is characterized by the construction of numerous hotels and guest houses to accommodate the rising number of tourists. However, this fast development has led to challenges in terms of water consumption and waste management (Khanday et al., 2024).

The high water demand, coupled with inadequate drainage systems, has caused water pollution. This pollution, directly or indirectly, has a detrimental impact on public health, particularly on waterborne diseases like diarrhoea. The study carried out in Leh Town shows that the incidence of diarrhoea has been increasing in Leh over the previous decade, which might be related to water pollution: Additionally, the author highlighted that fast-growing water usage and a lack of adequate water and sanitation infrastructure is triggering grave water pollution (Gondhalekar et al., 2014).

## 9.5 Sustainable Solutions for Public Health

### *9.5.1 Mitigating Indoor and Outdoor Pollution*

In Ladakh, the implementation of clean cooking technologies is vital in mitigating indoor air pollution, which poses significant health risks to the local population especially in winter. Clean cooking technologies, such as improved cookstoves, solar heaters, biogas systems, and proper ventilation systems in households, play a vibrant role in diminishing the emission of harmful pollutants and improving indoor air quality which in turn reduces morbidity of respiratory diseases. Shifting from traditional biomass-based cooking methods to clean cooking technologies, which can significantly decrease the levels of particulate matter, and other toxic substances not only improves indoor air quality but also has constructive environmental implications. It reduces emissions of greenhouse gases, promoting sustainable practices and addressing climate change concerns.

Government policy and support are essential for the effective adoption of clean cooking technologies. Household adoption of these technologies can be encouraged by incentives and subsidies, which will increase their availability and affordability. Protecting and enhancing the well-being and health outcomes of its residents can help pave the way for a sustainable and healthier future by prioritizing the mitigation of indoor and outdoor air pollution through clean cooking technologies and conducting community educational programs to promote consciousness about the health risks associated with dust storms and providing information on preventive measures, such as wearing masks and staying indoors during severe dust storms, etc.

### *9.5.2 Improving Access to Sanitary Facilities, Hygienic Practices, and Clean Drinking Water*

Clean water, hygienic practices, and access to proper sanitation facilities are fundamental in promoting public health and preventing the spread of disease. Promoting proper waste management techniques, such as the safe disposal of solid waste and recycling activities, along with better access to clean water sources and adequate sanitation facilities, can greatly raise overall hygiene standards and lower the incidence of water-related diseases like ADD in Ladakh. Hygiene education programs play a great role in raising awareness about the importance of personal hygiene practices, including regular handwashing, proper food handling, and safe waste disposal. Effective wastewater management systems, such as appropriate sewage treatment and recycling techniques, are vital for protecting water sources, reducing environmental contamination, and promoting healthy living conditions. Campaigns to promote hygiene that emphasize behavioural change and community involvement can enable people to adopt and maintain sanitary habits in their daily lives.

### *9.5.3 Raising Awareness About Environmental Sustainability and Public Health Linkages*

Increasing public awareness about the interrelation of environmental sustainability and public health is vital for promoting responsible and conscious decision-making in Ladakh.

- Educating individuals about the direct impact of environmental degradation on public health can encourage behavioural changes that lead to more sustainable practices in Ladakh.
- Public awareness programs highlighting the links between waste management, water quality, air pollution, and human health can endorse a sense of responsibility and motivate people to take preventative actions.
- Govt. as well as NGOs, should encourage people to adopt sustainable practices and improve their own and their community's quality of life by highlighting the health hazards linked to environmental pollution and degradation.
- For awareness campaigns that emphasize the significance of environmental sustainability as well as sustainable practices for public health, collaboration between governmental entities, educational institutions, and non-governmental organizations is crucial.

### *9.5.4 Empowering Local Communities for Sustainable Practices*

Providing training and resources in sustainable agriculture, waste management, renewable energy, and water conservation to local communities in Ladakh empowers them to take ownership of sustainable initiatives and make positive changes. Encouraging community-led projects and supporting local entrepreneurship in sustainable practices.

## 9.6 Conclusion

The interplay between environmental sustainability and public health in Ladakh is undeniable. The region's unique confrontations, like water scarcity, climate change, and habitat degradation, directly impact the well-being of its inhabitants. By adopting sustainable practices and promoting environmental stewardship, Ladakh can safeguard public health, preserve its natural resources, and safeguard a sustainable future for generations to come. Ladakh is a region that faces many challenges in public health due to its geographical features and lack of infrastructure. However, some sustainable as well as healthcare solutions can be implemented to improve public health in the region.

In addition to the Ayushman Bharat PMJAY, which is a fully financed health insurance program for all Ladakh residents, the union territory of Ladakh also implemented the Universal Health Care Scheme Which ensures the provision of high-quality healthcare to the people of Ladakh who live in the most isolated hilly regions and challenging terrain (Ladakh.nic.in). Another option is the Sowa-Rigpa medical system, which has been popular in Ladakh for many years and has enabled Amchis to manage at least 60% of the public health in this challenging environment (Sexena et al., 2018), especially during winters when roads get blocked due to snowfall. Ladakh Environment and Health Organisation (LEHO) works on environmental, ecological, and health-related issues. This was a response to the shifting agricultural practices, eating habits, social norms, and cultural practices of the various Ladakhi societies (Leho). The Aims and Objectives of LEHO are as under:

- To preserve and enhance Ladakh's ecological farming system, which is the foundation of the economy.
- To increase the public's awareness of health issues, particularly those about environmental cleanliness, nutrition, and steps to prevent the use of narcotics, smoking, alcohol, pollution, etc.
- Encourage afforestation, biodiversity, and ecological restoration vigorously.
- To put traditional, indigenous knowledge and technologies into practice and develop them.
- To promote and preserve the diversity of culture, tradition, and language.

The Ladakh Environment and Health Organisation (LEHO) plays a vital role in promoting sustainable solutions for public health in Ladakh. Their commitment to understanding and reducing the influence of environmental factors on health is demonstrated by their focus on modifying agricultural practices, dietary habits, social norms, and cultural practices in Ladakhi populations. LEHO contributes to the well-being of the local population and supports the broader goal of developing a healthier and more resilient community by working towards sustainable environmental practises.

In addition to this, The Department of Health is working with the main program to spread health awareness, improve treatment options, control communicable diseases, and reduce the morbidity of various diseases that are prevalent in the area.

Thus, tackling issues with public health and environmental sustainability in Ladakh necessitates a multidimensional approach that integrates several sustainable solutions. A healthier and more sustainable future for Ladakh will need moving to renewable energy, encouraging waste management and water conservation, implementing sustainable agriculture practices, protecting biodiversity, sustainable tourism practices, and raising awareness. Overall, ensuring environmental sustainability, preserving public health, and fostering a healthy economy in Ladakh requires a holistic approach that incorporates sustainable solutions, community empowerment, education, and responsible tourist practices as well as developing and implementing robust environmental policies, regulations, and enforcement mechanisms can provide a framework for sustainable development.

## References


Ahmad T, Gupta SK, Singh SK, Meraj G, Kumar P, Kanga S (2023) Unveiling nature's resilience: exploring vegetation dynamics during the COVID-19 era in Jharkhand, India, with the Google Earth engine. Climate 11(9):187

Alexander A (2005) Leh old town, Ladakh: a participatory approach to urban conservation, community-based upgrading and capacity-building. International Tibet Heritage Fund

Barrett K, Bosak K (2018) The role of place in adapting to climate change: a case study from Ladakh, Western Himalayas. Sustain 10(4):898

Behera MD, Matin S, Roy PS (2014) Biodiversity of Kargil cold desert in the Ladakh Himalaya. Integr Obs Assess 253–274

Behera MD, Chitale VS, Matin S, Pujar GS, Malik AH, Pasha SV (2022) Spatiotemporal dynamics of land and vegetation cover in cold desert region of the Ladakh Himalaya. Geospatial Technol Land Water Resour Manag 85–102

Bhutiyani MR, Kale VS, Pawar NJ (2010) Climate change and the precipitation variations in the northwestern Himalaya: 1866–2006. Int J Climatol 30:535–548

Carnero MC (2015) Assessment of environmental sustainability in health care organizations. Sustain 7(7):8270–8291

Charoenwat B, Suwannaying K, Paibool W, Laoaroon N, Sutra S, Thepsuthammarat K (2022) Burden and pattern of acute diarrhea in Thai children under 5 years of age: a 5-year descriptive analysis based on Thailand National Health Coverage (NHC) data. BMC Public Health 22(1):1–10

Chevuturi A, Dimri AP, Thayyen RJ (2018) Climate change over Leh (Ladakh), India. Theoret Appl Climatol 131:531–545

Commission B (1987) Our common future: report of the world commission on environment and development. https://doi.org/10.1080/07488008808408783

Dar SN, Wani MA, Shah SA, Skinder S (2019) Identification of suitable landfill site based on GIS in Leh, Ladakh Region. GeoJournal 84:1499–1513

Dimri AP, Dash SK (2012) Wintertime climatic trends in the western Himalayas. Clim Chang 111(3–4):775–800

Dolezal J, Dvorsky M, Kopecky M, Liancourt P, Hiiesalu I, Macek M, Schweingruber F (2016) Vegetation dynamics at the upper elevational limit of vascular plants in Himalaya. Sci Rep 6(1):24881

Dong Y, Hauschild MZ (2017) Indicators for environmental sustainability. Procedia Cirp 61:697–702

Farooq FN, Farooq M, Mushtaq F, Meraj G, Singh SK, Kanga S, Gupta A, Kumar P, Singh D, Avtar R et al (2022) Strategic slum upgrading and redevelopment action plan for Jammu City. Resources 11(12):120

Farooq M, Meraj G, Rishabh, Kanga S, Nathawat R, Singh SK, Ranga V (2022) Slum categorization for efficient development plan—a case study of Udhampur City, Jammu and Kashmir using remote sensing and GIS. In: Geospatial technology for landscape and environmental management: sustainable assessment and planning, pp 283–299

Geneletti D, Dawa D (2009) Environmental impact assessment of mountain tourism in developing regions: a study in Ladakh, India Himalaya. Environ Impact Assess Rev 29(4):229–242

Gondhalekar D, Akhtar A, Keilmann P, Kebschull J, Nussbaum S, Dawa S, Namgail P (2014) Drops and hot stones: towards integrated urban planning in terms of water scarcity and health issues in Leh Town, Ladakh, India. In: Ecological health: society, ecology and health. Emerald Group Publishing Limited, pp 173–193

Gondhalekar D, Nussbaum S, Akhtar A, Kebschull J (2015) Planning under uncertainty: climate change, water scarcity and health issues in Leh Town, Ladakh, India. In: Leal Filho W, Sümer V (eds) Sustainable water use and management. Green Energy and Technology. Springer, Cham. https://doi.org/10.1007/978-3-319-12394-3_16

Gondhalekara D, Nussbaumb S, Akhtara A, Kebschullb J, Keilmannc P, Dawad S et al (2013) Water-related health risks in rapidly developing towns: the potential of integrated GIS-based urban planning. Water Int 38(7):902–920

Goodland R (1995) The concept of environmental sustainability. Annu Rev Ecol Syst 26:1–24

Gulati FN, Gulati B, Sharma R, Kanga S, Singh SK, Sajan B, Meraj G, Kumar P, Ramanathan AL et al (2023) Unraveling the relationship between stubble burning and air quality degradation in Punjab: a temporal and spatial analysis (2019–2022). J Clim Chang 9(2):43−53

Hasan MM, Saha KK, Yunus RM, Alam K (2022) Prevalence of acute respiratory infections among children in India: regional inequalities and risk factors. Matern Child Health J 26(7):1594–1602

Husaini AM, Bhat MA, Kamili AN, Mir MA (2013) Is Ladakh a 'cold desert'? Curr Sci 104(6):687

Hussain S, Sharma S, Singh AN (2022) Evaluation of ecosystem supply services and calculation of economic value in Ladakh, India. Reg Sustain 3(2):157–169

Jeelani P, Shah SA, Dar SN, Rashid H (2023) Sustainability constructs of mountain tourism development: the evaluation of stakeholders' perception using SUS-TAS. Environ Devel Sustain 25(8):8299–8317

Joshi PK, Rawat GS, Padilya H, Roy PS (2006) Biodiversity characterization in Nubra Valley, Ladakh with special reference to plant resource conservation and bioprospecting. Biodivers Conserv 15:4253–4270

Khanday AA, Rather GM, Kumar MA, Lone AH (2024) Exploring disease dynamics of acute diarrhoeal disease in cold desert Ladakh-India: a geographical analysis. Clin Epidemiol Glob Health 101741

Kidd CV (1992) The evolution of sustainability. J Agric Environ Ethics 5:1–26

Kjaergard B, Land B, Bransholm Pedersen K (2014) Health and sustainability. Health Promot Int 29(3):558–568

Kumar MA, Rather GM, Khanday AA (2023) Mapping regional disparities in socio-economic dynamics in the mountain environs of Ladakh, India

Kumar P, Srivastava S, Banerjee A, Banerjee S (2022) Prevalence and predictors of water-borne diseases among elderly people in India: evidence from longitudinal ageing study in India, 2017–18. BMC Public Health 22(1):993

Li ZJ, Zhang HY, Ren LL, Lu QB, Ren X, Zhang CH, Yang WZ (2021) Etiological and epidemiological features of acute respiratory infections in China. Nat Commun 12(1):5026

Meraj FN, Meraj G, Farooq M, Singh SK, Islam MN, Kanga S et al (2022) Modeling the sediment retention and ecosystem provisioning services in the Kashmir valley, India, Western Himalayas. Model Earth Syst Environ 8(3):3859−3884

Meraj FN, Meraj G, Kanga S, Kranjčić N, Đurin B, Singh SK et al (2021) Role of natural capital economics for sustainable management of earth resources. Earth 2(3):622−634

Moldan B, Janoušková S, Hák T (2012) How to understand and measure environmental sustainability: indicators and targets. Ecol. Ind. 17:4–13

Morelli J (2011) Environmental sustainability: a definition for environmental professionals. J Environ Sustain 1(1):2

Müller J, Dame J, Nusser M (2020) Urban mountain waterscapes: the transformation of hydro-social relations in the trans-Himalayan Town Leh, Ladakh, India. Water 12(6):1698

Namgail T, Rawat GS, Mishra C, van Wieren SE, Prins HH (2012) Biomass and diversity of dry alpine plant communities along altitudinal gradients in the Himalayas. J Plant Res 125:93–101

Nusser M, Dame J, Kraus B, Baghel R, Schmidt S (2019) Socio-hydrology of "artificial glaciers" in Ladakh, India: assessing adaptive strategies in a changing cryosphere. Reg Environ Change 19(5):1327–1337

Rajashekariah K, Chandan P (2013) Value chain mapping of tourism in Ladakh. WWF-India, New Delhi, India

Saha J, Mondal S, Chouhan P, Hussain M, Yang J, Bibi A (2022) Occurrence of diarrheal disease among under-five children and associated sociodemographic and household environmental factors: an investigation based on national family health survey-4 in rural India. Children 9(5):658

Saxena S, Bhardwaj AK, Kumar V, Patel MK, Kumar R, Chaurasia OP (2018) Sustainable utilisation of medicinal plants of Ladakh and Lahaul-Spiti of Trans-Himalaya. Def Life Sci J 3:120–125

Schmidt S, Nüsser M (2017) Changes of high-altitude glaciers in the trans-Himalaya of Ladakh over the past five decades (1969–2016). Geosciences 7(2):27

Shekhar MS, Chand H, Kumar S, Srinivasan K, Ganju A (2010) Climate-change studies in the western Himalaya. Ann Glaciol 51(54):105–112

Sherman JD, Thiel C, MacNeill A, Eckelman MJ, Dubrow R, Hopf H, Bilec MM (2020) The green print: advancement of environmental sustainability in healthcare. Resour Conserv Recycl 161:104882

Sidhu AK, Chandra K, Palot J (2012) Observations on the status and diversity of butterflies in the fragile ecosystem of Ladakh (J & K). Rec Zool Surv India 112(3):101–112

Singh FN, Singh S, Singh H, Sharma V, Shrivastava V, Kumar P, Kanga S, Sahu N, Meraj G, Farooq M, Singh SK et al (2021) Impact of forest fires on air quality in Wolgan valley, New South Wales, Australia—a mapping and monitoring study using google earth engine. Forests 13(1):4

Singh FN, Singh H, Meraj G, Singh S, Shrivastava V, Sharma V, Farooq M, Kanga S, Singh SK, Kumar P et al (2022) Status of air pollution during COVID-19-induced lockdown in Delhi, India. Atmosphere 13(12):2090

Soheb M, Ramanathan A, Angchuk T, Mandal A, Kumar N, Lotus S (2020) Mass-balance observation, reconstruction, and sensitivity of Stok glacier, Ladakh region, India, between 1978 and 2019. J Glaciol 66(258):627–642

Sutton P (2004) A perspective on environmental sustainability. *Paper on the Victorian Commissioner for Environmental Sustainability*, 1, 32

Tuladhar S, Hussain A, Baig S, Ali A, Soheb M, Angchuk T, Shrestha AB (2023) Climate change, water, and agriculture linkages in the upper Indus basin: a field study from Gilgit-Baltistan and Leh-Ladakh. Front Sustain Food Syst 6:1012363

# Chapter 10
# Sustainable Agriculture and Food Security in India

**Sunita Singh, Abhay Kumar Chaubey, and Tarun Kumar**

## 10.1 Introduction

Agriculture holds a substantial share, approximately 18.8%, of India's gross value added (GVA) and stands as the largest employer of the workforce, as per the 2021–22 Economic Survey. Agricultural sustainability entails the effective management of agricultural resources to meet evolving human needs while simultaneously preserving or enhancing environmental quality and conserving natural resources, as articulated (Gregory, 1989). Sustainable agriculture encompasses practices that protect the environment, conserve natural resources, reduce chemical usage, and foster economic independence. In countries like India, characterized by a significant population below the poverty line and high production technology expenses, sustainable agriculture becomes both an economic and environmental imperative (Meraj et al. 2022). Therefore, progress should be assessed using both qualitative and quantitative criteria, considering the impacts of agricultural development on social, environmental, and health aspects. Attaining food security is of paramount importance for achieving the Sustainable Development Goals, particularly in a nation like India where over 20% of the population lives below the poverty line. According to the Global Food Security Index (2022), India grapples with challenges such as a 16.3% prevalence of undernutrition, 30.9% of children experiencing stunted growth, 33.4% being underweight, and 3.8% suffering from obesity. The International Food Security assessment for 2022–23 notes that India, owing to its vast population, significantly influences food insecurity trends, accounting for 65% of the food-insecure population in the same period. Over the next decade, the number of food-insecure individuals in India is projected to decline to 24.7 million, constituting 1.7%. National

S. Singh (✉) · A. K. Chaubey · T. Kumar
Department of Geography, Central University of South Bihar, Gaya, Bihar 824236, India
e-mail: sunitasingh@cusb.ac.in

© The Author(s), under exclusive license to Springer Nature Singapore Pte Ltd. 2024

G. Tripathi et al. (eds.), *Sustainability and Health Informatics*, Advances in Geographical and Environmental Sciences,
https://doi.org/10.1007/978-981-97-6706-9_10

Food Security plays a pivotal role in economic development and poverty alleviation by providing nutritional support and enhancing livelihoods (Singh et al. 2021).

Food security is a concept that has undergone a gradual evolution in its meaning and scope over time. Traditional notions of food security primarily revolved around straightforward measures like domestic food production, grain storage, national self-sufficiency in food, and food utility (Singh et al. 2021). With the burgeoning global population, there has been a surge in food consumption, posing substantial challenges for agriculture (Sajan et al. 2022). These challenges are further compounded by the adverse impacts of climate variations and climate change, soil and land degradation, and the depletion of vital resources such as freshwater, phosphates, fossil fuels, and fertile topsoil. Projections indicate that annual crop yields are expected to increase by less than 1% in the coming decades, and the availability of arable land for expansion is exceedingly limited (Alexandratos and Bruinsma, 2012). As the global population continues to grow, the overall demand for food is on the rise. Additionally, increased income and economic development have prompted dietary shifts toward meat and animal products, which are more resource-intensive to produce (OECD-FAO, 2010). To address the challenges of food scarcity and increasing food demand, technology plays a crucial role. This encompasses the development and utilization of nutrients, pest control solutions, crop cultivation techniques, and advanced farm equipment. It also encompasses the concept of genetically modified crops designed to offer enhanced nutritional efficiency, such as increased calorie yield or greater overall yield. Furthermore, it involves the manipulation of natural pest control agents and the implementation of farm management practices that emphasize long-term whole farm productivity, rather than focusing solely on annual production per hectare (Stone and Pedigo, 1972). Achieving sustainable agricultural growth requires the adoption of economically and environmentally sustainable agricultural practices, offering a viable approach to address the challenges of food security resulting from population growth and environmental degradation (Muhie 2022).

## 10.2 Sustainable Agriculture

The term "sustain," derived from the Latin sustinere (combining "sus-" meaning from below and 'tenere' meaning to hold or maintain), embodies the idea of long-term support and permanence. In context of agriculture, sustainable agriculture refers to farming systems that can consistently maintain their productivity and societal utility over an extended period (Meraj et al. 2021). These systems should effectively conserve resources, provide social benefits, remain competitive economically, and exhibit environmental responsibility. Sustainable agricultural systems are strategically developed to make the most of existing soil nutrients and water cycles, energy dynamics, beneficial soil microorganisms, and natural method of pest control. This approach helps minimize the potential for environmental harm by harnessing existing cycles and processes (Chel and Kaushik, 2011). Sustainable agricultural practices have gained significant recognition due to concerns about the Environmental consequences of modern agriculture, agriculture's reliance on finite resources, and the long-term productivity of Agricultural systems heavily dependent on substantial external inputs (Filho, 2000).

Sustainable Agriculture presents an alternative approach to addressing both fundamental and practical issues, as highlighted (Lal, 2008). While Green Revolution agriculture predominantly focused on productivity concerns, Sustainable Agriculture goes a step further, placing a stronger emphasis on productivity while also considering the multidimensional aspects of sustainability, including economic, environmental, and social dimensions (Rao and Roger, 2006). Sustainable agriculture encompasses environmentally responsible, economically viable, socially equitable, culturally appropriate practices rooted in a comprehensive scientific approach. *India Agriculture Policy 2000* defines sustainable agriculture as an approach that integrates technical comprehensiveness, economic feasibility, environmental non-degradation, and social suitability. It involves farming systems and methodologies that maintain or enhance the Economic sustainability of agricultural production, safeguard natural resources, and consider various factors influenced by Agricultural activities (Chouhan and Gaur, 2022).

Indian government started the National Mission for Sustainable Agriculture (NMSA) in 2014–15.

Sustainable Agriculture pursues to integrate three main objectives into their work:

- A healthy environment
- Economic profitability
- Social and economic equity

To ensure a sustainable agriculture system, stakeholders in the food chain, such as producers, food processors, suppliers, retailers, consumers, and waste management, can all contribute. The foundation of sustainability is the idea that we must satisfy our current needs without compromising the capacity of future generations to satisfy their own (Brodt et al., 2011).

Agriculture has undergone significant transformations since the conclusion of World War II. Advances in technique, machinery, increased chemical utilization, and implementation of various government policies aimed at boosting food production and reducing prices have led to a substantial increase in food and fiber productivity. Consequently, numerous farmers, now able to produce more food and fiber at lower costs (SAREP, 2021).

Sustainable Agriculture is not a singular, precisely defined objective. The scientific understanding of what constitutes sustainability in terms of the environment, society, and the economy is continually evolving and influenced by a multiple factors (Sinha et al. 2022). For instance, agriculture's capacity to adapt to climate change wasn't regarded as a critical concern two decades ago, but it has now garnered increasing attention. Furthermore, the specific criteria for what defines a sustainable system may vary from one set of conditions, such as soil types, climate, and labor costs, to another. They can also differ from one cultural and ideological perspective to another, leading to the term "sustainable" being a matter of ongoing debate. Hence,

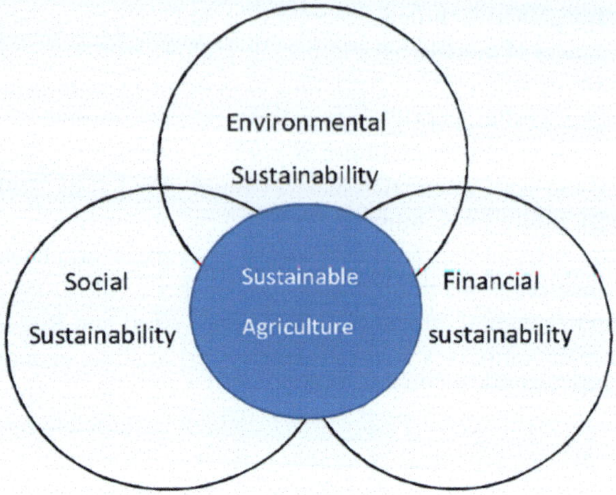


**Fig. 10.1** Components of sustainable agriculture

it is more constructive and pragmatic to consider agricultural systems as an existing with continuum from unsustainable to highly Sustainable, rather than forcing them into a simple sustainable/unsustainable categorization (Velten, 2015) (Fig. 10.1).

Goal oriented methods can be implemented and assessed for sustainability in agriculture (Lehr, 2001). The objective of sustainable development is to eliminate hunger, achieve food security, and improved nutrition. Some of the inter relationships that supports sustainable agriculture are to handle climate change, endowing small farmers, promoting gender equality, removing poverty, and ensuring healthy lifestyles (UN, 2023).

Sustainable agricultural practices and food systems encompass both the production and consumption of food products, and they should be approached from a holistic and integrated perspective. Key inputs for food production, such as land, soil, water, and plant resources, are of utmost importance. However, the increasing global demand for food has led to growing scarcity of these resources in many parts of the world, emphasizing the need to use and manage them sustainably (Farooq et al. 2022).

Boosting yields on existing agricultural land involves not only enhancing the productivity of fertile areas but also restoring degraded land through sustainable agricultural practices. Effective management of limited water resources, along with the adoption of improved irrigation and storage technologies, can contribute significantly to sustaining the productivity of arid and semi-arid regions. Additionally, the development of new crop varieties that are more resilient to drought conditions plays a crucial role in achieving these sustainability goals.

## 10.3 Food Security

Food security pertains to the continuous availability, accessibility, and affordability of food for all individuals at all times, underscoring its status as a fundamental human right. The right to food is intrinsically linked to the right to life, as stipulated in Article 21 of the Constitution of India, 1950. The concept of 'food security' encompasses both physical and economic access to food, with accessibility encompasses both the presence and affordability of food. Food security can defined as ensuring that every person, at all times, has the means to obtain enough safe and nutritious food to sustain a healthy and active life, as articulated by the World Food Summit (1996). The World Health Organization (WHO, 2022) recognizes that food security comprises three critical components:- "food availability," "food access" and "food use." Notably, the Food and Agriculture Organization of the United Nations (OECD-FAO, 2009) introduced a fourth dimension, which is "the dependability of the first three aspects of food security over time." As such, food security is a multi-dimensional and intricate concept.

Top of Form

It has the following dimensions:

- Availability: It means food production within the country, food imports, and the stock stored in government granaries.
- Accessibility: It means food is within range of every person without any discrimination.
- Affordability: It implies having adequate money to buy enough, safe and healthy food to meet dietary needs (Fig. 10.2).

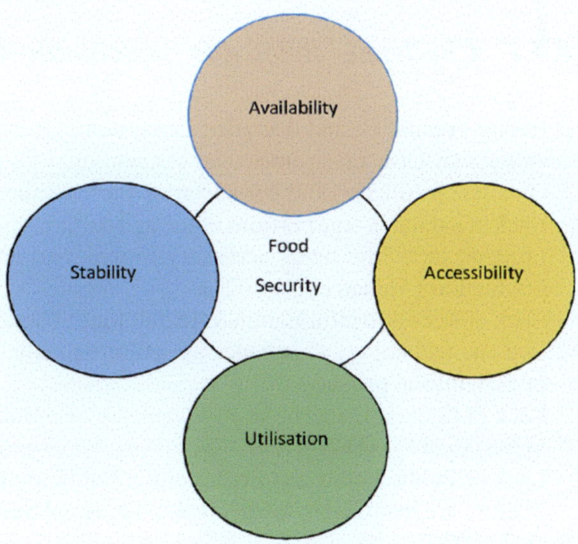


**Fig. 10.2** Dimensions of food security

Thus, a nation can only be guaranteed food security if there is enough food for each and every person, if everyone makes enough money to buy food of a reasonable calibre, and if there are no obstacles in the way.

India aspires to attain self-sufficiency in food production, which is integral to upholding national food security. Simultaneously, to bolster domestic food security, it is imperative to foster economic growth that elevates the income and purchasing capacity of disadvantaged populations. Furthermore, agricultural regulations play a pivotal role in securing food access within the country by establishing fixed procurement prices for food grains and implementing public distribution systems. These measures remain essential even when achieving self-sufficiency at the national level. Governments play a significant role in governing and stabilizing a substantial portion of the food supply through these interventions. However, inefficient pricing mechanisms can lead to undesirable consequences and hinder access to food supplies (FAO, 2023). Despite India's remarkable economic growth in recent years, it has yet to overcome persistent challenges related to poverty and food security. Food represents a fundamental and indispensable necessity for an individual's sustenance, development, and growth. While the Indian government has proactively addressed household food security through initiatives such as the Public Distribution System (PDS) and the National Food Security Act (NFSA, 2013), concerns about food security persist, particularly in the face of increasing population, climate change, and global supply disruptions like the Russia-Ukraine War. These issues demand continued attention and resolution (Dristi Sept 2022).

## 10.4 Issues and Challenges Related to Sustainable Agriculture and Food Security in India

### *10.4.1 Challenges Related to Sustainable Agriculture*

1. Rising Population and degraded Ecosystem: Growing population and damaged ecosystem have made intensive, conventional farming (using HYV seeds and chemical fertilizers) and deforestation more resilient (Ray, 2011).
2. Lack of capital: Large portions of the agricultural community (Small and medium farmers) lack 2the funds necessary to make switch to Sustainable Agriculture production (Ahmed et al., 2022).
3. Lack of Access to information & Technology: There is a lack of access to information and technology to enhance agriculture practices, processing and marketing of agricultural products.
4. Lack of Economic Incentives: Farmers are uncertain of the benefits of switching to Sustainable farming.
5. Lack of Public Policy & Infrastructure: Public policies and fundamental infrastructure are insufficient to encourage the use of Sustainable Agriculture (Edeme et al., 2020).

### 10.4.2 Current Framework for Sustainable Agriculture in India

1. **National Mission on Sustainable Agriculture**: Its goal is to increase Agricultural productivity, particularly in rain-fed regions, by emphasizing integrated farming, managing soil health, and coordinating resource conservation (Patel, 2016).
2. **Pradhan Mantri Krishi Vikas Yojna (PMKVY)**: The PMKVY program intends to encourage commercial organic production by involving a group of farmers in organic farming.
3. **Network Project on Organic Farming of ICAR**: Its goal is to compare the performance of important agricultural systems that are peculiar to a given place under organic and conventional farming and estimate the agronomic efficiency of various production systems (ICAR, 2023).

### 10.4.3 Challenges Related to Food Security

1. **Declining Soil Health**: An important element responsible for food production is healthy soil so good quality soil is required and its quality is declining due to various climatic and anthropogenic factors excessive, unsuitable use of agrochemicals and deforestation as well as natural disasters are the main challenges of the sustain food production. About 33% of the soil of the globe is already degraded (Brahmanand et al., 2013).
2. **Climate change**: The monsoon contributes to approximately 70% of India's annual precipitation and serves as the primary water source for irrigating 60% of the country's net sown area. Agricultural land utilization and food production are significantly influenced by the limited availability of water for irrigation and various other contributing factors. India is experiencing the effects of climate change, including rising temperatures, irregular and inadequate rainfall, and an increasing occurrence of extreme weather events (Fayaz et al., 2022). These factors collectively exert adverse impacts on food production systems and agricultural development. Several reports have indicated that climate change plays a substantial role in the heightened frequency of extreme events, such as floods and droughts. Alterations in precipitation patterns and the escalating frequency and severity of extreme weather events, such as heat waves and floods, are already diminishing agricultural productivity in India, posing a significant threat to food security (Brahmanand et al., 2013).
3. **Crop Diversification**: In recent times, agricultural scientists have placed significant emphasis on the adoption of crop diversification. This shift is primarily driven by the stagnant prices of food grains, such as rice and wheat, resulting in low returns for farmers. By encouraging farmers to focus on cultivating different crops, they stand to attain more substantial profits. Another proposed solution is the establishment of decentralized food grain banks at the block or district

level, which would enable rural communities to access subsidized food (Gulati et al., 2023). This innovative concept aims to enhance the distribution of food grains while simultaneously reducing instances of corruption (Brahmanand et al., 2013).

4. **Supply Chain Disruption due to Unstable Global Order**: In 2022, the Russia-Ukraine War disrupted the global supply of food grains, leading to food scarcity and price hikes, during a period when the Covid-19 epidemic had already affected food supplies worldwide in 2020 (Hassen and Bilali, 2022).

### *10.4.4 Current Framework for Food Security in India*

1. **Buffer Stock**: Buffer Stock is the stock of food grains, particularly wheat and rice. The public sector of India maintains Buffer Stocks to ensure food security and prepare for unexpected situations like drought, starvation, conflicts, etc. Food Corporation of India (FCI) stored stock of food grains at MSP (Kumar et al., 2007). The stocking of food grains was first implemented in 1969 during the 4th Five Year Plan (FYP). The objectives of Buffer Stock are:
   - Better return for farmers
   - Food Security
   - Price Stability
   - Food-grains for welfare programs
2. **Public Distribution System**: The Public Distribution System (PDS) in India is a government-run program that aims to distribute essential food items to the economically vulnerable population at subsidized rates. PDS was a general entitlement scheme for all consumers till 1992 but in 1992, it became Revamped PDS (RPDS) which focused the poor families. RPDS became Targeted PDS (TPDS) in 1997 and established Fair Price Shops for the distribution of food grains at subsidized rates (NFSA 2023). The objectives of the PDS are:
   - To provide essential food items at affordable prices.
   - To hunger and malnutrition among economically weaker section.
   - To prevent black marketing of food grains.
   - To stabilize prices of essential commodities in the market.
3. **National Food Security Act. 2013 (NFSA2013)**: Through ensuring access to a sufficient amount of high-quality food at reasonable rates, the National Food Security Act of 2013 seeks to ensure people's food and nutritional security. 75% of India's rural population and 50% of its urban population are eligible for subsidized food grains under the NFSA (2013) through the Targeted Public Distribution System (TPDS), which had a value of around 81.35 crores as per the 2011 Census.

## 10.5 Conclusion

Sustainable agriculture is an approach that seeks to uphold equilibrium between the escalating food demand and future production. As population growth, shifts in income demographics, and evolving food preferences occur, the demands for the food needs of the future population also undergo changes. It is crucial to remember that the majority of people living in severe poverty and hunger are from rural areas, and smallholder farmers and their families make up a sizable share of these populations. The situation leads to several health issues. Therefore, raising food production and agricultural productivity is essential to ending hunger and poverty. Global agricultural systems must aim to become more productive and waste-free. To do this, it is critical to implement sustainable agriculture methods and food systems that incorporate both production and consumption in a holistic and integrated manner.

Sustainable agriculture must include social, economic, and environmental sustainability. Without a transition to sustainable food and agriculture systems that will ensure food security in India, we will not be able to fulfill future needs as the country's population is growing day by day, its resources are exhausting, and the threat of climate change is growing. Therefore, active and sustainable agricultural research, based on the demand for food by the increasing population, is needed to combat hunger and maintain food security.

## References

Ahmed M, Asim M, Ahmad S, Aslam M (2022) Climate change, agricultural productivity, and food security. In Springer eBooks (pp 31–72). https://doi.org/10.1007/978-3-031-14973-3_2

Alexandratos N, Bruinsma (2012) World agriculture towards 2030/2050: the 2012 revision. Res Agric & Appl Econ

Bandla G, Kundu A, Abirami K, Swain S, Subramani T, Ahmad S KZ (2017) Organic farming in tropical Islands of India, ICAR-central Island agriculture research institute. Port Blair. pp 1–250. https://www.researchgate.net/publication/330837484

Brahmanand PS, Kumar A, Ghosh S, Chowdhury SR, Singandhupe RB, Singh R, Nanda P, Chakraborthy H, Srivastava SK, Behera MS (2013) Challenges to food security in India. Curr Sci 104(7):841–846. https://about.jstor.org/terms

Brodt S, Six J, Feenstra G, Ingels C, Campbell D (2011) Sustainable agriculture. Nat Educ Knowl 3(1)

Chel A, Kaushik G (2011) Renewable energy for sustainable agriculture. Agron Sustain Dev 31(1):91–118. https://doi.org/10.1051/agro/2010029

Chouhan AS, Gaur G (2022) Indian agriculture policy 2000: a birds eye view. Issue 5 Indian JL & Legal Rsch 4:1

Coulibaly TP, Du J, Diakité D (2021) Sustainable agricultural practices adoption. Agric. (Pol'nohospodarstvo) 67(4):166–176. De Gruyter Open Ltd. https://doi.org/10.2478/agri-2021-0015

Dristi IAS (2022) Food security in India.https://www.drishtiias.com/daily-updates/daily-news-editorials/food-security-in-india

Edeme RK, Nkalu NC, Idenyi JC, Arazu WO (2020) Infrastructural development, sustainable agricultural output and employment in ECOWAS countries. Sustain Futur 2. https://doi.org/10.1016/j.sftr.2020.1000

Farooq M, Meraj G, Rishabh, Kanga S, Nathawat R, Singh SK, Ranga V (2022) Slum categorization for efficient development plan—a case study of Udhampur City, Jammu and Kashmir using remote sensing and GIS. In: Geospatial technology for landscape and environmental management: sustainable assessment and planning, pp 283–299

Fayaz FN, Fayaz M, Meraj G, Khader SA, Farooq M, Kanga S, Singh SK, Kumar P, Sahu N et al (2022) Management of landslides in a rural–urban transition zone using machine learning algorithms—a case study of a National Highway (NH-44), India, in the Rugged Himalayan Terrains. Land 11(6):884

Filho WL (2000) Dealing with misconceptions on the concept of sustainability. Int J Sustain High Educ 1(1):9–19. https://doi.org/10.1108/1467630010307066

Food and Agriculture organization of the United Nation (2023) The state of food security and nutrition in the World 2023. https://www.fao.org/documents/card/en?details=cc3017en

Ganesh-Kumar A, Gulati A, Cummings R (2007) Foodgrains policy and management in India responding to today's challenges and opportunities. International Food Policy Research Institute. https://pdf.usaid.gov/pdf_docs/pnadk225.pdf

Gregory PJ (1989) Crop water use efficiency. Soil, water and crop/livestock management systems: rain-fed agriculture in the near east region. USAID, USDA, and ICARDA, Washington, DC, pp 162–172

Saini Ashok Gulati S (2015) The national food security act (NFSA) 2013: challenges, buffer stocking and the way forward. Indian Council for Research on International Economic Relations. http://hdl.handle.net/10419/176315

Gulati FN, Gulati B, Sharma R, Kanga S, Singh SK, Sajan B, Meraj G, Kumar P, Ramanathan AL et al (2023) Unraveling the relationship between stubble burning and air quality degradation in Punjab: a temporal and spatial analysis (2019–2022). J Clim Chang 9(2):43−53

ICAR (2023) Annual report 2022–23. In ICAR. Indian council of agricultural research. Retrieved 2024, from https://icar.org.in/sites/default/files/2023-07/ICAR-Annual-Report-2022-23-English.pdf

Hassen TB, Bilali HE (2022) Impacts of the Russia-Ukraine war on global food security: towards more sustainable and resilient food systems? Foods 11(15). https://doi.org/10.3390/foods11152301

Lal R (2008) Soils and sustainable agriculture. a review. Agron Sustain Dev 28(1):57–64. https://doi.org/10.1051/agro:2007025

Singh SK, Meraj G, Mondal N, Bera A, Verma MK, Tomar JS, Kanga S (2021) Assessment of seasonal vegetation dynamics over parts of thar desert using geospatial techniques. ANGRAU 105

Liu KC, Racherla R (2019) Innovation, economic development, and intellectual property in india and china comparing six economic sectors. Myth, Rities Oppor. http://www.springer.com/series/15958

Meraj FN, Meraj G, Farooq M, Singh SK, Islam MN, Kanga S et al (2022) Modeling the sediment retention and ecosystem provisioning services in the Kashmir valley, India, Western Himalayas. Model Earth Syst Environ 8(3):3859−3884

Meraj FN, Meraj G, Kanga S, Kranjčić N, Đurin B, Singh SK et al (2021) Role of natural capital economics for sustainable management of earth resources. Earth 2(3):622−634

Muhie SH (2022) Novel approaches and practices to sustainable agriculture. J Agric Food Res 10:100446. https://doi.org/10.1016/j.jafr.2022.100446

NFSA (2013) The national food security act. In the gazette of India (DL—(N)04/0007/2003—13). The controller of publications, Delhi. Retrieved 2024, from https://nfsa.gov.in/portal/nfsa-act

OECD-FAO (2009) Agricultural outlook 2009. In OECD agricultural outlook . . ./OECD-FAO agricultural outlook. https://doi.org/10.1787/agr_outlook-2009-en

OECD-FAO (2010) OECD-FAO agricultural outlook 2010. In OECD agricultural outlook . . ./ OECD-FAO agricultural outlook. OECD. https://doi.org/10.1787/agr_outlook-2010-en

Patel A (2016) Climate change & agriculture in india effective implementation of national mission for sustainable agriculture. Int J Agric Innov Res 5(3):341–352

Rao NH, Rogers PP (2006) Assessment of agricultural sustainability. Curr Sci 91(4):439–448

Ray S, Ray IA (2011) Impact of population growth on environmental degradation: case of India. J Econ Sustain Dev 2(8). www.iiste.org

Sajan FN, Sajan B, Mishra VN, Kanga S, Meraj G, Singh SK, Kumar P et al (2022) Cellular automata-based artificial neural network model for assessing past, present, and future land use/ land cover dynamics. Agronomy 12(11):2772

SAREP (2021) What is sustainable agriculture? Sustainable agriculture research and education program. Retrieved 2024, from https://sarep.ucdavis.edu/sustainable-ag

Singh FN, Singh S, Singh H, Sharma V, Shrivastava V, Kumar P, Kanga S, Sahu N, Meraj G, Farooq M, Singh SK et al (2021) Impact of forest fires on air quality in Wolgan valley, New South wales, Australia—a mapping and monitoring study using google earth engine. Forests 13(1):4

Sinha FN, Sinha KK, Gupta MK, Banerjee MK, Meraj G, Singh SK, Kanga S, Farooq M, Kumar P, Sahu N et al (2022) Neural network-based modeling of water quality in Jodhpur, India. Hydrol 9(5):92

Stoneand JD, Pedigo L (1972) Development and economic-injury level of the green cloverworm. J Econ Entomol 65(1):197–201. https://doi.org/10.1093/jee/65.1.197

United Nations (2023) Food security and nutrition and sustainable agriculture. https://sdgs.un.org/ topics/food-security-and-nutrition-and-sustainable-agriculture

Velten S, Leventon J, Jager N, Newig J (2015) What is sustainable agriculture? A systematic review. Sustain (Switzerland) 7(6):7833–7865. MDPI. https://doi.org/10.3390/su7067833

Von Wirén-Lehr S (2001) Sustainability in agriculture-an evaluation of principal goal-oriented concepts to close the gap between theory and practice. Ecosyst Environ 84

World Health Organization (2022) The state of food security and nutrition in the world 2022, https:// www.who.int/publications/m/item/the-state-of-food-security-and-nutrition-in-the-world-2022

# Chapter 11
# Assessing Interdependency of Climate Change and Food Security: An Overview

**Sneha Verma, Anurag Rawat, Kanchan Awasthi, Ramakant, and Akash Mishra**

## 11.1 Introduction

A person needs "sufficient, safe, and nutritious food that meets individual dietary needs and food preferences" to have an active and healthy life, and this food must always be readily available on a social, economic, and physical level (FAO 2002). There are four main aspects to it: availability, accessibility, use, and stability (FAO 2008; FAO 2009). Food insecurity and security have long been associated with issues in the "developing world". Research has so far focused on food security from the perspective of poor nations because historically, food shortages have predominantly affected the developing globe.

The observed effects of climate change on crop suitability have altered the production of the primary crops, as discussed in the current research. Despite the recent increase in interest in climate change adaptation consequences, the sectoral focus has remained the same (Ahmad et al. 2023). By reducing capacity or raising hazards in another location or sector, sectoral adaptation techniques can lead to maladaptation (Barnett and O'Neill 2010; Lele et al. 2013; Urwin and Jordan 2008; Walker et al. 2006). An estimated US$1.4 billion is wasted annually due to Pesticide overuse in China's food production has a detrimental effect on biodiversity both on and off farms (Norse et al. 2001). As a result, food production has been impacted. excessive groundwater use, waste of limited water supplies, and an increase in energy demand have resulted from groundwater extraction subsidies given by several nations in the

S. Verma · K. Awasthi · Ramakant · A. Mishra
Department of Science, Maharishi School of Science and Humanities, Maharishi University of Information Technology, Lucknow, India

A. Rawat (✉)
Department of Zoology, Shri Vishambhar Dayal Tripathi Government Degree College, Unnao, Chhatrapati Shahu Ji Maharaj University, Kanpur, India
e-mail: anurag11684@gmail.com

© The Author(s), under exclusive license to Springer Nature Singapore Pte Ltd. 2024 
G. Tripathi et al. (eds.), *Sustainability and Health Informatics*, Advances in Geographical and Environmental Sciences,
https://doi.org/10.1007/978-981-97-6706-9_11

Hindu Kush Himalayan (HKH) region in response to surface-water scarcity and unpredictable water availability, which in turn has compromised food and energy security. Pakistan's desert Balochistan province is currently using groundwater irrigation to cultivate apples and other fruits, a process that consumes a tremendous amount of energy while the nation struggles with severe energy shortages (Khair 2013; Mustafa and Qazi 2007).

Food, energy, and water are essential for human existence and long-term well-being. According to Bazilian et al. (2011), billions of individuals do not have access to all three (food, energy, and water), and they are all vulnerable to rapidly rising global demand as well as resource constraints. Addressing these pressing demands is the biggest problem society is currently facing. Climate change imposes a multitude of stresses or causes environmental depletion, therefore adaption necessitates thorough and coordinated solutions, as well as collaboration across several sectors and sizes (local, national, and regional) (Singh et al. 2021a, b). An evaluation of local vulnerabilities is necessary for effective planning for climate change adaptation to address the gap between local community demands and priorities and higher-level policy practices (Gulati et al. 2023).

Extreme weather occurrences, temperature, precipitation patterns, sea levels, land erosion, flooding, and other environmental aspects are all significantly impacted by climate change. It is mostly brought on by the increase in human population, anthropogenic activities including deforestation and fossil fuel combustion, and environmental deterioration (Singh et al. 2022). These modifications thus have a substantial influence on food security worldwide (Cottrell et al. 2019; Eriksson et al. 2009; FAO 2020; Rockstro¨m et al. 2009; Shrestha and Aryal 2011). One of humanity's biggest concerns is how to control global warming and lessen its negative consequences on social and ecological systems (Meraj et al. 2022). The adaptation of Climate change is a primary priority for the overall growth of human and humanity, however, developing countries—whose populations are largely concentrated in coastal and mountainous areas, have limited resources and capacity, and rely on climate-sensitive industries like forestry, agriculture, and fisheries—need to take note of this (WRI 2011).

## 11.2 Interdependency of Climate Change and Food Security

It is a complicated and important subject that necessitates a multifaceted analysis and has substantial consequences for global sustainability, social well-being, and economic stability. Climate change and food security are interdependent.

Climate change-related variations in temperature, precipitation, and extreme weather patterns have a direct impact on crop growth and productivity. Given that certain crops can find it challenging to adapt to shifting climate conditions, rising temperatures, altered rainfall patterns, droughts, floods, and shifting growing seasons impair agricultural productivity in certain areas (Debnath et al. 2023; Tomar et al.

2021). Food shortages, higher pricing, and less access to nutrition are all results of declining agricultural output. Existing disparities and food insecurity are made worse by these difficulties, particularly in disadvantaged groups and susceptible geographic areas. The main greenhouse gases produced by agricultural operations are methane and nitrous oxide emissions from farming practices and livestock rearing. The traditional cropping cycles and the suitability of locations for particular crops are further disrupted by these emissions, which also contribute to climate change (Meraj et al. 2021). Additionally, it aids in the transmission of illnesses and pests that are harmful to livestock and crops. More often storms, floods, and droughts can have disastrous effects on agriculture and livestock productivity. These occurrences undermine food security by causing crop failures, soil erosion, infrastructure damage, and loss of livelihood. Despite the prevalence of studies on the impact of climate change on crops, fisheries and animal productivity are also significantly impacted (Creighton et al. 2016; Herrero et al. 2015). The frequency and intensity of catastrophic weather events have increased, and farmers now have to contend with altering weather patterns, which makes farming much riskier (IPCC 2012).

Climate change affects water supplies because of warming, which is crucial for agriculture. Water shortages in some areas may result from altered precipitation patterns and higher evaporation rates, which influence the amount of water available for irrigation. Agriculture demands a lot of water, and alterations in water availability can affect irrigation, as a result, crop yields have been reduced, and endanger food supply in places with scarce water resources. Climate change is predicted to lower crop yield potential, according to modelling studies, especially in many tropical and mid-tropical nations (Cline 2007; Jarvis et al. 2001; Knox et al. 2012; Rosenzweig et al. 2014). As harsh weather conditions worsen, the price and yield volatility will increase further, affecting the livelihoods of vast populations (Wheeler and von Braun 2013). Climate change disproportionately affects vulnerable groups, including poor people, indigenous communities, and small-scale farmers. These populations often lack the resources to adapt to changing conditions and are more prone to food insecurity.

Most of the evidence points to decreased quality because of declines in protein, macro- and micronutrient (Fe, Zn, Mn, Cu) concentrations, leaf and grain N, and CO2 concentrations as well as more unpredictable and warm climates (DaMatta et al. 2010). Livestock systems are susceptible to several factors such as decreased feed availability and quality, shifts in the frequency of pests and diseases, and a decline in output due to physiological consequences.

Growth, meat, egg, and milk productivity and quality all decline as temperatures rise above 30 °C because less food is consumed (Thronton and Gerber 2010). Fish and plankton distribution are expected to alter along with suitable habitats because of rising ocean temperatures, shifting winds, and modifications to ice thickness, pH, and nutrient availability (Brander 2010). Changes in pest prevalence and an increase in the frequency of shock pest outbreaks brought on by climate change will also make agricultural systems more vulnerable in the twenty-first century (Bebber et al. 2013). According to White et al. (2010), rising food prices are anticipated and may be influenced by climatic changes. Large-scale changes in crop or livestock production suitability, for instance, are predicted to result from climate change (Havlik et al. 2014), and prices, trade patterns, and the availability of food could all be significantly impacted by these developments. The change in climate may also affect transportation and physical health, which in turn may affect physical access to food (White et al. 2010). In addition, there are issues regarding the distribution of food among family members, specifically women and children, and the potential effects of a more variable climate on this distribution.

Climate change primarily impacts food consumption through two main channels: the food supply chain and health issues that impact nutritional outcomes. The change in climate generally has an impact on food safety (Hammond et al. 2015), especially in supply chains for agricultural goods (Liu et al. 2013) and fisheries (Marques et al. 2010), because higher temperatures lead to increased rates of microbial growth. Numerous factors, such as heat stress, natural catastrophes, and vector-borne diseases all have an impact on health, and these factors also affect people's capacity to provide for the food security of their dependents and children (Costello et al. 2009). Consequences of climate change on water include limited access to water for sanitation (McDonald et al. 2011) and increased water contamination from more frequent and intense floods (Uyttendaele et al. 2014), which may likewise pose a threat to food safety and health. Rising disease incidence has raised worries that pesticides and veterinary medications will be overused, especially in fisheries (Tirado et al. 2010). Indigenous people and those who are already impoverished may be disproportionately affected by the indirect health effects of climate change (Costello et al. 2009; Ford 2012). Food security will suffer as a result of this. These outcomes may include migration, the loss of one's livelihood and employment, or the interruption of public health services.

The Millennium Ecosystem Assessment (2005) claims that ecosystems support, regulate, and provide services related to food security, including soil formation, water cycling, nutrient cycling, and flood control. These services include food, water, timber, and genetic resources. Ecosystems are under more stress due to the effects of climate change (Cabell and Oelofse 2012). For example, increased temperatures and intense weather are changing community assemblage interactions and decreasing biodiversity (Khoury et al. 2014; Oppenheimer et al. 2014), which puts the production and resilience of the world's present food systems at risk. Furthermore, it is thought that the social and economic underpinnings of agricultural systems are weakened by climate change and unpredictability. Because these groups are the most vulnerable to climate change, threats to them have the potential to erode community resource systems and intensify conflict. Marginalized and resource-poor populations are particularly vulnerable to the effects of climate change, and challenges to these groups have the potential to erode community resource systems and intensify conflict (Oppenheimer et al. 2014).

## 11.3 Adaptive Strategies and Technological Innovation

It needs both adaptation and mitigation methods, certain technology advancements, comprehensive policies, and international cooperation. Implementing sustainable water management and agricultural methods, creating climate-resilient crop types, and diversifying livelihoods are some examples of adaptation measures. Lowering greenhouse gas emissions is the primary goal of mitigation techniques for climate change. The immediate dangers that climate change poses to food security and the worldwide need to reduce greenhouse gas emissions from all sources, including agriculture, call for action-oriented research. The issues of food security and climate change require the implementation of effective policies and governance systems. Promoting sustainable farming methods, improving infrastructure, and strengthening social safety nets are included in these international agreements, national programs, and local efforts. Research and innovation must continue to comprehend how the dynamics of food security and climate change are changing. This involves developing novel agricultural technologies, improving climate modelling, and researching the socioeconomic effects on communities. Raising output has always been, and still is, a crucial strategy for lowering food poverty. Thus, there remains a strong idea that the needs of society may be met by increasing food supply. This is also theoretically true: everyone will be fed if there is sufficient food produced. 925 million people go hungry every day even though there is already more than enough food produced per person to sustain the world's population (FAO 2010). To establish future agricultural outlooks, crop-climate modelling is crucial modelling that can guide decision-making at the field level or in policy-making processes (Porter et al. 2014). Even though model-based assessments of future agricultural productivity can yield robust conclusions in some cases, they are susceptible to uncertainty that could reduce the predictability of the system under study and impede decisions about adaptation (Weaver et al. 2013). Based on these impacts, it appears that global food production will decrease due to climate change.

## 11.4 Role of Geospatial Technology in Climate change and Food Security

A major contributor to climate change is agriculture, which is influenced by climatic conditions and the physical features of the landscape in addition to significant seasonal patterns connected to the biological life cycle of crops (Rembold et al. 2016; Zampieri et al. 2017). Monitoring and providing precise estimates of the effects of climate change is critical for maintaining food security, early warning system design, and efficient use of limited resources. The development of new technologies for data acquisition (such as proximal and remote sensing) and the creation of reliable, validated models that are based on data from multiple sources is essential if the goal is to reduce the trade-offs associated with the effects of climate change on the food supply (Feizizadeh et al. 2021). The use of innovative technologies for developing strategies, creating policies for faster agricultural growth, and evaluating agricultural resources—including their biophysical and socioeconomic complexity—is becoming more and more important (Gommes et al. 2009). Reliable databases containing data on phenology, soil, agronomy, climate, and meteorology are necessary for these assessments, as are associated with innovative techniques and equipment (Ramasamy and Bernardi 2010).

Geospatial (GS) technology is essential to the realization of Climate Smart Agriculture (CSA) through its fundamental applications in agriculture, which include mapping, measuring, modelling, and monitoring. It is a rapidly evolving field of research that includes data management, information technology, global positioning systems (GPS), geographic information systems (GIS), and remote sensing (Sinha et al. 2022). In India, there are several initiatives have been proposed under national missions on climate change. Utilizing remotely sensed imagery from satellites, aerial platforms, and GPS-tagged drones, geospatial technology plays a major part in improving agricultural practices, crop development, and yields. Geospatial technology allows us to map and track crop and soil conditions as well as efficiently manage, safeguard, and conserve biodiversity worldwide (Sajan et al. 2022). Geospatial information offers promising methods for gathering, evaluating, understanding, and displaying spatial data for making evidence-based decisions. GS techniques reduce cost, time, and resources (Farooq et al. 2022; Singh et al. 2021a, b). They are very helpful in evaluating and monitoring land, as well as evaluating and managing weather anomalies like hailstorms, floods, and droughts for the development of agriculture. Additionally, GS technology aids in the evaluation of nutrient demands to improve effective regional fertilizer management and the establishment of indicators for traditional indigenous crops. Remote sensing supports risk assessment and sustainable land management by assisting in the identification of geographic hotspots such as crop loss, land degradation, and changes in land use. Furthermore, the assessment of several agricultural methodologies, such as sustainable agriculture, bio-economy, and technology utilization, can augment crop yields, profitability, and biodiversity preservation as pivotal constituents of these methodologies (Kanga et al. 2022). To maximize yields and agricultural productivity, remote sensing applications take into account ecological and environmental characteristics, soil variables, crop conditions, and plant-soil variety.

## 11.5 Conclusion

Climate change and food security are inextricably linked, posing a multifaceted challenge that necessitates a comprehensive and coordinated solution. It is a global challenge that aligns with many sustainable developmental goals in favour of good human health and an increasing Indian economy. To develop and execute solutions that enhance both climate resilience and food security for current and future generations, collaboration across government sectors, international organizations, research institutions, communities, regions, and disciplines is required. There will be significant climatic implications on food security, notwithstanding the uncertainty surrounding climate impact research and the limitations of crop and climate models. Consequently, we are in favour of additional research-oriented studies that contribute to the efforts needed to address issues related to food security. In order to develop adaptation and mitigation measures for the food security concern at global levels, science must, at the very least, work with practitioners and policy-makers. To provide climate-based advisories, it is crucial to combine important indigenous knowledge resources with additional scientific and technical expertise for disaster relief connected to climate change. Geospatial technologies like remote sensing, GIS, and GPS have become indispensable and immense in practicing climate-smart agriculture. These advanced technologies provide a mechanism for detailed inventory, mapping, monitoring as well and generation of reliable spatial databases on agriculture resources, especially in climate-vulnerable areas. This study highlights the role that food security and agriculture play in attaining the goals of the Sustainable Development Goals and maintaining good health. By facilitating agricultural practices, ensuring food security, protecting biodiversity, tracking soil and crop conditions, and fostering sustainable development, geospatial technology is essential to achieving the goals.

## References

Ahmad T, Gupta SK, Singh SK, Meraj G, Kumar P, Kanga S (2023) Unveiling nature's resilience: exploring vegetation dynamics during the COVID-19 Era in Jharkhand, India, with the Google Earth Engine. Climate 11(9):187

Barnett J, O'Neill S (2010) Maladaptation. Glob Environ Chang 20(2):211–213. https://doi.org/10.1016/j.gloenvcha.2009.11.004

Bazilian M, Rogner H, Howells M, Hermann S, Arent D, Gielen D, Steduto P, Mueller A, Komor P, Tol RSJ, Yumkella KK (2011) Considering the energy, water and food nexus: towards an integrated modelling approach. Energy Policy 39:7896–7906

Bebber DP, Ramotowski MAT, Gurr SJ (2013) Crop pests and pathogens move polewards in a warming world. Nat Clim Chang 3(11):985–988. https://doi.org/10.1038/nclimate1990

Brander K (2010) Impacts of climate change on fisheries. J Marine Syst 79:389–402. https://doi.org/10.1016/j.jmarsys.2008.12.015

Cabell JF, Oelofse M (2012) An indicator framework for assessing agroecosystem resilience. Ecol Soc 17:18. https://hdl.handle.net/10535/8331

Cline WR (2007) Global warming and agriculture: impact estimates by country. Center for Global Development. Washington D, USA

Costello A, Abbas M, Allen A, Ball S, Bell S, Bellamy R, Friel S, Groce N, Johnson A, Kett M, Lee M, Levy C, Maslin M, McCoy D, McGuire B, Montgomery H, Napier D, Pagel C, Patel J, de Oliveira JA, Redclift N, Rees H, Rogger D, Scott J, Stephenson J, Twigg J, Wolff J, Patterson C (2009) Managing the health effects of climate change. Lancet 373(9676):1693–1733. https://doi.org/10.1016/S0140-6736(09)60935-1

Cottrell RS, Nash KL, Halpern BS, Remenyi TA, Corney SP, Fleming A, Fulton EA, Hornborg S, Johne AS, Watson R (2019) Food production shocks across land and sea. Nat Sustain 2:130–137. https://doi.org/10.1038/s41893-018-0210-1

Creighton C, Hobday AJ, Lockwood M, Pecl G (2016) Adapting management of marine environments to a changing climate: a checklist to guide reform and assess progress. Ecosystem 19(2):187–219. https://doi.org/10.1007/s10021-015-9925-2

DaMatta FM, Grandis A, Arenque BC, Buckeridge M (2010) Impacts of climate changes on crop physiology and food quality. Food Res Int 43:1814–1823. https://doi.org/10.1016/j.foodres.2009.11.001

Debnath J, Sahariah D, Nath N, Saikia A, Lahon D, Islam MN, Hashimoto S, Meraj G, Kumar P, Singh SK, Kanga S (2023) Modelling on assessment of flood risk susceptibility at the Jia Bharali River basin in Eastern Himalayas by integrating multicollinearity tests and geospatial techniques. In: Modeling earth systems and environment, pp 1–27

Eriksson M, Xu J, Shrestha AB, Vaidya RA, Nepal S, Sandstrom K (2009) The changing Himalayas: Impact of climate change on water resources and livelihoods in the greater Himalayas. International Centre for Integrated Mountain Development, Kathmandu. https://doi.org/10.53055/ICIMOD.516

FAO (2010) The State of Food Insecurity in the World (SOFI): Addressing food insecurity in protracted crises. Rome

Farooq M, Mushtaq F, Meraj G, Singh SK, Kanga S, Gupta A, Kumar P, Singh D, Avtar R (2022) Strategic slum upgrading and redevelopment action plan for Jammu City. Resources 11(12):120

Feizizadeh B, Garajeh MK, Blaschke T, Lakes T (2021) An object-based image analysis applied for volcanic and glacial landforms mapping in Sahand Mountain. Iran. Catena 198:105073

Food and Agriculture Organization of the United Nations (FAO) (2008) Food security information for action: practical guides. EC - FAO Food Security Programme, Rome

Food and Agriculture Organization of the United Nations (FAO) (2002) The state of food insecurity in the world 2001. FAO, Rome, Italy

Food and Agriculture Organization of the United Nations (FAO) (2009) Declaration of the World Food Summit on Food Security, Rome

Food and Agriculture Organization of the United Nations (FAO). (2020). International Fund for Agricultural Development (IFAD), United Nations Children's Fund (UNICEF), World Food Programme (WFP) and World Health Organisation (WHO). 2020. The State of Food Security and Nutrition in the World (2020) Transforming food systems for affordable healthy diets. FAO, Rome, Italy. https://doi.org/10.4060/ca9692en

Ford JD (2012) Indigenous health and climate change. Am J Public Health 102(7):1260–1266. https://doi.org/10.2105/AJPH.2012.300752

Gommes R, Hairech ET, Rosillon D, Balaghi R, Kanamaru H (2009) World Bank- Morocco study on the impact of climate change on the agricultural sector. Impact of climate change on agricultural yields in Morocco. FAO, Rome: Government of Morocco, and World Bank

Gulati B, Sharma R, Kanga S, Singh SK, Sajan B, Meraj G, Kumar P, Ramanathan AL (2023) Unraveling the relationship between stubble burning and air quality degradation in Punjab: a temporal and spatial analysis (2019–2022). J Clim Change 9(2):43–53

Hammond ST, Brown JH, Burger JR, Flanagan TP, Fristoe TS, Mercado-Silva N, Nekola JC, Okie JG (2015) Food spoilage, storage, and transport: implications for a sustainable future. Biosciences 65(8):758–768. https://doi.org/10.1093/biosci/biv081

Havlík P, Valin H, Herrero M, Obersteiner M, Schmid E, Rufinno MC, Mosnier A, Thornton PK, Böttcher H, Conant RT, Frank S, Fritz S, Fuss S, Kraxner F, Notenbaert A (2014) Climate change mitigation through livestock system transitions. Proc Nat Acad Sci USA 111:3709–3714. https://doi.org/10.1073/pnas.1308044111

Herrero M, Wirsenius S, Henderson B, Rigolot C, Thornton PK, Havlik P, de Boer I, Gerber P (2015) Livestock and the environment: what have we learned in the past decade? Ann Rev Environ Resour 40:177–202. https://doi.org/10.1146/annurev-environ-031113-09350310.1016/j.enpol.2011.09.039

IPCC (2012) Managing the risks of extreme events and disasters to advance climate change adaptation. In: Field CB, Barros V, Stocker TF, Qin D, Dokken DJ, Ebi KL, Mastrandrea MD, Mach KJ, Plattner GK, Allen SK, Tignor M, Midgley PM (eds) A special report of working Groups I and

II of the inter-governmental panel on climate change. Cambridge University Press, Cambridge, UK and New York, NY, USA, 582 pp. https://doi.org/10.13140/2.1.3117.9529
Jarvis A, Lau C, Cook S, Wollenberg E, Hansen J, Bonilla O, Challinor A (2001) An integrated adaptation and mitigation framework for developing agricultural research: synergies and trade-offs. Exp Agric 47:185–203
Kanga S, Meraj G, Farooq M, Singh SK, Nathawat MS (2022) Disasters in the complex Himalayan Terrains. Disaster Management in the Complex Himalayan Terrains: Natural Hazard Management, Methodologies and Policy Implications. Springer International Publishing, Cham, pp 3–10
Khair MS (2013) The efficacy of groundwater markets on agricultural productivity and resource use sustainability: Evidence from the upland Balochistan region of Pakistan (Unpublished PhD thesis). Wagga Wagga, NSW: Charles Sturt University
Khoury CK, Bjorkman AD, Dempewolf H, Ramirez-Villegas J, Guarino L, Jarvis A, Rieseberg LH, Struik PC (2014) Increasing homogeneity in global food supplies and the implications for food security. Proc Nat Acad Sci USA 111(11):4001–4006. https://doi.org/10.1073/pnas.1313490111
Knox J, Hess T, Daccache A, Wheeler T (2012) Climate change impacts on crop productivity in Africa and South Asia. Environ Res Lett 7:034–132
Lele U, Klousia-Marquis M, Goswami S (2013) Good governance for food, water and energy security. Aquat Proc 1:44–63. https://doi.org/10.1016/j.aqpro.2013.07.005
Liu C, Hofstra N, Franz E (2013) Impacts of climate change on the microbial safety of pre-harvest leafy green vegetables as indicated by Escherichia coli O157 and Salmonella spp. Int J Food Microbiol 163(2–3):119–128. https://doi.org/10.1016/j.ijfoodmicro.2013.02.026
Marques A, Nunes ML, Moore SK, Strom MS (2010) Climate change and sea-food safety: human health implications. Food Res Int 43:1766–1779. https://doi.org/10.1016/j.foodres.2010.02.010
McDonald RI, Green P, Balk D, Fekete BM, Revenga C, Todd M, Montgomery M (2011) Urban growth, climate change, and freshwater availability. Proc Nat Acad Sci 108(15):6312–6317. https://doi.org/10.1073/pnas.1011615108
Meraj G, Kanga S, Kranjčić N, Đurin B, Singh SK (2021) Role of natural capital economics for sustainable management of earth resources. Earth 2(3):622–634
Meraj G, Farooq M, Singh SK, Islam MN, Kanga S (2022) Modeling the sediment retention and ecosystem provisioning services in the Kashmir valley, India, Western Himalayas. Model Earth Syst Environ 8(3):3859–3884
Millennium Ecosystem Assessment (2005) Ecosystems and human well-being: synthesis. Island Press, Washington
Mustafa D, Qazi MU (2007) Transition from karez to tubewell irrigation: development, modernization and social capital in Balochistan, Pakistan. World Dev 35:1796–1813. https://doi.org/10.1016/j.worlddev.2007.06.002
Norse D, Li J, Jin L, Zhang Z (2001) Environmental costs of rice production in China, lessons from Hunan and Hubei. Aileen International Press, Bethesda, MD
Oppenheimer M, Campos M, Warren R, Birkmann J, Luber G, O'Neill B, Takahashi K (2014) Emergent risks and key vulnerabilities. In: Field CB, Barros VR, Dokken DJ, Mach KJ, Mastrandrea MD, Bilir TE, Chatterjee M, Ebi KL, Estrada YO, Genova RC, Girma B, Kissel ES, Levy AN, MacCracken S, Mastrandrea PR, White LL (eds) Climate change 2014: impacts, adaptation, and vulnerability. Part A: global and sectoral aspects. Contribution of working group II to the fifth assessment report of the intergovernmental panel on climate change. Cambridge University Press, Cambridge, pp 1039–1099
Porter JR, Xie L, Challinor AJ, Cochrane K, Howden M, Iqbal MM, Lobell DB, Travasso MI (2014) Chapter 7. Food security and food production systems. climate change 2014: impacts, adaptation and vulnerability. In: Working group II contribution to the IPCC 5th assessment report, Geneva, Switzerland
Ramasamy S, Bernardi M (2010) Climate change impacts on agriculture in Africa: current assessments and the way forward. Nature and Faune 1(25):29–34

Rembold F, Kerdiles H, Lemoine G, Perez-Hoyos A (2016) Impact of El Niño on agriculture in Southern Africa for the 2015/2016 main season. Joint Research Centre (JRC) MARS Bulletin–Global Outlook Series. European Commission, Brussels

Rockström J, Steffen W, Noone K, Persson A, Chapin FS, Lambin EF, Lenton TM, Scheffer M, Folke C, Schellnhuber HJ, Nykvist B, de Wit CA, Hughes T, Leeuw SV, Rodhe H, Sorlin S, Snyder PK, Costanza R, Svedin U, Falkenmark M, Karlberg L, Corell RW, Fabry VJ, Hansen H, Walker B, Liverman D, Richardson K, Crutzen P, Foley JA (2009) A safe operating space for humanity. Nature 461:472–475. https://doi.org/10.1038/461472a

Rosenzweig C, Elliott J, Deryng D, Ruane AC, Muller C, Arneth A, Boote KJ, Folberth C, Glotter M, Khabarov N, Neumann K, Piontek F, Pugh TAM, Schmid E, Stehfest E, Yang H, Jones JW (2014) Assessing agricultural risks of climate change in the 21st century in a global gridded crop model intercomparison. Proc Nat Acad Sci 111(9):3268–3273. https://doi.org/10.1073/pnas.1222463110

Sajan B, Mishra VN, Kanga S, Meraj G, Singh SK, Kumar P (2022) Cellular automata-based artificial neural network model for assessing past, present, and future land use/land cover dynamics. Agronomy 12(11):2772

Shrestha A, Aryal R (2011) Climate change in Nepal and its impact on Himalayan glaciers. Reg Environ Change 11(1):565–577. https://doi.org/10.1007/s10113-010-0174-9

Singh S, Singh H, Sharma V, Shrivastava V, Kumar P, Kanga S, Sahu N, Meraj G, Farooq M, Singh SK (2021a) Impact of forest fires on air quality in Wolgan valley, New South Wales, Australia—A mapping and monitoring study using Google Earth Engine. Forests 13(1):4

Singh H, Meraj G, Singh S, Shrivastava V, Sharma V, Farooq M, Kanga S, Singh SK, Kumar P (2022) Status of air pollution during COVID-19-induced lockdown in Delhi India. Atmosphere 13(12):2090

Singh SK, Meraj G, Mondal N, Bera A, Verma MK, Tomar JS, Kanga S (2021) Assessment of seasonal vegetation dynamics over parts of thar desert using geospatial techniques. ANGRAU, p. 105

Sinha KK, Gupta MK, Banerjee MK, Meraj G, Singh SK, Kanga S, Farooq M, Kumar P, Sahu N (2022) Neural network-based modeling of water quality in Jodhpur India. Hydrology 9(5):92

Thornton PK, Gerber PJ (2010) Climate change and the growth of the livestock sector in developing countries. Mitig Adapt Strat Glob Change 15:169–184. https://doi.org/10.1007/s11027-009-9210-9

Tirado MC, Clarke R, Jaykus LA, McQuatters-Gollop A, Franke JM (2010) Climate change and food safety: a review. Food Res Int 43:1745–1765. https://doi.org/10.1016/j.foodres.2010.07.003

Tomar P, Singh SK, Kanga S, Meraj G, Kranjčić N, Đurin B, Pattanaik A (2021) GIS-based urban flood risk assessment and management—a case study of Delhi National Capital Territory (NCT). India. Sustainability 13(22):12850

Urwin K, Jordan A (2008) Does public policy support or undermine climate change adaptation? Exploring policy interplay across different scales of governance. Glob Environ Chang 18:180–191. https://doi.org/10.1016/j.gloenvcha.2007.08.002

Uyttendaele M, Liu C, Hofstra N (2014) Special issue on the impacts of climate change on food safety. Food Res Int 68:1–6. https://doi.org/10.1016/j.foodres.2014.09.001

Walker B, Salt D, Reid W (2006) Resilience thinking: Sustaining people and ecosystems in a changing world. Island Press, Washington, DC. http://hdl.handle.net/20.500.11822/2141

Weaver CP, Lempert RJ, Brown C, Hall JA, Revell D, Sarewitz D (2013) Improving the contribution of climate model information to decision making: the value and demands of robust decision frameworks. Wires Clim Change 4(1):39–60. https://doi.org/10.1002/wcc.202

Wen W, Timmermans J, Chen Q, van Bodegom PM (2020) A review of remote sensing challenges for food security with respect to salinity and drought threats. Remote Sens 13(1):6. https://doi.org/10.3390/rs13010006

Wheeler T, VonBraun J (2013) Climate change impacts on global food security. Science 341:508–513. https://doi.org/10.1126/science.1239402

White R, Stewart B, O'Neill P (2010) Access to food in a changing climate. Environmental Change Institute, School of Geography and the Environment, Oxford University. http://www.oxfordmartin.ox.ac.uk/downloads/reports/2011-OIA-ECI-report.pdf⟩

WRI (2011) World resources Report 2010–2011: Decision making in a changing climate – adaptation challenges and choices. Washington, DC: World Resources Institute, United Nations Development Programme, United Nations Environment Programme, World Bank

Zampieri M, Ceglar A, Dentener F, Toreti A (2017) Wheat yield loss attributable to heat waves, drought and water excess at the global, national and subnational scales. Environ Res Lett 12(6):064008. https://doi.org/10.1088/1748-9326/aa723b

# Chapter 12
# Environmental Drivers of Vector-Borne Disease Utilizing Geospatial Techniques in Jaipur

**Priyanka Roy, Shruti Kanga, and Suraj Kumar Singh**

## 12.1 Introduction

Vector-borne diseases are a significant public health concern worldwide, causing a substantial burden on human populations and ecosystems. These diseases, such as malaria, dengue fever, Zika virus, Lyme disease, and others, are transmitted by various vectors, including mosquitoes, ticks, fleas, and sandflies. Environmental factors play a crucial role in the transmission and spread of vector-borne diseases (Singh et al. 2021). This chapter explores the key environmental drivers that influence the dynamics of these diseases and their vectors. The dynamics of vector-borne diseases entail intricate interactions between the pathogen, vector, host (humans or animals), and the environment. The most important aspects of vector-borne disease dynamics are (a) the vector life cycle and the disease transmission cycle, which consists of -pathogen transmission -vector reproduction and dispersal-host immunity (Ahmad et al. 2023). (b) climate and weather variables (c) vector habitat (d) land use and land cover (e) density and distribution of the population (f) water management and irrigation practises (g) healthcare infrastructure and surveillance (h) globalization and travel (i) Socioeconomic elements.

Understanding and mitigating the environmental factors that contribute to the prevalence of these diseases requires geospatial techniques. Researchers can discover

P. Roy (✉) · S. K. Singh
Centre for Climate Change and Water Research, Suresh Gyan Vihar University, Jaipur, India
e-mail: priyanka.roy@mygyanvihar.com

S. K. Singh
e-mail: suraj.kumar@mygyanvihar.com

S. Kanga
Department of Geography, School of Environment and Earth Sciences, Central University of Punjab, Bathinda, India
e-mail: shruti.kanga@cup.edu.in

© The Author(s), under exclusive license to Springer Nature Singapore Pte Ltd. 2024 
G. Tripathi et al. (eds.), *Sustainability and Health Informatics*, Advances in Geographical and Environmental Sciences,
https://doi.org/10.1007/978-981-97-6706-9_12

patterns, anticipate epidemics, and devise targeted therapies by combining spatial data and diseases information. This discussion clarifies the environmental factors that contribute to the prevalence of vector-borne diseases and examines the application of geospatial methodologies to the study of these phenomena (Meraj et al. 2021). Climate and meteorological factors, such as temperature, precipitation, humidity, and seasonal variations, have a significant effect on vector populations and their ability to transmit disease. Geospatial techniques are capable of analysing historical climate data and subsequently developing climate models for predicting future climate changes (Farooq et al. 2022a, b). Consequently, scholars are able to establish a connection between these climate models and disease-related data in order to identify geographical regions vulnerable to disease transmission as a result of climatic changes.

**Climate Change and Temperature** has a profound impact on vector-borne diseases by altering the distribution and abundance of vectors and pathogens (Singh et al. 2022). Rising temperatures influence the development and reproduction rates of many vectors, leading to increased transmission potential (Meraj et al. 2022). Changes in precipitation patterns can create favourable breeding habitats for mosquitoes (Bomblies 2012), while droughts can concentrate human populations around limited water sources, increasing exposure to disease vectors (Caminade et al. 2019).

**Land Use and Urbanization**: Changes in land use, such as deforestation, urbanization, and agricultural expansion, can significantly affect vector-borne disease transmission. Deforestation, for example, can disrupt natural ecosystems, driving vectors to find alternative hosts, including humans, and potentially increasing disease transmission (Ferraguti et al. 2023). Urbanization creates environments that favour certain vectors, such as Aedes mosquitoes, leading to higher transmission rates of diseases like dengue and chikungunya.

**Water Management**: Water plays a crucial role in the life cycle of many disease vectors, particularly mosquitoes. Poor water management practices, such as improper drainage and water storage, can create breeding sites for mosquitoes and increase the risk of disease transmission (Sinha et al. 2022; Narisetty et al. 2023). The construction of dams and irrigation systems can also alter vector habitats, affecting disease patterns in nearby communities (Olagunju 2023).

**Human Behaviours and Migration**: Human behaviours and migration patterns can influence the spread of vector-borne diseases. People moving from endemic areas to new regions can introduce diseases to susceptible populations, causing outbreaks in non-immune individuals. Additionally, human behaviour, such as outdoor activities and housing conditions, can influence exposure to disease vectors (Hainesa et al. 2006; Duval 2022).

**Ecological Disruptions**: Environmental disturbances, such as natural disasters, can lead to temporary or long-term changes in vector habitats and behaviours (Farooq et al. 2022a, b). After events like hurricanes or floods, stagnant water and disrupted ecosystems can create ideal breeding conditions for disease vectors, leading to outbreaks of diseases like West Nile virus or malaria (Bennett et al. 2003; Brett and Wilcox 2009).

**Wildlife Reservoirs**: Wildlife can serve as reservoirs for various vector-borne pathogens, acting as a source of infection for vectors and, potentially, humans. Changes in wildlife populations due to factors like deforestation or habitat fragmentation can alter the dynamics of diseases by affecting vector-host interactions (Brett and Wilcox 2009).

**Socioeconomic Factors**: Socioeconomic factors, including access to healthcare, education, and infrastructure, influence the vulnerability and resilience of communities to vector-borne diseases. Impoverished communities often lack resources for vector control measures and timely medical treatment, exacerbating the impact of these diseases (Ramirez 2017).

**Vector Control and Environmental Management**: Understanding the environmental drivers of vector-borne diseases is essential for designing effective control strategies. Integrated vector management approaches that consider environmental factors, such as climate, land use, and water management, alongside human behaviour, can help reduce disease transmission. Environmental management practices, such as the restoration of natural habitats and improved water resource management, can contribute to controlling vector populations and disease prevalence.

Environmental drivers play a central role in shaping the dynamics of vector-borne diseases. Climate change, land use changes, water management practices, human behaviour, and wildlife reservoirs all influence disease transmission. Developing a comprehensive understanding of these factors is crucial for designing effective prevention and control strategies to mitigate the impact of vector-borne diseases on human populations and ecosystems.

## 12.2 Vector Life Cycles

The biology of vector life cycles is an essential aspect of understanding how certain diseases are transmitted from one host to another. Vectors are organisms, typically arthropods like mosquitoes, ticks, fleas, and flies, that can transmit pathogens (such as viruses, bacteria, or parasites) from one host to another, often causing diseases (Basurko et al. 2018).

### *12.2.1 Life Cycle of Vectors*

Understanding the life cycle of vectors and the pathogens they carry is crucial in designing effective strategies for disease prevention and control. By targeting specific stages of the vector life cycle or interrupting the transmission process, we can reduce the incidence of vector-borne diseases.

Eggs: The life cycle of vectors usually begins with the female laying eggs in an appropriate environment. The location and conditions for egg-laying vary among different vector species. For example, mosquitoes may lay their eggs in standing water, while ticks and fleas often lay eggs in leaf litter or soil.

Larva: After the eggs hatch, the immature form of the vector, called the larva, emerges. Larvae typically have several developmental stages during which they undergo molting before progressing to the next stage. Larvae are usually less mobile than their adult counterparts and mostly focus on feeding and growing.

Pupal: Some vectors, like mosquitoes, go through a pupal stage. During this stage, they undergo significant internal changes and do not feed. Instead, they are relatively inactive as they transform into the adult stage.

Adult: After completing the pupal stage (if present), the vector emerges as an adult. This is the stage when they are capable of transmitting diseases to hosts. The adult vector's primary activities involve finding a host, feeding, mating, and laying eggs to continue the life cycle.

The vector's life cycle is often intertwined with the life cycle of the pathogen it carries. Let's take the example of mosquitoes, which are vectors for diseases such as malaria, dengue fever, and Zika virus.

A female mosquito lays her eggs in stagnant water or other suitable breeding places. The eggs hatch into larvae known as "wigglers", which live in water and feed on organic matter. The eggs hatch into larvae, which moult many times, becoming larger with each moult. Once they have finished the larval stage, they grow into pupae, which are also aquatic but do not ingest. After emerging from the pupal stage, the adult mosquito remains on the water's surface to dry and harden its exoskeleton. Once the adult mosquito's exoskeleton is grown, it takes flight and begins its search for a blood meal in order to obtain the nutrients needed for egg production (Fig. 12.1). If a mosquito bites an infected host during this blood feast, the disease is ingested along with the blood. The pathogen grows and multiplies within the mosquito's body. Following an incubation period, the virus or parasite migrates to the mosquito's salivary glands. The infected mosquito can now transmit this disease to a new victim during its next blood meal.

### *12.2.2 Pathogen Transmission in Vector-Borne Diseases*

Pathogen transmission in vector-borne diseases refers to the process by which disease-causing microorganisms, known as pathogens, are transmitted from one host to another through the bite of a vector, typically an arthropod such as mosquitoes, ticks, fleas, or sandflies. The vector acts as an intermediary in the transmission cycle, facilitating the transfer of the pathogen between different hosts, which may include humans, animals, or other organisms. pathogen transmission process in vector-borne diseases begins from Acquisition of the Pathogen, upon feeding on an infected host,

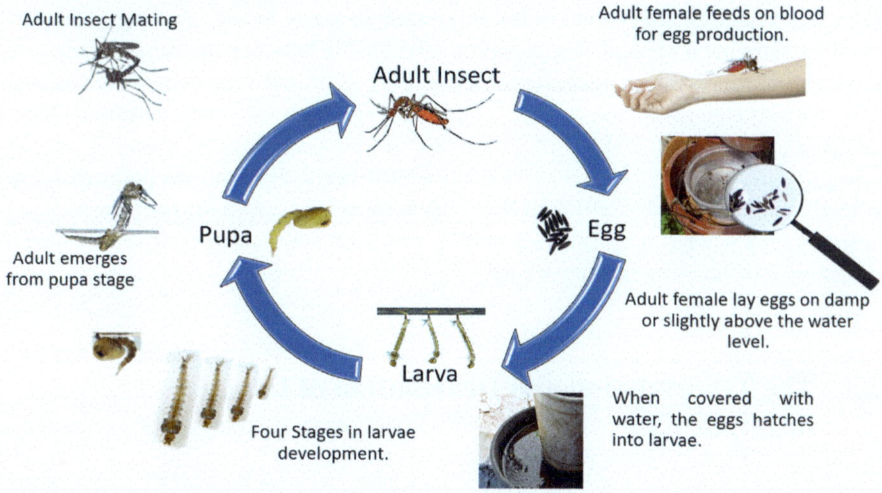


**Fig. 12.1** Life cycle of Aedes mosquitoes

the vector becomes infected with the pathogen (Pani et al. 2008). During the blood meal, the vector ingests the pathogen, which multiplies and replicates within its body, frequently residing in particular tissues or organs (Gubler and Rosen 1976; WHO 2009, 2019, 2022).

After acquiring the pathogen, the vector endures an incubation period during which the pathogen replicates and reaches transmission levels. Once the pathogen has reached a transmissible stage within the vector, it searches for a new host to prey on. This is frequently determined by the natural feeding behaviour of the vector and can be influenced by environmental factors and host availability. During the feeding process, the vector injects the pathogen into the host's circulation or tissue when feasting on a new host. Afterwards, the pathogen establishes an infection in the new host (Deen et al. 2006).

Pathogen-Induced Immune Response and Disease Progression: The pathogen induces an immune response in the newly afflicted host, which results in the development of clinical symptoms and signs of the disease. Recent virology and immunology research by Fernandes-Santos and Azeredo (2022) has enhanced our understanding of how immune responses impact DENV infection from both viral and host perspectives. These observations clarify the relationship's intricacy. Understanding infection and sickness dynamics, innate and adaptive immunity, and their interaction helps scientists create and test vaccines. Prospective longitudinal investigations indicate changing virus-host immune responses in acute dengue fever patients. Dengue fever starts when viral levels are maximum. When severe patients haemorrhage plasma, viral burden lowers significantly. Virulence increases sickness severity but does not cause plasma leakage. This implies additional strategies increase vascular permeability. Dengue stimulates innate and adaptive immunity. Multiple signalling pathways are activated, including pro- and anti-inflammatory substances (Shukla et al.

2020). Immunological responses are connected to virus levels, making causation difficult when time is critical. The complex relationship between the dengue virus and the host's immune system generates dengue fever. Extreme symptoms and immune responses may determine the disease's fate, although high viremia is required (King et al. 2020; Póvoa et al. 2014).

*Amplification and Maintenance*: In many vector-borne diseases, the pathogen may replicate within the newly afflicted host, reaching levels sufficient for transmission to other vectors. This amplification within the host population contributes to the maintenance of the transmission cycle.

## 12.3 The Transmission and Distribution of Dengue in Jaipur City

The intricate relationship between climate and vector-borne diseases underscores the importance of understanding climate patterns in disease surveillance, prevention, and control efforts. Public health measures that consider the influence of climate on vector-borne diseases can be more effective in mitigating the impact of these diseases on human and animal populations. Climate plays a significant role in the transmission and distribution of vector-borne diseases. Vector-borne diseases encompass the transmission of pathogens to animals or people through arthropod vectors, such as sandflies, parasites, mosquitoes, and fleas. These vectors act as carriers of pathogens, such as viruses, bacteria, and parasites, and can transmit the diseases to susceptible hosts.

**Temperature** has a direct influence on the growth and survival of vectors and pathogens. Warmer temperatures can hasten the pathogen's development within the vector, resulting in a shortened incubation period. This, in turn, can increase the frequency of disease transmission and vector-borne bites. Furthermore, milder climates may expand the geographical range of certain vectors, allowing them to persist in areas where they previously could not, thereby expanding the distribution of diseases (Brady et al. 2012).

Over the past 50 years, eastern city winters have been warmer. Hot to moderate temperatures were at Bagru Ansik and Adarsh Nagar. Temperatures were 19–22 °C. This temperature range limits viral growth by immobilizing Aedes mosquito larvae and adults. Spring temperatures in the east and southeast are higher than winter temperatures. Bagru Ansik, Adarsh Nagar, Malviya Nagar, Hawamahal, and Sanganagar saw hot to moderate temperatures. Despite temperatures between 31 and 34 °C, vectors cannot thrive. Eastern and southeastern summers are hotter. Summer temperatures in Bagru Ansik, Adarsh Nagar, Malviya Nagar, Hawamahal, and Sanganagar are 32–35 °C. Vectors also suffer from the temperature range. Egg development is possible this season. However, the Aedes mosquito, which causes this process, cannot survive summer temperatures and dies. Eastern and southeastern fall temperatures are high. Bagru Ansik, Adarsh Nagar, Malviya Nagar, Hawamahal,

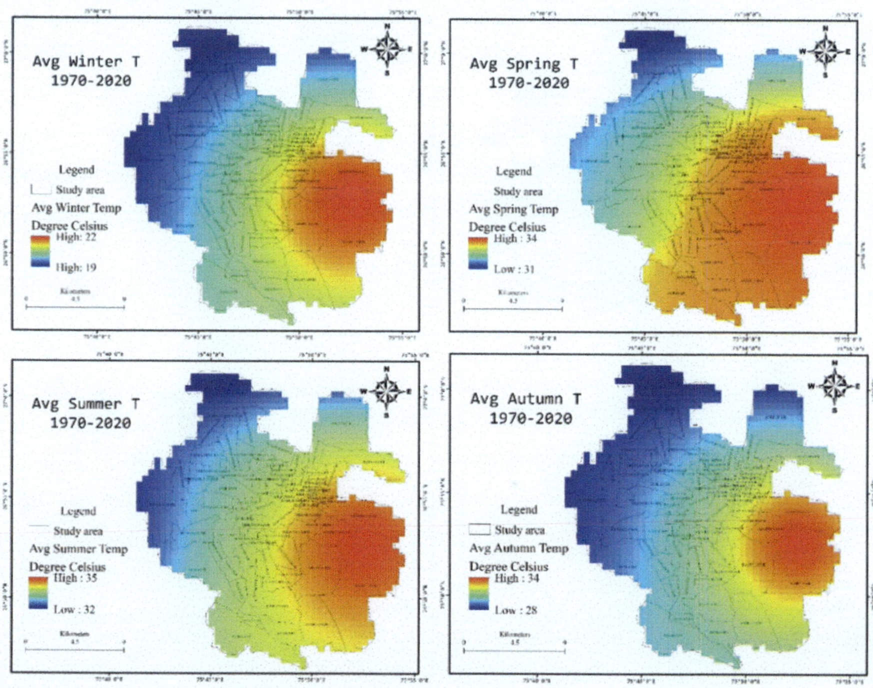


**Fig. 12.2** Average five-decade seasonal temperature Jaipur city

and Sanganagar average 28–34 °C (Fig. 12.2). Ades species' adult vectors and larvae thrive in this temperature range because it maintains wet breeding sites. Aedes mosquitoes peak at these temperatures. Aedes mosquitoes reproduce and eat in low and high temperatures. Dengue transmission increases due to this.

Rainfall and humidity are essential for the reproduction and survival of many vector species. Standing water is a mosquito breeding ground, and an increase in precipitation can create more breeding sites, leading to a rise in mosquito populations. Nevertheless, excessive precipitation can also flood away reproductive sites and disrupt the life cycle of certain vectors, thereby impacting disease transmission.

The studied regions average 92–96 mm of winter precipitation. Vidhyadhar Nagar and Jhotwara get less rain. Most study areas have high precipitation, excellent for vector density. Spring precipitation averages 71–75 mm. Bhagru Ansik reported 75 mm of rain in Jhalan Panther Park's forested hill zone, whereas the other study sites got less. The vector's life cycle is hampered by insufficient rainfall. Summer precipitation averages 61–65 mm, which is low. Insufficient rainfall makes vector survival difficult throughout their life cycle.

In the studied region, autumn precipitation averaged 71–76 mm. Eastern wards including Jhotwars Ansik, Civil Lines, and Vidgyadhar Nagar receive more rainfall than the west. In the research region, the Jhalan Panther Park forest receives the maximum precipitation, suggesting it may be optimal for vector density (Fig. 12.3).

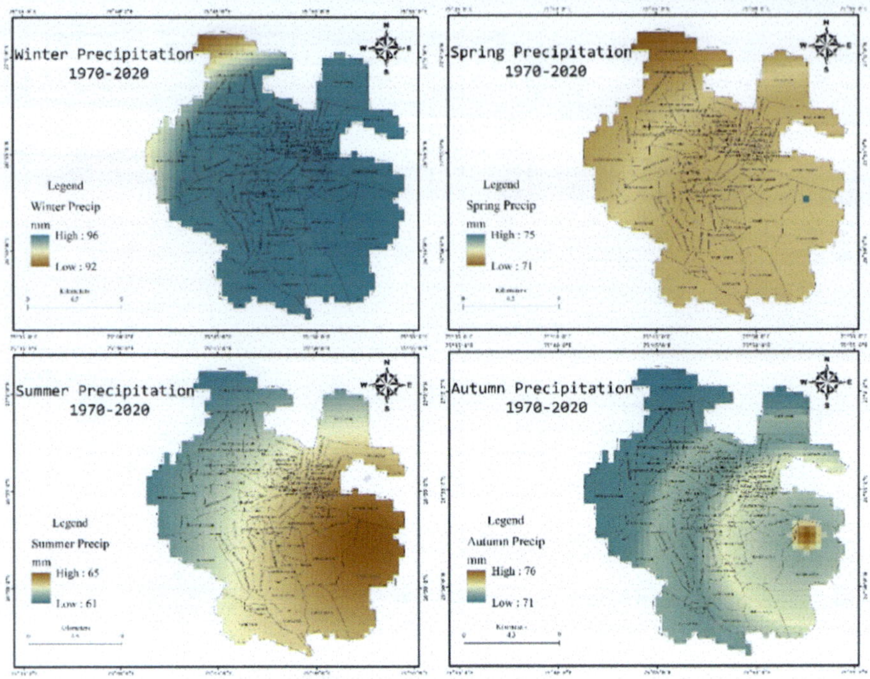


**Fig. 12.3** Average five-decade seasonal precipitation Jaipur city

Seasonality in vector-borne diseases often reflects patterns in the climate. In temperate regions, for instance, mosquito-borne maladies such as West Nile virus are more prevalent during the summer when mosquito populations are at their peak. In contrast, some diseases may be more prevalent during the rainy season as a result of the increased reproductive opportunities for vectors (Dhiman et al. 2010). Extreme weather events, such as hurricanes, floods, and droughts, can have a substantial effect on vector populations and disease transmission. Inundations can generate new reproductive grounds and contribute to an increase in vector populations, whereas droughts can concentrate vectors in remaining water sources, thereby increasing the risk of disease transmission (Pani et al. 2008).

Changes in climate can alter the distribution and abundance of vectors and the pathogens they transmit. As the climate warms, certain vectors may expand their range to higher latitudes and altitudes, putting at risk populations that were previously safe. Climate change can also cause alterations in the timing of seasonal patterns, which may result in discrepancies between vector activity and human activity, thereby increasing disease risk and exposure (Lafferty 2009).

The relationship between altitude and temperature and humidity is close. As one ascends to greater altitudes, the climate becomes colder, which can restrict the survival and spread of certain vectors and pathogens (Gubler et al. 2001). Consequently, vector-borne diseases may be more prevalent at lower altitudes and decrease as one ascends in elevation (Purse et al. 2012). How will climate change pathways and mitigation options alter the incidence of vector-borne diseases. A framework for leishmaniasis in South and Meso-America.

## 12.4 Role of Land Use and Urbanization in Vector-Borne Disease

Urbanization and changes in land use frequently involve the modification of natural habitats, resulting in the creation of novel environments suitable for vector species. Deforestation and urbanization can create mosquito breeding grounds, such as stagnant water on construction sites, discarded tyres, and receptacles. This increases the likelihood of mosquito-borne illnesses such as dengue, Zika, and chikungunya (Correia et al. 2007).

Urbanization results in a greater concentration of human populations in urban areas. Dense populations increase the likelihood of vectors coming into contact with susceptible individuals, thereby increasing disease transmission. Diseases such as malaria and dengue, which require direct contact between vectors and humans, thrive in densely populated urban areas (Pani et al. 2008).

Migration and Travel: Urban centres frequently serve as transportation and commercial nodes, facilitating the movement of people and products across regions and countries. This can result in the introduction and dissemination of vector-borne diseases in areas where the vector or pathogen was not previously found. Infected individuals who travel to new regions have the potential to spread the disease as a result of the heat island effect caused by human activities and structures, urban areas can be warmer than surrounding rural regions. Warmer temperatures can hasten the development of vectors and pathogens, which could increase the transmission rate of vector-borne diseases in urban areas (Tatem et al. 2012).

Access to Healthcare and Resources: Access to healthcare facilities and resources is typically greater in urban areas, which may influence the detection and reporting of vector-borne diseases. This could result in earlier diagnosis and treatment, thereby reducing the severity and spread of the diseases (Campbell-Lendrum et al. 2015). The management of water resources in urban areas can have an effect on vector reproductive sites. Inadequate drainage systems and water storage methods can result in stagnant water bodies that are optimal for mosquito reproduction. Alternatively, well-maintained water management systems can reduce vector reproductive sites and reduce disease risk. Some vector species can adapt to urban environments by using man-made structures as reproductive locations and by taking advantage of the abundant food supply (i.e. human and animal hosts). This adaptation may result in the persistence and spread of vector-borne diseases in urban environments (Baker et al. 2022).

Therefore, integrated urban planning and public health measures are required to address the challenges posed by land use and urbanization in relation to vector-borne diseases. This includes vector control strategies, appropriate waste management to prevent reproductive sites, early disease surveillance and monitoring, public education, and the promotion of green spaces to reduce the impact of urbanization on disease transmission. By taking into account the interactions between land use, urbanization, and vector-borne diseases, policymakers, and health authorities can create urban environments that are healthier and more resilient.

### *12.4.1 Ecology and Habitat-Vector Borne Disease*

The transmission and proliferation of vector-borne diseases are heavily reliant on ecology and habitat. Vector-borne diseases are those transmitted to humans and animals by infected arthropods such as mosquitoes, parasites, and fleas. Key considerations regarding the ecology and habitat of vector-borne diseases include the following.

**Habitat and distribution of vectors**: Their habitat requirements have a significant impact on the presence and abundance of vectors. Different vector species have distinct environmental preferences that influence their distribution and abundance. Some mosquitoes flourish in urban areas, whereas others favour rural or forested areas. It is essential to comprehend the habitat preferences of vectors in order to predict disease transmission patterns (Hinne et al. 2021). The climate and atmospheric conditions play an important influence in the ecology of vector-borne diseases. Temperature, humidity, and precipitation can affect the development, survival, and reproduction of vectors. Changes in climate patterns can increase the geographical distribution of vectors and the diseases they transmit.

Vectors require blood feeds from susceptible hosts in order to conclude their life cycles. The presence of these hosts (e.g., humans, animals, and mammals) in a specific habitat influences the abundance and behaviour of vectors. Human activities that alter host populations or increase contact between humans and vectors can increase the risk of disease transmission.

Life Cycle and Reproduction of Vectors The ecology of vector-borne diseases is intricately linked to the life cycle and reproductive characteristics of vector species. Understanding the reproductive sites, development periods, and behaviour of vectors is essential for the implementation of effective control strategies. Changes in land use, including deforestation, urbanization, and agriculture, can alter the landscape and establish new vector habitats. Certain vector species can proliferate in urban areas characterized by artificial water containers, inadequate sanitation, and a high population density. When infected individuals or animals migrate from one location to another, they can introduce vector-borne diseases to new regions. Travel and migration can facilitate the spread of diseases to non-endemic regions where vectors may find suitable habitats. Human behaviour can also influence the transmission of diseases transmitted by vectors. For instance, improper waste disposal can lead to the accumulation of stagnant water, which provides mosquito reproductive grounds. In order to reduce disease transmission, public health interventions such as vector control programmes and community education are essential (2009, 2019, 2022).

**Ecological Resilience and Climate Change**: Environmental changes, such as deforestation, habitat devastation, and climate change, can have an effect on the distribution and abundance of vectors and their associated pathogens. Certain regions may experience the emergence or reappearance of vector-borne diseases as a result of these changes (Zou et al. 2006). For design effective control and prevention strategies for vector-borne diseases, it is essential to comprehend the intricate interactions between vectors, pathogens, and their habitats. To mitigate the impact of these maladies on human and animal populations, integrated approaches that incorporate ecological, environmental, and socio-economic factors are necessary.

In order to comprehend the intricacies of illness episodes and their spatial distribution, the implementation of a vector surveillance system becomes imperative. Various larval biomarkers have been utilized in the field of entomological monitoring. The validity of these indicators has been called into doubt due to their correlation with adult Aedes populations. The utilization of the Breteau Index is now observed in contemporary contexts. Entomological data pertaining to Aedes mosquito species was generated by the use of the Home Index, Container Index, and Breteau Index during vector surveillance and larva presence and survival studies. The residential dwelling or property serves as the principal unit of analysis, subject to a comprehensive examination to identify any water-holding containers. The containers undergo inspection to determine the presence of mosquito larvae and pupae. The termination of the search might occur upon the detection of Aedes larvae, or alternatively, it may be prolonged until all containers have been thoroughly examined, contingent upon the specific objectives of the survey. The process of data collection was employed to ensure the inclusion of Aedes mosquito species in this particular study. Aedes species indicators are utilized as a means to address the challenge of comprehending their distribution.

(a) Container Index (Cl): percentage of larvae or pupae infested water holding vessels.

$$\mathrm{CI} = \frac{\text{Number of positive containers}}{\text{Number of containers inspected}} \times 100$$

(b) Breteau Index (Bl): per 100 residences assessed, the number of positive containers.

$$\mathrm{BI} = \frac{\text{Total Number of positive containers}}{\text{Total Number of containers inspected}} \times 100$$

To ascertain the locations with high vector abundance and density The larval indices were derived from the Container Index (Cl) and Breteau Index (Bl), which were acquired from a survey done by the Department of Public Health and Family Welfare and the Vector Borne Disease Control Programme (VBDCP) Department in Jaipur. Subsequently, the Indexes were geographically mapped onto the boundaries of the municipal wards.

## 12.5 Conclusion

The study delves into the intricate relationship between environmental factors and the transmission of vector-borne diseases in Jaipur. Geospatial techniques prove crucial in understanding and mitigating these diseases, considering factors like climate change, land use, water management, human behaviours, wildlife reservoirs, and socioeconomic elements. The vector life cycle, from eggs to adults, plays a pivotal role in disease transmission, with the pathogen transmission process adding complexity. The immune response in the host and the subsequent disease progression, along with amplification and maintenance, contribute to the dynamics of vector-borne diseases.

This study of Dengue transmission in Jaipur highlights the impact of climate on vectors, with temperature, rainfall, and humidity influencing their life cycles. Seasonal variations and extreme weather events play a role, emphasizing the need for climate-informed public health measures. The role of land use and urbanization in disease spread is evident, as changes in habitats and population density affect vector-borne disease transmission. Urban areas become hotspots due to human activities, migration, and travel. Access to healthcare resources and water management further shape the urban landscape's impact on disease prevalence.

The study underscores the importance of ecology and habitat in understanding vector-borne diseases. The distribution of vectors, habitat preferences, and the influence of climate on their ecology are critical factors. Environmental changes, including deforestation and climate change, can reshape disease patterns, necessitating integrated approaches for control and prevention. The comprehensive analysis provides valuable insights into the environmental drivers of vector-borne diseases, emphasizing the need for geospatial techniques and integrated strategies for effective prevention and control.

## References

Ahmad T, Gupta SK, Singh SK, Meraj G, Kumar P, Kanga S (2023) Unveiling nature's resilience: exploring vegetation dynamics during the COVID-19 era in Jharkhand, India, with the Google Earth engine. Climate 11(9):187

Baker RE, Mahmud AS, Miller IF et al (2022) Infectious disease in an era of global change. Nat Rev Microbiol 20:193–205. https://doi.org/10.1038/s41579-021-00639-z

Basurko C et al (2018) Estimating the risk of vertical transmission of dengue: a prospective study. Am J Trop Med Hyg 98(6):1826–1832

Bennett SN, Holmes EC, Chirivella M, Rodriguez DM, Beltran M, Vorndam V, Gubler DJ, McMillan WO (2003) Selection-driven evolution of emergent dengue virus. Mol Biol Evol 20:1650–1658
Bomblies A (2012) Modelling the role of rainfall patterns in seasonal malaria transmission. 112(3/4):673–685
Brady OJ et al (2012) Refining the global spatial limits of dengue virus transmission by evidence-based consensus. PLoS Negl Trop Dis 6(8):1760
Brett RE, Wilcox BA (2009) The ecological dimensions of vector-borne disease research and control. Cad Saudepublica 25:155–167
Caminade C, McIntyre KM, Jones AE (2019) Impact of recent and future climate change on vector-borne diseases. Ann N Y Acad Sci 1436(1):157–173. https://doi.org/10.1111/nyas.13950
Campbell-Lendrum D, Manga L, Bagayoko M, Sommerfeld J (2015) Climate change and vector-borne diseases: what are the implications for public health research and policy? Philos Trans R Soc Lond Ser B Biol Sci 370(1665):20130552. https://doi.org/10.1098/rstb.2013.0552
Deen JL, Harris E, Wills B, Balmaseda A, Hammond SN, Rocha C, Dung NM, Hung NT, Hien TT, Farrar JJ (2006) Viewpoint the WHO dengue classification and case definitions: time for a reassessment. Popul Engl Ed 368:170–173
Dhiman RC, Pahwa S, Dhillon GPS, Dash AP (2010) Climate change and threat of vector borne diseases in India: are we prepared? Parasitol Res 106:763
Duval L (2022) Climate change, vector-borne diseases, and migration. In: Zimmermann KF (eds) Handbook of labor, human resources and population economics. Springer, Cham. https://doi.org/10.1007/978-3-319-57365-6_247-1
Farooq M, Meraj G, Rishabh Kanga S, Nathawat R, Singh SK, Ranga V (2022a) Slum categorization for efficient development plan—a case study of Udhampur City, Jammu and Kashmir using remote sensing and GIS. Geospatial technology for landscape and environmental management: sustainable assessment and planning, pp 283–299
Farooq M, Mushtaq F, Meraj G, Singh SK, Kanga S, Gupta A, Kumar P, Singh D, Avtar R (2022b) Strategic slum upgrading and redevelopment action plan for Jammu City. Resources 11(12):120
Fernandes-Santos C, Azeredo EL (2022) Innate immune response to dengue virus: toll-like receptors and antiviral response. Viruses 14(5):992. https://doi.org/10.3390/v14050992
Ferraguti M, Magallanes S, Suarez-Rubio M et al (2023) Does land-use and land cover affect vector-borne diseases? A systematic review and meta-analysis. Landsc Ecol 38:2433–2451. https://doi.org/10.1007/s10980-023-01746-3
Gubler DJ, Rosen L (1976) A simple technique for demonstrating transmission of dengue virus by mosquitoes without the use of vertebrate hosts. Am J Trop Med Hyg 25:146–150. https://doi.org/10.4269/ajtmh.1976.25.146
Gubler DJ, Reiter P, Ebi KL, Yap W, Nasci R, Patz JA (2001) Climate variability and change in the United States: potential impacts on vector and rodent-borne diseases. Environ Health Perspect 109(2):223–233
Haines A, Kovats RS, Campbell-Lendrum D, Corvalan C (2006) Climate change and human health: impacts, vulnerability and public health. Public Health 120(7):585–96. https://doi.org/10.1016/j.puhe.2006.01.002. Epub 2006 Mar 20. PMID: 16542689
Hinne IA, Attah SK, Mensah BA et al (2021) Larval habitat diversity and *Anopheles* mosquito species distribution in different ecological zones in Ghana. Parasites Vectors 14:193. https://doi.org/10.1186/s13071-021-04701-w
King CA, Wegman AD, Endy TP (2020) Mobilization and activation of the innate immune response to dengue virus. Front Cell Infect Microbiol 10:1–16. https://doi.org/10.3389/fcimb.2020.574417
Lafferty KD (2009) The ecology of climate change and infectious diseases. Ecology 90:888–900. https://doi.org/10.1890/08-0079.1
Meraj G, Kanga S, Kranjčić N, Đurin B, Singh SK (2021) Role of natural capital economics for sustainable management of earth resources. Earth 2(3):622–634

Meraj G, Farooq M, Singh SK, Islam MN, Kanga S (2022) Modeling the sediment retention and ecosystem provisioning services in the Kashmir valley, India, Western Himalayas. Model Earth Syst Environ 8(3):3859–3884

Narisetty NG, Tripathi G, Kanga S, Singh SK, Meraj G, Kumar P, Đurin B, Matijević H (2023) Integrated multi-model approach for assessing groundwater vulnerability in Rajasthan's semi-arid zone: incorporating DRASTIC and SINTACS variants. Hydrology 10(12):231

Olagunju EA (2023) Is the presence of mosquitoes an indicator of poor environmental sanitation? J Water Health 21(3):385–401. https://doi.org/10.2166/wh.2023.280

Pani DB, Katyal R, Abhay S, Raina VK, Saxena VK, Lal S (2008) Natural vertical transmission of dengue virus in peak summer collections of Aedes aegypti (Diptera: Culicidae) from urban areas of Jaipur (Rajasthan) and Delhi. 40(2):155–157

Póvoa TF, Alves AMB, Oliveira CAB, Nuovo GJ, Chagas VLA, Paes MV (2014) The pathology of severe dengue in multiple organs of human fatal cases: histopathology, ultrastructure and virus replication. PLoS One 9:e83386. https://doi.org/10.1371/journal.pone.0083386

Purse BV, Masante D, Golding N, Pigott D, Day JC, Ibañez-Bernal S, Kolb M, Jones L (2012) How will climate change pathways and mitigation options alter incidence of vector borne diseases. A framework for leishmaniasis in South and Meso-America

Ramirez B (2017) Support for research towards understanding the population health vulnerabilities to vector-borne diseases: increasing resilience under climate change conditions in Africa, Ramirez, TDR-IDRC Research Initiative. 6(1):164

Shukla R, Ramasamy V, Shanmugam RK, Ahuja R, Khanna N (2020) Antibody-dependent enhancement: a challenge for developing a safe dengue vaccine. Front Cell Infect Microbiol 10:572681. https://doi.org/10.3389/fcimb.2020.572681

Singh S, Singh H, Sharma V, Shrivastava V, Kumar P, Kanga S, Sahu N, Meraj G, Farooq M, Singh SK (2021) Impact of forest fires on air quality in Wolgan valley, New South Wales, Australia—a mapping and monitoring study using Google Earth Engine. Forests 13(1):4

Singh H, Meraj G, Singh S, Shrivastava V, Sharma V, Farooq M, Kanga S, Singh SK, Kumar P (2022) Status of air pollution during COVID-19-induced lockdown in Delhi, India. Atmosphere 13(12):2090

Sinha S, Kumarasamy N, Suram VK, Chary SS, Naik S, Singh VB, Jain MK, Suthar CP, Borthakur S, Sawardekar V, Noushadali Sk, Reddy N, Talluri L, Thakur P, Reddy M, Panapakam M, Vattipalli R (2022) Efficacy and safety of molnupiravir in mild COVID-19 patients in India. Cureus 14(11):e31508. https://doi.org/10.7759/cureus.31508. PMID: 36532902; PMCID: PMC9751243

Tatem A, Huang Z, Das A, Qi Q, Roth J, Qiu Y (2012) Air travel and vector-borne disease movement. Parasitology 139(14):1816–1830. https://doi.org/10.1017/S0031182012000352

WHO (2009) Dengue guidelines for diagnosis, treatment, prevention and control. WHO, Geneva, Switzerland

World Health Organization (2019) Dengue: guidelines for diagnosis, treatment, prevention and control. World Health Organization, Geneva, pp 1–147

World Health Organization (2022) World health statistics 2022. World Health Organization. https://www.who.int/news/item/20-05-2022-world-health-statistics-2022

Zou L, Miller SN, Schmidtmann ET (2006) Mosquito larval habitat mapping using remote sensing and GIS: implications of coalbed methane development and West Nile virus. J Med Entomol 43(5):1034–1041

# Chapter 13
# Health Informatics: An Emerging Strategy for Clinical Care

**Kulshrestha Himani, Jadoun Singh Pratap Surya, Asthana Shobhit, and Singh Akhand Pratap**

## 13.1 Introduction

### *13.1.1 What is Health Informatics*

An Evolutionary Paradigm in Healthcare Transformation. In today's era, the healthcare landscape is undergoing a profound transformation, primarily driven by the integration of information technology and data science. At the forefront of this evolution lies the field of health informatics, which synergizes healthcare, technology, and data analytics (Jones et al. 2013). This chapter delves into a scholarly exploration of health informatics, unveiling its essence, significance, and its influence on modern healthcare.

### *13.1.2 Defining the Genesis*

Health informatics weaves a complex tapestry from threads of clinical care, computer science, information technology, and management (Staggers and Thompson 2002). Succinctly put, it's the scientific amalgamation and operational deployment of multidisciplinary knowledge to leverage information technology for advancing healthcare processes and patient outcomes (Jones et al. 2013). The International Medical Informatics Association (IMIA) crystallizes it as "The science and the practice that integrates knowledge from several disciplines and fields, including clinical care,

K. Himani (✉) · J. S. P. Surya · S. A. Pratap
Department of Biotechnology, Maharishi University of Information Technology, Lucknow, India
e-mail: himanikul01@gmail.com

A. Shobhit
School of Engineering and Applied Sciences, Bennett University, Greater Noida, India

© The Author(s), under exclusive license to Springer Nature Singapore Pte Ltd. 2024

G. Tripathi et al. (eds.), *Sustainability and Health Informatics*, Advances in Geographical and Environmental Sciences,
https://doi.org/10.1007/978-981-97-6706-9_13

computer and information science, biology, social science, engineering, and management" (Staggers and Thompson 2002). It emerges as a dynamic entity fostering the seamless convergence of diverse domains to generate transformative insights and tangible benefits in healthcare.

### *13.1.3 Significance in the Healthcare Landscape*

Health informatics, symbolic of a paradigm shift, holds immense significance across the healthcare spectrum. Among its epochal contributions is the establishment of Electronic Health Records (EHRs), replacing traditional patient records with a sophisticated digital narrative. EHRs transcend basic documentation; they evolve into holistic repositories containing patient histories, medical conditions, treatments, and diagnostics. Beyond mere digitization, EHRs drive a leap forward in patient care. They empower healthcare providers to access a comprehensive patient narrative in real time, eliminating uncertainties, optimizing diagnoses, and ensuring well-informed interventions.

The pivotal sphere of Health Information Exchange (HIE) embraces health informatics as a conduit for seamless data orchestration across various healthcare silos (Jones et al. 2013). HIE forms the foundation of care coordination, particularly as patients navigate multiple healthcare entities (Ahmad et al. 2023). This leads to a reduction in redundant diagnostics, improved care outcomes, and a sense of continuity for the patient (Singh et al. 2022).

Undoubtedly, the vanguard of health informatics is the Clinical Decision Support (CDS) systems, symbolizing the synergy between technology and evidence-based practice. This digital coordination empowers healthcare professionals with cognitive augmentation, integrating patient data with clinical knowledge to facilitate informed clinical decisions (Jones et al. 2013). CDS emerges as a guardian of patient safety, warding off errors, reinforcing best practices, and aligning interventions with the latest research.

## 13.2 Importance of Health Informatics

### *13.2.1 Illuminating the Nexus of Healthcare and Technology*

The modern healthcare landscape stands on the cusp of a paradigm shift, where the integration of health informatics and technological innovation takes center stage. The fusion of clinical expertise and technological prowess heralds a new era, reshaping healthcare delivery, research methodologies, and administrative processes. This chapter embarks on an enlightening journey to uncover the multifaceted importance of health informatics, unveiling its pivotal role in shaping the trajectory of global healthcare.

### 13.2.2 Enhanced Clinical Decision-Making Through Data-Driven Insights

Health informatics, with its data-rich ecosystem, plays a pivotal role in augmenting clinical decision-making (Ammenwerth et al. 2011). By harnessing expansive datasets from electronic health records (EHRs) and wearable devices, healthcare professionals gain access to comprehensive patient profiles. This holistic view enables timely and informed decisions, leading to more accurate diagnoses, tailored treatment plans, and optimal interventions. Moreover, data analytics unveil intricate patterns that empower clinicians with evidence-based insights, propelling medical practice towards higher standards of care.

### 13.2.3 Empowering Patient-Centered Care

The patient lies at the core of healthcare, and health informatics reinforces this central relationship. Patient engagement is elevated through health informatics platforms offering personalized portals, enabling individuals to access medical records, appointment schedules, and educational resources (Bates et al. 2014a, b). This empowerment fosters a sense of ownership over one's health and nurtures collaborative patient-provider relationships. Consequently, healthcare becomes a joint journey, enhancing adherence to treatment plans and catalyzing positive health outcomes.

### 13.2.4 Optimized Resource Allocation and Population Health Management

In the realm of public health, health informatics assumes a strategic role in optimizing resource allocation and population health management (Terry et al. 2010). By analyzing aggregated data, healthcare administrators forecast disease trends, allocate resources efficiently, and design targeted interventions. This data-driven approach enhances health promotion, disease prevention, and response strategies, ultimately fostering healthier communities and reducing healthcare costs.

### *13.2.5 Elevating Research and Innovation*

Health informatics serves as a fulcrum for transformative medical research and innovation. Vast datasets, aggregated from diverse sources, fuel research endeavors that pave the way for breakthroughs in disease understanding, treatment modalities, and drug discovery (Jha et al. 2008). These insights catalyze the evolution of evidence-based medicine, advancing clinical protocols and propelling the medical community toward evidence-driven decisions.

### *13.2.6 Streamlined Healthcare Administration*

Efficiency in healthcare administration is instrumental in delivering timely and effective patient care. Health informatics streamlines administrative processes through automated workflows, digitized records, and seamless communication systems (Bates and Gawande 2003). This optimization liberates healthcare professionals from administrative burdens, enabling them to allocate more time to patient interactions and clinical pursuits.

The significance of health informatics lies not merely in its technological dimensions, but also in its potential to humanize and elevate healthcare. It embodies empowerment, enables data-driven decision-making, and heralds equitable and efficient healthcare delivery. As the journey of health informatics unfolds, its importance resonates across clinical, administrative, and research domains, igniting a transformation that transcends boundaries and impacts lives.

## 13.3 Key Components of Health Informatics

Medical informatics is a field that combines technology and healthcare to transform patient data management, healthcare decision-making, and medical service delivery. As a dynamic and evolving field, health informatics encompasses several key components that collectively shape the landscape of modern healthcare. This chapter delves into the fundamental components of health informatics, exploring their significance and impact on the healthcare ecosystem.

### *13.3.1 Electronic Health Records (EHRs)*

Electronic Health Records (EHRs) are the central component of health informatics, which digitize patient health information. Accessible to authorized healthcare practitioners, electronic health records (EHRs) serve as complete stores of patient data, including medical history, diagnosis, prescriptions, and treatment plans. These electronic records replace the traditional paper-based approach, offering a centralized and real-time view of patients' health journeys. EHRs not only improve patient safety by reducing medication errors and redundant tests but also empower clinicians to make informed decisions by providing a holistic perspective on patient health. The study underscores how EHRs, particularly within the framework of meaningful use, have become integral to healthcare quality enhancement (Mars 2015).

### 13.3.2 Health Information Exchange (HIE)

The interchange of health information between various organizations has become increasingly important in an era where healthcare systems are integrated. Systems for Health Information Exchange (HIE) make it possible for patient data to be securely shared between various healthcare facilities. This element improves the coordination of treatment, especially when patients see several different physicians. Seamless information exchange through HIE mitigates duplications in tests and procedures, minimizes data silos, and fosters comprehensive patient care. Emphasize the role of HIE in achieving efficient and collaborative healthcare across diverse national contexts (Ahn et al. 2020).

### 13.3.3 Clinical Decision Support (CDS)

Technology-enabled clinical decision support systems (CDS) have emerged as a critical tool in guiding healthcare professionals' decision-making processes. By leveraging patient data and evidence-based guidelines, CDS provides clinicians with real-time recommendations, alerts, and reminders, enhancing clinical outcomes. This component serves as a safeguard against errors, improves adherence to best practices, and ensures that treatment plans align with current medical knowledge. underscores how CDS contributes to patient safety, reducing the likelihood of adverse events (Adler-Milstein and Pfeifer 2017a, b).

### 13.3.4 Telemedicine and Telehealth

The advent of telemedicine and telehealth has ushered in a new era of healthcare accessibility and convenience. These components leverage technology to provide remote medical consultations, diagnostics, and monitoring. Telemedicine bridges geographical barriers, enabling patients to receive expert care regardless of their location. Beyond primary care, telehealth applications extend to mental health services, specialist consultations, and postoperative follow-ups exploring the transformative potential of telemedicine in improving healthcare delivery, especially in underserved and rural areas (Lluch 2011).

### *13.3.5 Health Data Analytics*

In an age characterized by vast amounts of healthcare data, health informatics embraces the power of data analytics. Health data analytics involves processing and analyzing large datasets to unearth insights, trends, and patterns that inform clinical decisions and research endeavors. This component is pivotal in evidence-based medicine, predictive modeling, and public health management. Data analytics facilitates the identification of disease outbreaks, optimization of treatment strategies, and formulation of research hypotheses for medical advancements.

## 13.4 Challenges and Opportunities of Health Informatics

In the ever-evolving intersection of healthcare and technology, the realm of health informatics presents a nuanced landscape marked by a tapestry of challenges and boundless opportunities. The synergy between medical expertise and technological innovation has not only reshaped the healthcare industry but has also introduced complex challenges that necessitate strategic addressing. Concurrently, these challenges underscore the latent potential for transformative advancements that have the capacity to revolutionize patient care, medical research, and healthcare administration on a global scale.

**Challenges**

**1. Data Security and Privacy Concerns**

The monumental shift towards digitization has magnified concerns over data security and patient privacy (Wootton 2012). The electronic storage and exchange of sensitive patient information create vulnerabilities that demand fortified cybersecurity measures. Breaches, unauthorized access, or data leaks could severely compromise patient confidentiality and erode public trust in healthcare systems.

**2. Interoperability Complexities**

The diversified landscape of health information systems often leads to interoperability challenges (Adler-Milstein and Pfeifer 2017a, b). Siloed data repositories hinder seamless data exchange between healthcare providers and across different systems. This lack of interoperability disrupts the flow of crucial patient information and impedes effective care coordination.

**3. Usability and User Adoption**

The integration of intricate health informatics solutions into clinical workflows has encountered resistance due to usability concerns (Lluch 2011). Complex interfaces and a steep learning curve may deter healthcare professionals from embracing new technologies. Consequently, successful user adoption hinges on designing user-friendly interfaces and delivering comprehensive training.

**Opportunities**

**1. Data-Driven Insights and Precision Medicine**

Health informatics harnesses a vast repository of data that, when harnessed through advanced analytics, holds the potential to yield invaluable insights (Murdoch and Detsky 2013). From predictive modeling for disease outbreaks to personalized treatment plans based on genetic markers, health informatics can reshape medical decision-making and usher in an era of precision medicine.

**2. Telemedicine Revolution**

A prominent facet of health informatics, telemedicine offers a transformative opportunity to expand healthcare accessibility (Wootton 2012). Remote consultations, diagnostics, and real-time patient monitoring overcome geographic barriers, ensuring equitable healthcare distribution and enhancing patient experiences.

**3. Empowering Patient Engagement**

Patient portals and mobile applications serve as conduits for heightened patient engagement (Ammenwerth et al. 2011). These platforms grant individuals access to their health records, facilitate appointment scheduling, and provide medication reminders. This empowerment fosters greater patient adherence to treatment plans and encourages proactive involvement in personal health management.

**4. Healthcare Process Optimization**

Health informatics optimizes administrative processes, reducing inefficiencies and costs (Terry et al. 2010). Administrative tasks, including billing and scheduling, benefit from streamlined electronic systems, freeing up resources for improved patient care and enhancing operational efficiency.

## 13.5 Future Trends of Health Informatics

### *13.5.1 A Glimpse into Transformative Pathways.*

In the realm of healthcare, the trajectory of progress is marked by the evolving interplay of health informatics and technological innovation. As the digital landscape continues to expand, health informatics is poised to steer healthcare towards novel horizons. This chapter delves into the promising future trends of health informatics, elucidating their potential to reshape patient care, research methodologies, and healthcare administration.

### *13.5.2 AI-Powered Clinical Decision Support*

Through clinical decision support systems driven by artificial intelligence (AI), artificial intelligence (AI) has the potential to dramatically alter clinical decision-making (Topol 2019a, b). These systems will leverage vast datasets, decipher intricate patterns, and offer tailored insights to healthcare professionals, enabling more precise diagnoses, treatment plans, and prognostic assessments (Farooq et al. 2022). The fusion of clinical acumen with AI-driven algorithms has the potential to elevate patient outcomes to unprecedented levels (Sinha et al. 2022).

### *13.5.3 Predictive and Prescriptive Analytics*

Health informatics is poised to embrace a data-driven era where predictive and prescriptive analytics play a pivotal role in medical research and patient care. By analyzing historical data, these analytics will forecast disease trends, optimize resource allocation, and guide preventive interventions (Bates et al. 2014a, b). These insights will reshape public health strategies and catalyze proactive patient management.

### *13.5.4 Interconnected Healthcare Ecosystems*

The future of health informatics lies in the seamless connectivity of healthcare ecosystems (Adler-Milstein and Pfeifer 2017a, b). Interoperability will transition from a challenge to an industry standard, enabling patient data to flow effortlessly between providers, hospitals, and clinics. This interconnectedness will empower patients with comprehensive health records, enable data-driven research collaborations, and facilitate informed clinical decisions.

### *13.5.5 Telehealth Evolution*

The telehealth paradigm is set to evolve, driven by advancements in health informatics (Bashshur et al. 2018). High-definition video consultations, wearable monitoring devices, and real-time remote diagnostics will redefine patient-clinician interactions.

The seamless integration of telehealth into healthcare workflows will enhance access to care, especially in remote or underserved areas.

### *13.5.6 Blockchain for Data Security*

The integration of blockchain technology is poised to revolutionize data security in health informatics (Ekblaw et al. 2016). Decentralized data storage, immutability, and enhanced privacy protection will mitigate data breaches and establish trust in patient data management. The transparency and security offered by blockchain will reshape how patient information is stored, shared, and accessed.

In the vista of health informatics, these trends represent not only futuristic possibilities but also pathways to tangible transformation. They reflect a harmonious confluence of technological prowess and healthcare expertise, propelling the industry towards a data-driven, patient-centric, and seamlessly connected future.

## 13.6 Conclusion

The quickly developing field of health informatics is an essential combination of information systems and technology in the healthcare sector, and it is essential to improving clinical outcomes, patient care, and professional communication (Mantas 2019). This diverse field includes electronic health records and health information exchange in addition to telemedicine, remote patient monitoring, clinical decision support systems, and other essential elements (Ammenwerth et al. 2020). Health informatics makes it easier to make fast, correct decisions by digitizing patient data, which ultimately increases the efficacy and efficiency of healthcare delivery (Cresswell and Sheikh 2012). Personalized treatment plans, public health campaigns, and research have all benefited from the systematic gathering and analysis of enormous amounts of healthcare data (Dixon and Pfeifer 2017).

Two essential components of health informatics, telemedicine, and remote patient monitoring, have been vital in enabling access to healthcare services, especially in underprivileged or isolated locations (Wade et al. 2016). These technological advancements allow for remote diagnosis, monitoring, and consultation between patients and healthcare providers, expanding access to high-quality treatment (Bashshur et al. 2011). Health informatics is changing the healthcare sector by utilizing data analytics and technology to its fullest potential. Aside from improving patient care, healthcare organizations that adopt this discipline stand to advance medical research and impact the future of healthcare delivery as well as the general well-being of individuals and communities (Häyrinen et al. 2008).

## References

Jones SS, Rudin RS, Perry T, Shekelle PG (2013) Health information technology: an updated systematic review with a focus on meaningful use. Ann Intern Med 160(1):48–54

Staggers N, Thompson CB (2002) The evolution of definitions for nursing informatics: a critical analysis and revised definition. J Am Med Inform Assoc 9(3):255–261

Topol EJ (2019a) High-performance medicine: the convergence of human and artificial intelligence. Nat Med 25(1):44–56

Ammenwerth E, Schnell-Inderst P, Hoerbst A, Müller G (2011) The impact of electronic patient portals on patient care: a systematic review of controlled trials. J Med Internet Res 13(4):e99

Bates DW, Saria S, Ohno-Machado L, Shah A, Escobar G (2014a) Big data in health care: using analytics to identify and manage high-risk and high-cost patients. Health Aff 33(7):1123–1131

Murdoch TB, Detsky AS (2013) The inevitable application of big data to health care. JAMA 309(13):1351–1352

Terry AL, Thorpe CF, Giles G, Brown JB, Harris SB, Reid GJ (2010) Implementing electronic health records: key factors in primary care. Can Fam Physician 56(4):382–389

Jha AK, Doolan D, Grandt D, Scott T, Bates DW (2008) The use of health information technology in seven nations. Int J Med Inf 77(12):848–854

Bates DW, Gawande AA (2003) Improving safety with information technology. N Engl J Med 348(25):2526–2534

Mars M (2015) Telemedicine and advances in urban and rural healthcare delivery in Africa. Prog Cardiovasc Dis 58(3):328–337

Ahn E, Kang SH, Kim DK (2020) Healthcare information systems and data security. In: Healthcare data analytics and management. Springer, pp 105–122

Adler-Milstein J, Pfeifer E (2017a) Information blocking in health IT: a literature review. J Am Med Assoc 172(11):1096–1103

Lluch M (2011) Healthcare professionals' organisational barriers to health information technologies—a literature review. Int J Med Inf 80(12):849–862

Wootton R (2012) Telemedicine support for the developing world. J Telemed Telecare 18(5):251–253

Topol EJ (2019) High-performance medicine: the convergence of human and artificial intelligence. Nat Med 25(1):44–56

Bates DW, Saria S, Ohno-Machado L, Shah A, Escobar G (2014) Big data in health care: using analytics to identify and manage high-risk and high-cost patients. Health Affairs 33(7):1123–1131

Adler-Milstein J, Pfeifer E (2017b) Information blocking in health IT: a literature review. J Am Med Assoc 317(17):1774–1784

Bashshur RL, Shannon GW, Krupinski EA, Grigsby J, Kvedar JC (2018) The taxonomy of telemedicine. Telemed e-Health 24(6):484–494

Ekblaw A, Azaria A, Halamka JD, Lippman A (2016) A case study for blockchain in healthcare: "MedRec" prototype for electronic health records and medical research data. IEEE Open J Comput Soc 1:99–107

Mantas J (2019) Health informatics: the integration of technology and information systems in healthcare. Health Inf Sci Syst 7(1):1

Ammenwerth E, Duftschmid G, Gall W (2020) Clinical information systems for intensive care: applications and barriers. Int J Med Inf 63(2–3):119–131

Cresswell K, Sheikh A (2012) Organizational issues in the implementation and adoption of health information technology innovations: an interpretative review. Int J Med Inf 81(2):73–88

Dixon BE, Pfeifer N (2017) New horizons in patient safety and quality: an analysis of key themes in research and practice. J Patient Saf 13(3):151–157

Wade VA, Karnon J, Elshaug AG, Hiller JE (2016) A systematic review of economic analyses of telehealth services using real-time video communication. BMC Health Serv Res 16(1):1–11

Bashshur RL, Shannon GW, Smith BR, Woodward MA (2011) The empirical foundations of telemedicine interventions in primary care. Telemed e-Health 17(3):237–245

Häyrinen K, Saranto K, Nykänen P (2008) Definition, structure, content, use and impacts of electronic health records: a review of the research literature. Int J Med Informatics 77(5):291–304

Sinha KK, Gupta MK, Banerjee MK, Meraj G, Singh SK, Kanga S, Farooq M, Kumar P, Sahu N (2022) Neural network-based modeling of water quality in Jodhpur India. Hydrology 9(5):92

Singh H, Meraj G, Singh S, Shrivastava V, Sharma V, Farooq M, Kanga S, Singh SK, Kumar P (2022) Status of air pollution during COVID-19-induced lockdown in Delhi India. Atmosphere 13(12):2090

Farooq M, Mushtaq F, Meraj G, Singh SK, Kanga S, Gupta A, Kumar P, Singh D, Avtar R (2022) Strategic slum upgrading and redevelopment action plan for Jammu City. Resources 11(12):120

Ahmad T, Gupta SK, Singh SK, Meraj G, Kumar P, Kanga S (2023) Unveiling nature's resilience: exploring vegetation dynamics during the COVID-19 Era in Jharkhand, India, with the Google Earth Engine. Climate 11(9):187

# Chapter 14
# Telemedicine: An Information Strategy for Health

**Kulshrestha Himani, Mishra Ankita, Asthana Shobhit, and Singh Akhand Pratap**

## 14.1 Introduction

Pediatrists can use the concept of telemedicine for a wide usage. It may be used for distance teaching, distance counseling, distance medicine, and distance research. Distance learning can be sent via live AV links, live video surfing, and watching of contained learning things. distance educating programmes enable doctors to maintain the status quo, reduce the frequency of travel for ongoing medical education (CME), receive free CME, facilitate terms between academic and category-based doctors, and create large pry groups to know from each other and academics (Smith et al. 2009; González-Espada et al. 2019).

Telemedicine involves setting up a contact link between a physician requesting to see a patient under care and a specialist at a telemedicine facility. Such consultation can be done via an interactive live active link or via accumulating and sending technology. For example, storing echocardiographic images, but not in real time, for fast reading by a remote cardiologist. Telemedicine is effective in the management of acute and chronic diseases (Doolittle et al. 2011; Hall-Barrow et al. 2009; Lim et al. 2012; Mahnke et al. 2011; Marcin et al. 2004; Munoz et al. 2012). The benefits of such visits include increased access for disadvantaged populations, and improved access for children in rural and urban areas (Ahmad et al. 2023).

Telemedicine implies communication among a doctor and a patient in a kindergarten, kindergarten, school or juvenile hall. Telemedicine does not replace face-to-face visits, but either adds to them. The benefits of such linkages comprises strengthening medicinal facilities where primary care physicians care for kids. Health care

K. Himani (✉) · M. Ankita · S. A. Pratap
Department of Biotechnology, Maharishi University of Information Technology, Lucknow, India
e-mail: himanikul01@gmail.com

A. Shobhit
School of Engineering and Applied Sciences, Bennett University, Greater Noida, India

© The Author(s), under exclusive license to Springer Nature Singapore Pte Ltd. 2024 
G. Tripathi et al. (eds.), *Sustainability and Health Informatics*, Advances in Geographical and Environmental Sciences,
https://doi.org/10.1007/978-981-97-6706-9_14

amounts are reduced and children are absent from school. Parents spend less money on travel. Parents spend less time away from work. Reducing crowding in emergency departments (Doolittle et al. 2011; Izquierdo et al. 2009).

## 14.2 Requirements

### *14.2.1 Hardware Requirements*

The main hardware requirements include various power burning, probably power evolution, and the number of workspaces embraced. Most needed are not replacements for telemedicine projects in developed countries, but the same requirements can be important in rural areas and some urban areas in developing countries.

### *14.2.2 Communication Requirements*

Any project can face serious blocks due to contact set ups. A huge communication needs to involve the bandwidth of the network buildings. Bandwidth (a data transmission rate) has a significant impact on the type of transmission. Smaller bandwidths can transmit simple information, mostly text, while higher bandwidths can transmit more complex information such as speech, audio (such as stethoscopes), still images, moving images, and video. When it comes to hardware, most transmission requirements are critical in progressing nations, both rural and urban (Debnath et al. 2023).

### *14.2.3 Software Requirements*

Basic software should appraise both primary locations (mainly large hospitals running as telemedicine servers) and major operated areas.

A major limiting requirement comes from the customer side. We differentiate between client-side tele services that involve a single remote workstation (eg, a PC) or multiple remote workstations (eg, a PC and other biomedical equipment such as biometric recorders or imaging devices).

## 14.3 Infrastructure

The technological needs of telemedicine depend on the type of telemedicine a corporation aims to implement. Simultaneous telemedicine, or ongoing telecare, involves live interactions between candidates in two or more facilities (Singh et al. 2022). Store-and-forward telemedicine involves recording medicinal queries at a facility and then transferring the recorded stored and noted information to another facility or facility for a change by a healthcare professional.

Telemonitoring is a key area in telemedicine that enables professionals to monitor patients in real time in hospitals and at home, enabling infected people to monitor hazardous conditions like breathing issues, diabetes type 1 and 2, and heart disease (Marcin 2013; Copeland et al. 2013; Health Information and Management Systems Society 2015).

### *14.3.1 Equipments*

The instruments required can vary from dedicated video conferencing and components of software video conferencing applications for computers or electronic gadgets. The innovation must be able to serve enough AV clarity for forbearing and the potential for healthcare suppliers to easily connect with others. Based on the needs of the telemedicine application, the technology must be capable of connecting hard-wired or portable peripheral medical devices (Marcin et al. 2009).

### *14.3.2 Connection*

To support the needs of the telehealth program goals the adequate bandwidth must be provided to the organization. Point-to-point connectivity from inside or outside the health care facility should be able to be provided to the organization. Occasionally, telemedicine interactions use an integrated services digital network connection when sites lack the infrastructure to support high-speed Internet connections. The minimum recommended speed for a telemedicine link is 384-kilobits-per-second, the general is bidirectional connection between sites.

### *14.3.3 Privacy and Security*

Health Insurance Portability must be combined with Telemedicine interactions with the Accountability Act of 1996 (HIPAA) and other executive needs. To protect the privacy of personal health information is the primary purpose of HIPAA. Most telemedicine technologies provide point-to-point encryption between interacting devices. Although the encryption and privacy of telemedicine are determined by Internet connection security of telemedicine interactions, each corporation should establish the security of health information overlapping to or in line with face-to-face interactions with patient by developing adequate privacy related to data assembly, exploring, and reporting (US Department of Health and Human Services 2003).

## 14.4 Method

Telemedicine mediations are reviewed organized. Interactions include all electronic health mediations for Healthcare communication, data, and technology connections, involvement for discovery and treatment, and community care in combination with Healthcare for the suffering person with critical and hazardous conditions. Every possible sequenced opinion was evaluated in the form of full text by members of an outer adept panel using the Cochrane Effective Practice and Organization Group (EPOC) checklist for quality assessment. It moreover highlights the significance of telemedicine in crisis and calamity reactions and giving get to to pediatric medication for farther and underserved populaces. Impediments to the development of telemedicine were depicted as lawful issues, deficient installment for administrations, innovation costs, and opportuneness, and the need for an innovative foundation at the national level. In spite of the fact that a few challenges constrain its more extensive usage, the current utilization of telemedicine demonstrates its viability and potential. Broad selection of telemedicine will affect compliance with key arrangements of the Understanding Security and Reasonable Care Act, mechanical progresses, and expanded understanding requests for virtual visits.

We looked into writing utilizing a few bibliographic databases. In expansion, we physically looked at major telemedicine diaries and recognized key articles from the reference records of diary articles. Within the Autonomous Prove Report on Telemedicine for the Medicare Population, we outlined an exploration for telemedicine distributions and utilized it to look at the MEDLINE, CINAHL, and HealthSTAR databases for all accessible a long time. Through this handle, we collected things about indirect home telemedicine with pediatricians, obstetricians, and doctors. In any case, they were excluded from most reports as they are exterior the scope of most reports. In this supplemental report, we checked on the first look that comes about and distinguished ponders important to this report. We distinguished extra ponders by hand looking at the reference records of included articles and two peer-reviewed telemedicine diaries: Diary of Telemedicine and Telecare and Telemedicine Diary. We fundamentally evaluated the included ponders for each investigated region and key address and talked about the qualities and impediments of the foremost imperative things about in week-after-week investigative group gatherings. We moreover investigated proposals to address information crevices in telemedicine. To oblige these crevices with the capabilities of particular inquiries about strategies, we categorized telemedicine administrations based on the sort of proof required to decide whether the specific objectives that telemedicine administrations cover. We highlighted the relationship between the sorts and levels of proof found in orderly audits of adequacy and the sorts of inquiries about that may well be supported to address information holes in this developing inquiry about the region.

### *14.4.1 Barriers to Telemedicine*

Telemedicine is leading a huge success, as subjected to the expanding rise in the application of this technique in pediatrics over these 10 years. It has been used beneficially in neonatology, in chronic diseases (Hall-Barrow et al. 2009), in ophthalmology for retinopathy of prematurity screening (Marcin et al. 2004), for following chronic diseases such as asthma and diabetes, in psychology, for outpatient care, in dermatology, and in education. However, barriers to its use exist. These include personal (provider and patient), technological, legal, administrative, and licensing barriers (Nelson et al. 2010; Warshaw et al. 2010; Merrell and Doarn 2012).

To provide acceptance, telemedicine is used to relate Personal barriers and candidates training. The most difficult hurdle has been the provider acceptance (Jameson et al. 2011; Martin et al. 2012; Whitten and Mair 2000).

Telemedicine can reduce their legal risk by having the steps witnessed by a counsel familiar with telemedical law. The key responsibility issues in telemedicine are as follows:

- Doctor–patient communication.
- Technical fails.
- Responsibility insurance.
- Wrongdoing actions
- Standard level of care.
- Informed problems.
- Safety.
- Unknown legal risks in context with telemedicine.

## 14.5 Observation

Telemedicine has several implementations in pediatrics. As the inventions follow to upgrade and degrade in cost, telemedicine will increase the quality of research, knowledge, way to care, quick response, and the drop of normal and specialized pediatrics in broad systems. This will also enlarge connections among family and medicinal caring of the youngsters, paediatrics in category-based operations. Greatest strength of telemedicine lies in its capability to cross over the problems of separation and time to get a population that is medically unreached. This power may automate

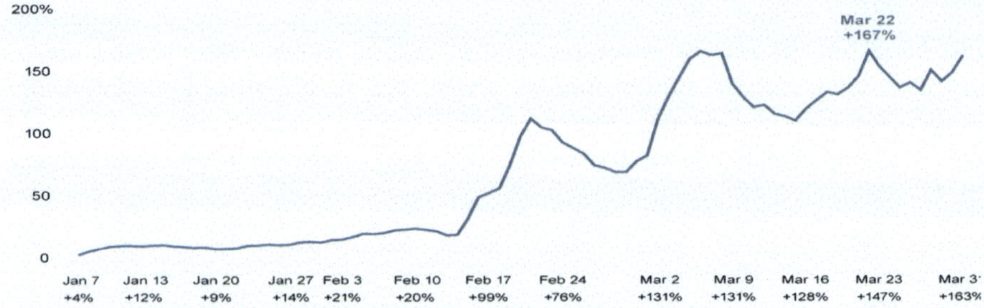


**Graph 14.1** This graph represents the adoption of telemedicine in Asia during Covid-19

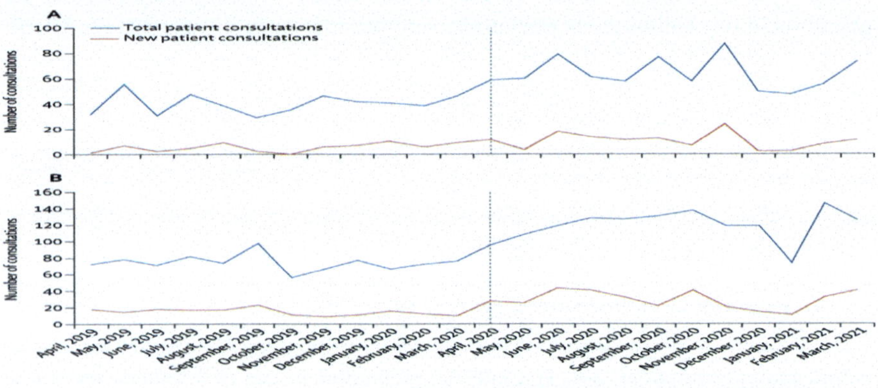


**Graph 14.2** It represents the telemedicine for rural and regional patient access to hematology services during Covid-19

the use of telemedicine to expand various paediatric operations (Graphs 14.1 and 14.2).

## 14.6 Result

Widespread adoption of telemedicine is stuck by the absence of proven survival. The trade model is important and shall be applied early in curriculum growth. Grant-funded projects may fail criteria for not meeting original program and stakeholder goals, which may require additional financial support, highlighting the importance of balancing costs and benefits for all stakeholders and patients.

After quality witness, the review included 80 identical sequential reviews. Up-grooming trends are issues specific to the economic analysis of telemedicine, the benefits of telecare to patients, and the collaborative nature of telecare's complex also unpredictable process.

## 14.7 Conclusion

This subsidiary report, which includes the areas of pediatricians and passive clinical home telemedicine, repeats the results of the first report for the Medicare domain, as the use of telemedicine is low but rising, and the proof of its effectiveness is incomplete. Most studies are short and limited, so the effectiveness of telemedicine cannot be determined.

In this dynamic field, new subject areas are emerging, and researchers are beginning to explore new questions beyond clinical and cost-effectiveness. Observers point to the need for greater research into telemedicine as a controlled intervention, focusing on the patient's perspective, economic analysis, and innovation in telemedicine as a more complex process and an ongoing collaborative effort. Formative assessment is emerging as an area of interest.

## References


Ahmad FN, Ahmad T, Gupta SK, Singh SK, Meraj G, Kumar P, Kanga S et al (2023) Unveiling nature's resilience: exploring vegetation dynamics during the COVID-19 era in Jharkhand, India, with the google earth engine. Climate 11(9):187

Cole SL, Grubbs JH, Din C, Nesbitt TS (2012) Rural inpatient telepharmacy consultation demonstration for after-hours medication review. Telemed J E Health 18(7):530–537

Copeland KC, Silverstein J, Moore KR et al (2013) American academy of pediatrics management of newly diagnosed type 2 diabetes mellitus (T2DM) in children and adolescents. Pediatrics 131(2):364–382

Doolittle GC, Spaulding AO, Williams AR (2011) The decreasing cost of telemedicine and telehealth. Telemed J E Health 17(9):671–675

Farooq FN, Farooq M, Mushtaq F, Meraj G, Singh SK, Kanga S, Gupta A, Kumar P, Singh D, Avtar R et al (2022) Strategic slum upgrading and redevelopment action plan for Jammu City. Resources 11(12):120

González-Espada WJ, Hall-Barrow J, Hall RW, Burke BL, Smith CE (2009) Achieving success connecting academic and practicing clinicians through telemedicine. Pediatrics 123

Hall-Barrow J, Hall RW, Burke BL Jr (2009) Telemedicine and neonatal regionalization of care-ensuring that the right baby gets to the right nursery. Pediatr Ann 38(10):557–561

Health Information and Management Systems Society. Definitions of mHealth. Accessed 18 May 2015

Izquierdo R, Morin PC, Bratt K et al (2009) School-centered telemedicine for children with type 1 diabetes mellitus. J Pediatr 155(3):374–379

Jameson JP, Farmer MS, Head KJ, Fortney J, Teal CR (2011) VA community mental health service providers' utilization of and attitudes toward telemental health care: the gatekeeper's perspective. J Rural Health 27(4):425–432

Debnath J, Sahariah D, Mazumdar M, Lahon D, Meraj G, Hashimoto S, Kumar P, Singh SK, Kanga S, Chand K, Saikia A (2023) Evaluating flood susceptibility in the Brahmaputra river basin: an insight into Asia's eastern Himalayan floodplains using machine learning and multi-criteria decision-making. Earth Syst Environ 1–28

Lim D, Oakley AM, Rademaker M (2012) Better, sooner, more convenient: a successful teledermoscopy service. Australas J Dermatol 53(1):22–25

Mahnke CB, Jordan CP, Bergvall E, Person DA, Pinsker JE (2017) The Pacific asynchronous TeleHealth (PATH) system: review of 1,000 pediatric teleconsultations. Telemed J E Health 17(1):35

Marcin JP (2013) Telemedicine in the pediatric intensive care unit. Pediatr Clin North Am 60(3):581–592

Marcin JP, Nesbitt TS, Kallas HJ, Struve SN, Traugott CA, Dimand RJ (2004) Use of telemedicine to provide pediatric critical care inpatient consultations to underserved rural northern California. J Pediatr 144(3):375–380

Marcin JP, Trujano J, Sadorra C, Dharmar M (2009) Telemedicine in rural pediatric care: the fundamentals. Pediatr Ann 38(4):224–226

Marcin JP, Nesbitt TS, Kallas HJ, Struve SN, Traugott CA, Dimand RJ (2004) Use of telemedicine to provide pediatric critical care inpatient consultations to underserved rural northern California. J Pediatr 144(3):375

Martin AB, Probst JC, Shah K, Chen Z, Garr D (2012) Differences in readiness between rural hospitals and primary care providers for telemedicine adoption and implementation: findings from a statewide telemedicine survey. J Rural Health 28(1):8–15

Merrell RC, Doarn CR (2012) Barriers or barricades. Telemed J E Health 18(2):79–80

Modifications to the HIPAA Privacy, Security, Enforcement, and Breach Notification rules under the Health Information Technology for Economic and Clinical Health Act and the Genetic Information Nondiscrimination Act; other modifications to the HIPAA rules. Fed Regist 78(17):5565–5702. Accessed March 31, 2014 (2013)

Munoz RA, Burbano NH, Motoa MV, Santiago G, Klevemann M, Casilli J (2012) Telemedicine in pediatric cardiac critical care. Telemed J E Health 18(2):132–136

Nelson EL, Miller EA, Larson KA (2010) Reliability associated with the Roter interaction analysis system (RIAS) adapted for the telemedicine context. Patient Educe Coins 78(1):72–78

Singh FN, Singh H, Meraj G, Singh S, Shrivastava V, Sharma V, Farooq M, Kanga S, Singh SK, Kumar P et al (2022) Status of air pollution during COVID-19-induced lockdown in Delhi, India. Atmosphere 13(12):2090

Smith CE, Fontana-Chow K, Boateng BA et al (2009) Tele-education: linking educators with learners via distance technology. Pediatr Ann 38(10):550–556

US Department of Health and Human Services (2014) Summary of the HIPAA privacy rule 2003

Warshaw EM, Gravely AA, Nelson DB (2010) Accuracy of teledermatology/teledermoscopy and clinic-based dermatology for specific categories of skin neoplasms. J Am Acad Dermatol 63(2):348–352

Whitten PS, Mair F (2000) Telemedicine and patient satisfaction: current status and future directions. Telemed J E Health 6(4):417–423

# Chapter 15
# Cryptanalysis and Improvement of Scheme on Wireless Sensor Networks for Agriculture Monitoring

**Sunil Prajapat, Anchana, and Pankaj Kumar**

## 15.1 Introduction

Agriculture has been the foundation stone in the economic growth from the ancient times to now. Agricultural methods must provide for the people's necessities to sustain the population's lifestyle and progress. Agriculture is India's primary economic activity, and 60–70% of the country's population relies on it. According to academic studies, it is projected that the world population would increase to 9.8 billion by the year 2050, indicating a minimum growth of 25% compared to the present population. It is tough to satisfy the needs of the deficient food supply with conventional farming practices since the industry has also been increasingly seizing agricultural land daily.

Hence, traditional agriculture has to be replaced with smart agriculture to maximize agricultural yield while protecting the harvest from natural disasters on the limited amount of available land (Sajan et al. 2022; Ahmad et al. 2023). Advanced techniques that enable computing, transmission, and control via devices make this conceivable. Accurate knowledge of the influencing elements needed in agriculture is equally important to achieve this goal as advances in agricultural equipment (Singh et al. 2022). Wireless sensor networks (WSN) are the most feasible approach and WSN technologies are becoming the cornerstone of modern agriculture. WSNs aid agriculture in scattered data collecting, monitoring of environmental conditions, precise irrigation, and fertilizer supply, and farmers gain from authentic data gathering. In the WSN system, sensor nodes are strategically placed across a farming field to collect meteorological and other relevant data on the variables that impact

S. Prajapat (✉) · Anchana · P. Kumar
Srinivasa Ramanujan Department of Mathematics, Central University of Himachal Pradesh, Dharamshala 176215, India
e-mail: sunilprajapat645@gmail.com

© The Author(s), under exclusive license to Springer Nature Singapore Pte Ltd. 2024 
G. Tripathi et al. (eds.), *Sustainability and Health Informatics*, Advances in Geographical and Environmental Sciences,
https://doi.org/10.1007/978-981-97-6706-9_15

crop development and field management, such as temperature, water quality, soil chemistry, air humidity (Singh et al. 2021).

A wireless sensor network (WSN) is a collection of different sensors that are dispersed around an area and are capable of detecting. The sensor nodes gather, transmit, and process analyzing data. The WSN has been made up of a variety of sensor nodes, including humidity, temperature, light, and soil pH sensors. In WSN sensors sense the data and transmit it and the gathered data is critical. The sensors' gathered data is transported across the network, increasing the security risk. User authentication is a crucial security requirement for WSN secure communication. For instance, fertilizer plays a significant role in agricultural development and boosts crop output, but it also has negative environmental effects and may upset the soil's nitrogen balance.

One of the key objectives of WSN in agriculture is to guarantee an accurate and trustworthy conveyance of information. So before transferring the real data, the sensor first verifies its legality, and then the system verifies the agriculture user's identity. Attackers can change the real data, which can negatively influence or harm the crop if the system cannot verify the authorized user and he attempts to connect with them. Therefore, authentication, integrity, and secrecy will eventually be affected without secure data transmission (Boujelben et al. 2011; Dodis et al. 2004; Le Mouël and Forslund 2017; Mehmood et al. 2017; Prajapat et al. 2022a, b). To achieve data integrity, authentication, and nonrepudiation, an authentication scheme assures that only a legitimate individual may access the data. The basic structure of the WSN-based agriculture monitoring is shown in Fig. The information is gathered by the sensor nodes and transmitted to the gateway via an access point. Also, the gateway node sends that information to the base station. Therefore, all users must register at the base station to access the base station's data.

### *15.1.1 Motivation and Contributions*

We reviewed Rifaqat Ali's scheme (2018) authentication and key agreement scheme using the WSNs for agriculture monitoring. We analyze that the scheme lacks anonymity and untraceability and is susceptible to insider assault, session key attack, and secret empirical key leaking attacks.

## 15.2 Related Work

The development of an authentication protocol that satisfies the requirements of being low-cost for sensor nodes while still being secure, lightweight, and reliable is of utmost importance. This is due to the limited processing and storage capabilities of sensor nodes in WSNs. In addition, the two essential security services, authentication, and access control, protect many networking environments, including wireless sensor networks and farm monitoring.

In 2011, Mazid et al. proposed an identity-based key management protocol for sensor networks. This protocol gives strong security by making use of ECC. The authors say their proposed scheme exhibits resilience against various assaults, with particular emphasis on the node compromise attack. This scheme's flaw is that it doesn't have message identification. Also, its energy use and computational expense efficiency could be more efficient. In the year 2014, a scholarly article authored by Turkanovic et al. (2014) introduced an authentication and key agreement mechanism designed specifically for WSN. A system was devised to provide mutual authentication among the user, sensor node, and gateway node (GWN), even in scenarios where the user does not directly communicate with the GWN. The authors said that the proposed system enables and offers features such as password security, unrestricted password selection, password modification, and the ability to dynamically add nodes. It offers high security because it can withstand attacks like impersonation, multiple users logged in with the same login ID, GWN bypassing, password change, and DoS assaults. But M. Turkanovic's scheme has some security weaknesses.

According to research done in 2015 by Farash et al. (2016), it is susceptible to cryptographic attacks. Farash discovered that the protocol (2016) is vulnerable to smart card theft attacks and man-in-the-middle attacks. Furthermore, the protocol lacks forward/backward secrecy and untraceability. Also, the wireless communications and mobile computing side of this scheme's transmission and storage overhead is higher. With the development of WSN technology, it has gained popularity. It is now being utilized extensively in various fields, such as environmental monitoring, greenhouse use, industrial and commercial applications, forest monitoring, vehicle ad-hoc networks, military, etc. In 2015, He et al. (2015) suggested an authentication scheme using WSN in medical applications. Their scheme provides security against replay, Man-in-the-middle, and impersonation attacks.

Extension, they asserted that the cost of communication and computation for their protocol is low. Later, it was noted that He et al.'s protocol had security weaknesses, such as an insufficient authentication process, a denial of services attack, no password verification, and a bad session key agreement. Kumari and Om (2016) designed an authentication protocol to reduce the security issues in WSN and used AVISPA for the formal security proof. In 2016, Umar et al. suggested a secure key distribution scheme for the WSN. One notable benefit of this technique is that in the event that a cluster head depletes its energy reserves, and its battery life is nearing its end, it effectively transfers its management responsibilities to another node and subsequently disengages.

However, this scheme's flaw is that it is vulnerable to replay attacks, denial of service attacks, and numerous more. Therefore, its security still needs to be improved. Amin et al. presented an improved version of the Farash et al. scheme, the authenticated scheme using WSNs based on IoT. In 2021, Sadhukhan et al. (2021) developed an authentication scheme using the ECC, and they performed an informal security analysis of their scheme using the AVISPA tool. However, the protocol is vulnerable to user impersonation and privileged insider attacks. Ali et al. (2018) proposed an authentication scheme for WSN for agriculture monitoring using the fuzzy biometric extractor. They claimed their scheme withstands the known session key temporary information attack and sensor node impersonation attack. However, we found the scheme vulnerable to the session key attack, insider attack, and empirical secret key leakage attack. The scheme also does not support anonymity.

### *15.2.1 Structuring Your Paper*

The following sections of this paper are structured as follows: In Sect. 15.2, we provide an overview of the mathematical terminology and notations utilized in this study to describe the proposed methodology. Section 15.3 reviews Rifaqat Ali's scheme, followed by cryptanalysis in Sect. 15.4, Sect. 15.5 the improved scheme, Sect. 15.6 includes the informal security analysis, and Sect. 15.7 includes the conclusion.

## 15.3 Preliminaries

### *15.3.1 Fuzzy Extractor*

An (B, b, n, Δ, ϵ)-fuzzy extractor composed of the two algorithms Gen and Rep with the following properties:

- Gen: The Gen algorithms take the biometric input $b_a \epsilon B$ outputs an extracted secret output string $s_a \in \{0,1\}^n$ and the helper public string $p_a \in \{0,1\}^*$.
- Rep: The Rep algorithm takes b' ϵ B and bit string $p \in \{0,1\}^*$ as inputs. Then Rep $(b'_a, p_a) = s_a$ with the property that $d_s(b_a, b'_a) \leq \Delta$.

A fuzzy extractor is appropriate if Gen and Rep execute in the estimated polynomial time.

### *15.3.2 The Elliptic Curve Cryptography*

Elliptic Curve Cryptography ($ECC$) is often preferred public-key cryptography because it offers strong security with relatively small key sizes. It makes ECC an effective way to achieve robust security while minimizing computational resources and the risk of security breaches. Elliptic Curve Cryptography ($ECC$) utilizes the Elliptic Curve $E_Q(a, b)$ over the finite field $F$, which can be represented by the equation $t^2 = s^3 + as + b \ mod \ q$, where $q$ is a prime number of significant size and $4a^3 + 27b^2 \ mod \ q \neq 0$. The Additive Elliptic Curve Cryptography ($ECC$) group, denoted as G, is defined as $G = \{(t, s) : t, s \in F(t, s) \in E(a, b)\} \cup \{\varphi\}$, where $\varphi$ is the identity element of $G$. The group $G$ is cyclic. The scalar multiplication operation on $G$ is defined as $mS = S + S + \cdots + S(mtimes)$, where S is the base point on $E_Q(a, b)$, and $n$ is a positive integer in $F$.

## 15.4 Revisit of Rifaqat Ali's Designed Authentication Protocol

The scheme proposed by Rifaqat Ali (2018) involves a five-stage process that includes system setup, registration of users and agriculture professionals, login and authentication, session key agreement, and password updates. The scheme involves four types of participants: Users/Agriculture Professionals ($U_a$), Base Stations ($BS_t$), Gateway Nodes ($GTWN_j$), and Sensor Nodes ($SEN_j$). The meanings of all symbols used in the scheme are provided in Table 15.1.

**Table 15.1** Notation used in revisit of Rifaqat Ali's scheme

| Symbol | Description |
|---|---|
| $U_a$ | User/agriculture professional |
| $BS_t$ | Base Station |
| $GTWN_j$ | Gateway Node |
| $SEN_j$ | Sensor Node |
| $ID_a$ | Identity of User $U_a$ |
| $ID_{SEN_j}$ | Identity of sensor node |
| $ID_{GTWN_j}$ | Identity of gateway node |
| $pw_a$ | Password of $U_a$ |
| $b_a$ | Biometric of $U_a$ |
| $R_{U_a}$ | Random nonce of $U_a$ |
| $R_{SEN_j}$ | Random nonce of sensor node |
| $R_{BS_t}$ | Random nonce of base station |
| $R_{GTWN_j}$ | Random nonce of the gateway node |
| $\hat{A}\alpha$ | Adversary |
| $X_{BS t}$ | Secret key of base station |
| $X_{BS_t - GTWN_j}$ | Secret key shared between the $BS_t$ and $GTWN_j$ |
| SK | Session key |
| H(.) | Bio-hash function |
| h (.) | Hash function |
| $\oplus$ | XOR Operation |
| $\Vert$ | Concatenation Operation |

### 15.4.1 System Setup Phase

During this phase, the system administrator *SAR* selects unique identities for a total of m sensor nodes, denoted as $ID_{SEN_j}$ where $1 \leq j \leq m$. *SAR* also assigns individual identities for each gateway node, represented as $ID_{GTWN_j}$. Once these identities are selected, *SAR* calculates $RT_j$ by computing $RT_j = h\left(ID_{SEN_j}||X_{BS_t}\right)$, where $X_{BS_t}$ is the identity of the base station $BS_t$ and h is a one-way hash function. *SAR* then stores $\{RT_j, ID_{SEN_j}\}$ in the memory of each sensor node $SEN_j$ and deploys both the sensor nodes and gateway nodes in the target region.

### 15.4.2 User/Agriculture Professional Registration Phase

During this phase, the user, who is an agriculture professional $U_a$, first completes the registration process with the $BS_j$ platform. Subsequently, the subsequent procedures are carried out in order to finish this stage.

- ***Step 1:*** User $U_a$ chooses their own identity $ID_a$, password $pw_a$, and biometric imprint $Gen(b_a) = (s_a, p_a)$; here $(s_a, p_a)$ refer to the secret and public keys, respectively. Then $U_a$ obtained s. $U_a$ afterward securely sends the information $\{ID_a||RPW_a\}$ to $BS_t$.
- ***Step 2:*** When $BS_t$ receives a registration request from $U_a$, $BS_t$ begins the registration process by computing several values upon receiving the registration request from $U_a$. Firstly, it evaluates $A_a$ = h $(ID_a||X_{BS_t})$. Next, $B_a$ is obtained and $B_a = A_a \oplus h(RPW_a||ID_a)$. Then, $C_a$ is computed by $C_a = A_a \oplus h\left(B_a||X_{BS_j}\right)$, and finally, $D_a$ is $h(A_a||RPW_a||\text{ID}_a)$. Afterward, $BS_t$ issues smart-card gain parameters i.e., $\{B_a, C_a, D_a, h(.)\}$ and sends to $U_a$ via the secure channel.
- ***Step 3:*** After receiving the smartcard from$BS_t$, $U_a$ inserts $p_a$ and Gen(.) into the smartcard's memory, i.e., $B_a, C_a, D_a, h(.), Gen(.), p_a\}$.

### 15.4.3 Login and Authentication Phase

When user/agricultural professional $U_a$ requests environmental data like light, humidity, temperature, soil, etc., then, to access the information, they must log into the $BS_t$. To complete this phase, the subsequent actions are carried out:

- ***Step 1:*** The user $U_a$ inserts their own smartcard into the reader and enters their $ID_a$, $PW_a$, as well as imprinting $b_a$ on a sensor device. Card reader now calculates $Rep(b_a, p_a) = s_a$, $RPW_a^* = h(PW_a||s_a)$, $A_a^* = B_{(a)} \oplus h(RPW_a^*||ID_a)$, $D_a^* = h(A_a^*||RPW_a^*||ID_a)$, $\left[h\left(B_a||X_{BS_t}\right)\right]^* = C_a \oplus A_a^*$ and verifies $D_a^* = D_a$ or not. If the verification is successful, the process is continued; if not, the session is ended.
- ***Step 2:*** Then, $U_a$ create a random nonce $R_{U_a}$ and enumerates $DID_a = ID_a \oplus h(B_a||X_{BS_t})$, $M_1 = E_{A_a}\left(R_{U_a}||ID_{SEN_j}||ID_{GTWN_j}||T_1\right)$,

$M_2 = h(R_{U_a}||ID_a||T_1||h(B_a||X_{BS_t})$ and makes a public channel transmission of $\{B_a, DID_a, M_1, M_2\}$ to $BS_t$.

- ***Step 3:*** After receiving the message $\{B_a, DID_a, M_1, M_2\}$ from $U_a$, the $BS_t$ evaluates $ID_a^* = DID_a \oplus h(B_a||X_{BS_t})$, $\left(R_{U_a}^*||T_1||ID_{SEN_j}^*||ID_{GTWN_j}\right) = D_{A_a}(M_1)$ and checks $T_2 - T_1 \leq \Delta T$, the session expires if this is false. If this holds, $BS_t$ determines if $M_2^* = M_2$ or not; if it does, $U_a$ is valid and moves on to the next phase.
- ***Step 4:*** he $BS_t$ now generate a random nonce $R_{BS_t}$ and determines $M_3 = E_{h(X_{BS-GTWN_j})}(ID_a||RT_j||ID_{SEN_j}||R_{U_a}||R_{BS_t}||T_3)$, $M_5 = h(M_2||ID_{U_a}||T_3||R_{BS_t}||R_{U_a})$ and $\{M_3, M_2, M_5\}$ are then sent to $GTWN_j$ through public channels.
- ***Step 5:*** After receiving the message from $BS_t$, the $GTWN_j$ calculates $(ID_a||RT_j||ID_{SEN_j}||R_{U_a}||R_{BS_t}||T_3) = D_{h\left(X_{BS_j-GTWN_j}\right)}(M_3)$, and $M_5^* = h(M_2||ID_a||T_3||R_{BS_j}||R_{U_a})$ and then determines whether two conditions $T_4 - T_3 \leq \Delta T$ and $M_5^* = M_5$ are true. If both prerequisites are met, the next step is taken. Alternatively, end the session.
- ***Step 6:*** $GTWN_j$ now creates a random nonce $R_{GTWN_j}$, and calculates $M_6 = E_{RT_j}(R_{BS_t}||T_5||R_{U_a}||R_{GTWN_j}||ID_a)$, $M_7 = h(M_2||RT_j||R_{GTWN_j}||ID_a||R_{U_a})$ before sending $\{M_2, M_6, M_7\}$ to $SEN_j$.
- ***Step 7:*** When the $SN_j$ receives the message from GTWN $j$, it calculates $(R_{BS_t}||T_5||R_{U_a}||R_{GTWN_j}||ID_a) = D_{RT_j}(M_6)$, and check $T_6 - T_5 \leq \Delta T$ holds or not. If holds then compute $M_7^* = h(M_2||RT_j||R_{GTWN_j}||ID_a||R_{U_a})$ and check $M_7^* = M_7$ holds or not. The next step is taken if both conditions are true; else, the session is ended.
- ***Step 8:*** The $SEN_j$ creates a random nonce $R_{SEN_j}$ and enumerates $M_8 = R_{SEN_j} \oplus h(R_{GTWN_j})$, $SK = h(R_{GTWN_j}||R_{U_a}||R_{SEN_j}||RT_j||M_2)$, $M_9 = h(SK||ID_a)$ and before sending $\{M_8, M_9, T_7\}$ to $GTWN_j$ via the public channel.
- ***Step 9:*** After receipt of the message from $SEN_j$, $GTWN_j$ initially confirms $T_8 - T_7 \leq \Delta T$. Continue if this is true; else, end the session. Now, further evaluates $R_{SEN_j}^* = M_8 \oplus h(RT_j||R_{GTWN_j})$, $SK* = h\left(R_{GTWNj}||R_{U_a}||R_{SEN_j}^*||RT_j||M_2\right)$, computes $M_9^* = h(SK^*||ID_a)$, check $M_9^* = M_9$ or not. Do the subsequent action if this is the case; else, end the session.
- ***Step 10:*** The computes and sends via the same channel. $\{M_9, M_{10}\}$ sent to $U_a$ through the same public channel after the $GTWN_j$ calculates $M_{10} = E_{h(R_{U_a}||ID_a)}(RT_j||T_9||R_{GTWN_j}||R_{SEN_j}||M_2)$.
- ***Step 11:*** Upon receipt of the message from $GTWN_j$, $U_a$ calculates $(RT_j||T_9||R_{GTWN_j}||R_{SEN_j}||M_2) = D_{h(R_u||ID_a)}(M_{10})$ and confirms $T_{10} - T_9 \leq \Delta T$. If it does, move on to the next step. Alternatively, end the session.
- ***Step 12:*** In order to determine whether mutual authentication and session key agreement are valid, the $U_a$ calculates $SK^* = h(R_{GTWN_j}||R_{U_a}||R_{SEN_j}||RT_j||M_2)$, $M_9^* = h(SK^*||ID_a)$, and compares $M_9^* = M_9$. If this comparison is valid, then the U a can proceed.

### 15.4.4 Password Update Phase

At this stage, the $U_a$ modifies its password independently of the base station. For this, the steps listed below are used.

- ***Step 1:*** In this stage, user $U_a$ inserts their own smartcard into the reader and enters their $ID_a, PW_a$, as well as imprinting $b_a$ on a sensor device. Card reader now calculates $Rep(b_a, p_a) = s_a$, $RPW_a^* = h(PW_a||s_a)$, $A_a^* = B_a \oplus h(RPW_a^*||ID_a)$, $D_a^* = h(A_a^*||RPW_a^*||ID_a)$, $[h(B_a||X_{BS_t})]^* = C_a \oplus A_a^*$ and verifies $D_a^* = D_a$ or not. If the verification is successful, the process is continued; if not, the session is ended.
- ***Step 2:*** The $U_a$ inputs a fresh password $pw_a^{new}$ and computes $RPW_a^{new} = h(PW_a^{new}||s_a)$, $B_a^{new} = B_a \oplus h(RPW_a||ID_a) \oplus h(RPW_a^{new}||IDa)$, $A_a^{new} = B_a^{new} \oplus h(RPW_a^{new}||ID_a)$, $C_a^{new} = C_a \oplus A_a \oplus A_a^{new}$ and $D_a^{new} = h(A_a^{new}||RPW_a^{new}||ID_a)$ and then replaces $\{B_a^{new}, C_a^{new}, D_a^{new}\}$ in place of $\{B_a, C_a, D_a\}$ respectively.

## 15.5 Cryptanalysis of Rifaqat Ali et al. Scheme

In this section, we have done a cryptanalysis of Rifaqat Ali et al. proposed authentication protocol.

### 15.5.1 Password Update Phase

Insider attacks are legitimate and practicable in Ali et al.'s scheme because, at the user registration stage, the $U_a$ securely submits registration request $\{ID_a, RPW_a\}$ to $BS_t$. Then, a privileged insider or an adversary of the base station has this information, Then, an elite insider or an adversary $\hat{A}_\alpha$ of base station $BS_t$ possesses this information, i.e., $\{ID_a, RPW_a\}$, and from the Side channel attack, $\hat{A}_\alpha$ gets data stored in the smart card i.e., $\{B_a, C_a, D_a, h(.), Gen(.), p_a\}$. Adversary $\hat{A}_\alpha$ can compute the encryption and decryption key Aa as

$$A_a = B_a \oplus h(RPW_a||ID_a).$$

### 15.5.2 User Impersonation Attack

In this adversary, $\hat{A}_\alpha$ obstructs the login message i.e., $\{B_a, DID_a, M_1, M_2\}$ from the public channel and also extracts all the parameters from the smart card by the side-channel attacks and also from the privileged insider attack. In this attack, the adversary blocks the login message, which is $i\{B_a, DID_a, M_1, M_2\}$, from the public channel and also takes all the parameters from the smart card, i.e., $\{B_a, C_a, D_a, h(.), Gen(.), p_a\}$ through side-channel attacks and $\{ID_a, RPW_a\}$ a through privileged insider attacks. Then adversary tries to act as a legitimate user.

- ***Step 1:*** Firstly, generate a random nonce and time stamp $T_1$ and computes $DID_a = ID_a \oplus h(B_a||X_{BS_t})\ M1 = E_{A_a}(R_{U_a}||ID_{SEN_j}||ID_{GTWN_j}||T_1)$ $M_2 = h(R_{U_a}||ID_a||T_1||h(B_a||X_{BS_t})$ and finally, $\hat{A}_\alpha$ sends $\{DID_a, B_a, M_1, M_2\}$ via the public channel.

### 15.5.3 Attack on the Session Key

Adversary $\hat{A}_\alpha$ gets $\{ID_a, RPW_a\}$ from the privileged insider attack, and from the smart card stolen attack, he gets $\{B_a, C_a, D_a, h(.), Gen(.), p_a\}$. Also, the adversary $\hat{A}_\alpha$ eavesdrops on the messages $\{B_a, DID_a, M_1, M_2\}$, $\{M_2, M_6, M_7\}$, $\{M_8, M_9, T_7\}$, $\{M_9, M_{10}\}$ from the public channel.

- ***Step 1:*** Adversary computes $A_a = B_a \oplus h(RPW_a||ID_a, [h(B_a||X_{BS_t})] = C_a||A_a$

  And then decrypt message M1 using $A_a$ and get $(RU_a||ID_{SEN_j}||ID_{GTWN_j}||T_1) = D_{A_a}(M_1)$

- ***Step 2:*** Compute $h(RU_a||ID_a)$ and decrypt $M_{10}$ and get $(RU_a||ID_{SEN_j}||ID_{GTWN_j}||T_1) = D_{A_a}(M_1)$
- ***Step 3:*** Adversary $\hat{A}_\alpha$ can compute the session key as

$$SK = h(GTWN_j||R_{U_a}||R_{SEN_j}||RT_j||M_2).$$

## 15.6 Improved Scheme

### 15.6.1 Setup Phase

The system administrator *SYS* selects unique identities $IT_{SNNj}$ for each of the $m$ $SNN_i$ (sensor nodes), where $m$ is a positive integer. They also select distinct identities $IT_{GAWj}$ for each gateway node $GAW_i$. The *BA* chooses a non-singular elliptic curve $E_Q(a, b)$ and a base point $P$ in a finite field $F$, where $Q$ is a prime number. They also select a secret parameter $K$, compute the public key $PU = K.P$, and make Rep(.), Gen(.), h(.), and $PU$ as public parameter. Finally, *SYS* computes $RI_i = h(IT_{SNN_i}||K)$ and keeps $\{RI_i, IT_{SNN_i}\}$ into $SNN_i$'s memory, where h(.) is a cryptographic hash function, and || denotes concatenation.

### 15.6.2 Registration Phase

The user initially undergoes a registration process at the base station. The registration carried out in the following steps.

1. The user, $U_n$ , begins registration by selecting their own identity, $IT_n$, along with a password, $PWA_n$, and a random nonce, $r_n$. They then place their biometric data, $T_n$, on the sensor device. Next, they compute the function $Gen(T_n) = (X_T, P_T)$, where $(X_T, P_T)$ are the secret and public parameter respectively. Using these parameters, $U_n$.
2. Calculates $RPAW_n = h(PAW_n||X_T||r_n)$. Once the above steps are completed, $U_n$ securely transmits the pair $\{ID_n, RPAW_n\}$ to the base station, $BA$, through a trusted communication channel.
3. Upon receiving the registration request from $U_n$, the base station $BS$ proceeds to calculate $A_n = h(IT_n||K||C)$, where $C$ is the registration counter. Next, BS computes $B_n = A_n \oplus h(RPAW_n||IT_n)$, $C_n = A_n \oplus h(B_n||K)$, $D_n = h(A_n||RPAW_n||IT_n)$. $BS$ store $\{IT_n, C, B_n, C_n\}$ for the further use. The $BS$ issues a smartcard containing $\{D_n, h(.)\}$ and sends it, along with $\{B_n, C_n\}$, to $U_n$ through a trusted communication channel.
4. After getting the smart card from $BA$, $U_n$ calculates $N_n = B_n \oplus h(r_n||X_T)$, $M_n = C_n \oplus h(r_n||X_T)$, $r_n = h(X_T||PAW_n)$ and embeds $N_n, M_n, P_T, Rep(.)$ into the smart card i.e., $\{N_n, M_{n,}D_n, h(.), Gen(.), P_T, r_n\}$.

### 15.6.3 Login Phase

When user $U_n$ desires to access information regarding environmental conditions like temperature, humidity, soil pH, etc., they are required to initiate a login process with the base station (BA). The login procedure encompasses the following steps.

1. The $U_n$ insert own smart-card into the card reader and input $IT_n$, $PWA_n$, and also imprint $T_n$ on the sensor device. Then the card reader calculates the $Rep(T_n, P_T) = X_T^*$, $r_n^* = h\big(X_T^*||PAW_n\big)$, $RPWA_n^* = h(PAW_n||X_T^*||r_n^*)$, $B_n^* = N_n \oplus h\big(r_n^*||X_T^*\big)$, $C_n^* = M_n \oplus h\big(r_n^*||X_T^*\big)$, $A_n^* = B_n^* \oplus h\big(B_n^*||IT_n\big)$, $D_n^* = h\big(A_n^*||RPAW_n^*||IT_n\big)$, $\big[h\big(B_n^*||K\big)\big]^* = C_n^* \oplus A_n^*$ And then verifies $D_n^*? = D_n$, if holds then continue the process, otherwise the session.
2. Now $U_n$ generates a random nonce $R_{U_n}$ and computes pseudo identity $DID_n = IT_n \oplus \big[h\big(B_n^*||K\big)\big]^*$. Then computes message $M_1 = E_{A_n}\big(R_{U_n}||IT_{SNN_i}||IT_{GAW_i}||T_1\big)$, $M_2 = h\big(R_{U_n}||IT_n||T_1||h(B_n||K)\big)$ and sends $\{DID_n, M_1, M_2\}$ to the $BA$ by the public channel.

### 15.6.4 Authentication and Session Key Agreement Phase

1. etrieve $\{IT_n, C, B_n, C_n\}$ and computes $IT_n^* = DID_n \oplus h(B_n||K)$. Decrypt $M_1$ as $\left(R_{U_n}^*||T_1||IT_{SNN_i}^*||IT_{GAW_i}^*\right) = D_{A_n}(M_1)$. If $T_2 - T_1 \leq \Delta T$ holds, then further compute $M_2^* = h\left(R_{U_n}^*||IT_n^*||T_1||h(B_n||K)\right)$. If $M_2^*? = M_2$ holds and then $BA$ produces random nonce $R_{BA}$ and computes $M_3 = E_{h\left(K_{BA-GAW_i}\right)}\left(IT_n||RI_i||IT_{SNN_i}||R_{U_n}||R_{BA}||T_3\right)$, $\left(IT_n||RI_i||IT_{SNN_i}||R_{U_n}||R_{BA}||T_3\right)$, $M_4 = h\left(M_2||IT_n||T_3||R_{BA}||R_{U_n}\right)$. And then sends $\{M_3, M_2, M_4\}$ to the $GAW_i$ by public channel.
2. After getting the message from $BA$, the $GAW_i$ decrypt $\left(IT_n||RI_i||IT_{SNN_i}||R_{U_n}||R_{BA}||T_3\right) = D_{h(BA-GAW_i)}(M_3)$. If $T_4 - T_3 \leq \Delta T$ holds, then further compute $M_4^* = h\left(M_2||IT_n||T_3||R_{BA}||R_{U_n}\right)$. If $M_4^* = M_4$ holds and then take $R_{GAW\,i}$ and computes $M_5 = E_{RI_n}\left(R_{BA}||T_5||R_{U_n}||R_{GAW_i}||IT_n\right)$, $M_6 = h\left(M_2||RI_i||R_{GAW_i}||IT_n||R_{U_n}\right)$. Then $GAW_n$ sends message $\{M_2, M_5, M_6\}$ to the $SNN_i$ by the public channel.
3. After receiving message from $GAW_i$, the $SNN_i$ Decrypt $\left(R_{BA}||T_5||R_{U_n}||R_{GAW_i}||IT_n\right) = D_{RI_n}(M_5)$.
4. Verify $T_6 - T_5 \leq \Delta T$ if holds then compute $M_6^* = h\left(M_2||RI_i||R_{GAW_i}||IT_n||R_{U_n}\right)$. If $M_6^* =?M_6$ holds, then generates random nonce $R_{SNN\,i}$ and computes $M_7 = R_{SNN_i} \oplus h\left(RI_i||R_{GAW_i}\right)$, and the session key $SEK = h\left(R_{GAW_i}||R_{U_n}||R_{SNN_i}||RI_i||M_2\right)$, $M_8 = h(SEK||IT_i)$ and sends $\{M_7, M_8, T_7\}$ to $GAW_i$ by public channel.
5. After getting message from $SNN_i$, $GAW_i$ check $T_8 - T_7 \leq \Delta T$, if holds then compute $R_{SNN_i}^* = M_7 \oplus h\left(RI_i||R_{GAW_i}\right)$, $SEK^* = h\left(R_{GAW_i}||R_{U_n}||R_{SNN_i}^*||RI_i||M_2\right)$ and $M_8^* = h(SEK^*||IT_i)$. If $M_8^* = ?M_8$ holds, then compute$M_9 = E_{h\left(R_{U_n}||IT_n\right)}\left(RI_i||T_9||R_{GAW_i}||R_{SNN_i}||M_2\right)$. After that send message $\{M_8, M_9\}$ to the $U_n$ by the public channel.
6. After getting message from $GAW_i$, $U_n$ Decrypt$\left(RI_i||T_9||R_{GAW_i}||R_{SNN_i}||M_2\right) = D_{h\left(R_{U_n}||IT_n\right)}(M_9)$. Verify $T_{10} - T_9 \leq \Delta T$, if holds then compute $SEK^* = h\left(R_{GAW_i}||R_{U_n}||R_{SNN_i}||RI_i||M_2\right)$, $M_8^* = h(SEK^*||IT_i)$ and compare $M_8^*? = M_8$ if this holds then mutual authentication and session key agreement holds.

### 15.6.5 Password Update Phase

In this phase $U_n$ update their password. This phase is carried in the following step:

1. The $U_n$ insert own smart card into the card reader and input $IT_n$, $PWA_n$, and also imprint $T_n$ on the sensor device. Then the card reader calculates the $Rep(T_n, P_T) = X_T^*$, $r_n^* = h\left(X_T^*||PWA_n^*\right)$, $RPWA_n^* = h(PAW_n||X_T^*||r_n^*)$, $B_n^* = N_n \oplus h\left(r_n^*||X_T^*\right)$, $C_n^* = M_n \oplus h\left(r_n^*||X_T^*\right)$, $A_n^* = B_n^* \oplus h(B_n^*||IT_n)$, $D_n^* = h(A_n^*||RPAW_n^*||IT_n)$, $[h(B_n^*||K)]^* = C_n^* \oplus A_n^*$ And then verifies $D_n^*? = D_n$, if holds then continue the process, otherwise the session.

2. The user enters a new password $PAW_n^{NEW}$ and computes $r_n^{NEW} = h(X_T||PAW_n^{NEW})$, $RPAW_n h(PAW_n^{NEW}||X_T||r_n^{NEW})$,$B_n^{NEW} B_n \oplus h(RPAW_n||IT_n) \oplus h(RPAW_n^{NEW}||IT_n)$, $A_n^{NEW} B_n^{NEW} \oplus h(RPAW_n^{NEW}||IT_n)$, $C_n^{NEW} C_n \oplus A_n \oplus A_n^{NEW} D_n^{NEW} h(A_n^{NEW}||RPAW_n^{NEW}||IT_n)$, $N_n^{NEW} = B_n^{NEW} \oplus h(r_n^{NEW}||X_T)$, $M_n^{NEW} = C_n^{NEW} \oplus h(r_n^{NEW}||X_T)$, and then replaces $\{N_n, M_{n,} D_n\}$ by $\{N_n^{NEW}, M_n^{NEW}, D_n^{NEW}\}$.

## 15.7 Informal Security Analysis

- **Privacy preserving**

The proposed scheme ensures user anonymity by using a pseudo-identity $PID_n$ instead of the actual identity $IT_n$.

- **Three factor security**

The proposed scheme incorporates three-factor security, requiring the presence of three parameters: a password, a smart card, and biometric verification. An attacker attempting to log in to the base station must possess all three of these factors to be considered a legitimate user. If the attacker only has two of the three factors, they cannot successfully log in. Specifically, if the attacker possesses the biometric data and the password, they still require the smartcard information $\{N_n, M_n, D_n\}$ in order to proceed with the login process.

- **Replay attack**

In this attack scenario, the attacker intercepts a login message and forwards it to the intended system without making any changes. When the login message $\{DID_n, M_1, M_2\}$ intercepted and transmitted to the $BA$, the system first decrypts the message $M_1$ and verifies if the time difference $T_1 - T_2 \leq \Delta T$. Since the system has already verified this message, it rejects the request. If the attacker intercepts the message $\{M_3, M_2, M_4\}$ and retransmits it to the target system $GAW_i$, the system first decrypts the message $M_3$ and checks if the time difference $T_4 - T_3 \leq \Delta T$. If it is not, the system terminates the session. Similarly, if the attacker eavesdrops on the message $\{M_2, M_5, M_6\}$ and forwards it to the target system $SNN_i$ the system decrypts the message $M_5$ and checks if the time difference $T_6 - T_5 \leq \Delta T$. If the condition is not met, the system terminates the session.

- **Insider attack**

In this user tries to get the information of the user (password, etc.) as attackers get the message going through the trusted channel. Then he can't get user information from it, as $RPAW_n = h(PAW_n||X_T||r_n)$ involve the random nonce $r_n$ and the biometric secret parameter $X_T$. In this scenario, the user attempts to obtain sensitive user information such as passwords. The attacker intercepts a message $\{IT_n, RPAW_n\}$ that is transmitted through a trusted channel. Where $RPAW_n = h(PAW_n||X_T||r_n)$ But he can't get any helpful personal information from the $RPAW_n$.

- **Smart card stolen attack**

In this attack, the attacker's objective is to obtain personal information about the user by extracting data stored in their smart card through power analysis. In the proposed scheme if the user, $U_n$, extract all the information stored in the smart card $\{N_n, M_{n,}D_n, h(.), Gen(.), P_T, r_n\}$, Where, $N_n = B_n \oplus h(r_n||X_T), M_n = C_n \oplus h(r_n||X_T)$, $r_n = h(X_T||PAW_n)$ Then attacker needs the biometric $X_T$, and has to guess the password in polynomial time, after that, he gets the $B_n$ and $C_n$, where $B_n = A_n \oplus h(RPAW_n||IT_n)$, $C_n = A_n \oplus h(B_n||K)$ So, it is not possible for the attacker to get personal information about the $U_n$ using smart card.

- **Offline password guessing attack**

In this attack, the attacker's goal is to guess the password of the user $U_n$. However, in the proposed scheme, even if the attacker obtains the message $\{ID_n, RPAW_n\}$ through an insider attack, they cannot determine the user's password from this message alone. This is because the calculation of $RPAW_n = h(PAW_n||X_T||r_n)$ involves the biometric $X_T$ and random nonce $r_n$ that can't be guessed in polynomial time. This makes it significantly difficult for the attacker to deduce the user's password, providing an added layer of security to the system.

- **Empirical secret key**

In this attack, the attacker's objective is to acquire the encryption and decryption keys, enabling them to access the user's information. However, in the proposed scheme, the attacker is unable to obtain the encryption and decryption key $A_n$, where $A_n = h(IT_n||K||C)$ and $B_n = A_n \oplus h(RPAW_n||IT_n)$ If the attacker stole the Smart card and extracted all the information stored in it i.e. $\{N_n, M_{n,}D_n, h(.), Gen(.), P_T, r_n\}$, where, $N_n = B_n \oplus h(r_n||X_T)$,$M_n = C_n \oplus h(r_n||X_T)$, $r_n = h(X_T||PAW_n)$ Which involves the biometric secret parameter, so he can't be able to calculate the $A_n$.

- **User impersonation attack**

In this attack, the attacker assumes the identity of the user by impersonating them. The attacker intercepts the login message, modifies it, and subsequently sends it in an attempt to appear as a legitimate user. Our scheme withstands this attack because as attacker eavesdrops on the login message $\{DID_n, M_1, M_2\}$, where $DID_n = IT_n \oplus [h(B_n^*||K)]^*$, $M_1 = E_{A_n}(R_{U_n}||IT_{SNN_i}||IT_{GAW_i}||T_1), M_2 = h(R_{U_n}||IT_n||T_1||h(B_n||K))$ The attacker can generate the random nonce $R_{U_n}^*$ and time stamp $T_1^*$, but he can't be able to obtain the message $\{ M_1, M_2\}$ as he does know the values $A_n$, $RPAW_n$, and $IT_n$.

- **Message authentication**

The proposed scheme provides message authentication between the $U_n$, $BA$ at $M_2^*? = M_2$, between $BA$, $GAW_i$ at $M_4^*? = M_4$, and between $GAW_n$, $SNN_i$ at $M_6^*? = M_6$. Also provides mutual authentication at $M_8^*? = M_8$.

- **Withstand to the known session key temporary information attack**

In this attack attacker knows all short-term key secret information like random nonce, time stamps and tries to evaluate the session key. In the proposed scheme attacker fails to compute the session key as $SEK = h(R_{GAW_i}||R_{U_n}||R_{SNN_i}||RI_i||M_2)$ Where,$RI_i = h\big(IT_{SNN_i}||K\big)$, and $M_2 = h\big(R_{U_n}||IT_n||T_1||h(B_n||K)\big)$ involves the $\{IT_n, B_n, K\}$ which are not known.

## 15.8 Conclusion

After conducting a thorough analysis of Rifaqat Ali et al.'s authentication and key agreement protocol (2018), we identified several vulnerabilities including empirical secret key leakage, impersonation attacks, and compromised session keys. To address these weaknesses, we have proposed an enhanced scheme for the authentication and key agreement protocol proposed by Rifaqat Ali et al. Our improved scheme offers message authentication, three-factor security, protection against replay attacks, user impersonation attacks, offline password guessing attacks, empirical secret key attacks, smart card theft attacks, insider attacks, and resilience against known session key temporary information attacks.

## References

Ahmad T, Gupta SK, Singh SK, Meraj G, Kumar P, Kanga S (2023) Unveiling nature's resilience: exploring vegetation dynamics during the COVID-19 era in Jharkhand, India, with the Google earth engine. Climate 11(9):187

Ali R, Pal AK, Kumari S, Karuppiah M, Conti M (2018) A secure user authentication and key-agreement scheme using wireless sensor networks for agriculture monitoring. Futur Gener Comput Syst 84:200–215

Boujelben M, Youssef H, Mzid R, Abid M (2011) IKM—an identity based key management scheme for heterogeneous sensor networks. J Commun 6(2):185–197

Dodis Y, Reyzin L, Smith A (2004) Fuzzy extractors: How to generate strong keys from biometrics and other noisy data. In: Advances in cryptology-EUROCRYPT 2004: international conference on the theory and applications of cryptographic techniques, Interlaken, Switzerland, May 2–6, 2004. Proceedings, vol 23. Springer Berlin Heidelberg, pp 523–540

Farash MS, Turkanović M, Kumari S, Hölbl M (2016) An efficient user authentication and key agreement scheme for heterogeneous wireless sensor network tailored for the Internet of Things environment. Ad Hoc Netw 36:152–176

He D, Kumar N, Chen J, Lee CC, Chilamkurti N, Yeo SS (2015) Robust anonymous authentication protocol for health-care applications using wireless medical sensor networks. Multimedia Syst 21:49–60

Kumari S, Om H (2016) Authentication protocol for wireless sensor networks applications like safety monitoring in coal mines. Comput Netw 104:137–154

Le Mouël C, Forslund A (2017) How can we feed the world in 2050? A review of the responses from global scenario studies. Eur Rev Agric Econ 44(4):541–591

Mehmood A, Umar MM, Song H (2017) ICMDS: secure inter-cluster multiple-key distribution scheme for wireless sensor networks. Ad Hoc Netw 55:97–106

Prajapat S, Kumar P, Sharma V (2022a) A lightweight group authentication scheme over lattices. In: 2022 4th international conference on advances in computing, communication control and networking (ICAC3N). IEEE, pp 1215–1219

Prajapat S, Kumar P, Sharma V (2022b) An efficient CL-signature scheme over NTRU lattices. In: 2022 4th international conference on advances in computing, communication control and networking (ICAC3N). IEEE, pp 1220–1224

Sadhukhan D, Ray S, Biswas GP, Khan MK, Dasgupta M (2021) A lightweight remote user authentication scheme for IoT communication using elliptic curve cryptography. J Supercomput 77:1114–1151

Sajan B, Mishra VN, Kanga S, Meraj G, Singh SK, Kumar P (2022) Cellular automata-based artificial neural network model for assessing past, present, and future land use/land cover dynamics. Agronomy 12(11):2772

Singh S, Singh H, Sharma V, Shrivastava V, Kumar P, Kanga S, Sahu N, Meraj G, Farooq M, Singh SK (2021) Impact of forest fires on air quality in Wolgan valley, New South Wales, Australia—a mapping and monitoring study using Google earth engine. Forests 13(1):4

Singh H, Meraj G, Singh S, Shrivastava V, Sharma V, Farooq M, Kanga S, Singh SK, Kumar P (2022) Status of air pollution during COVID-19-induced lockdown in Delhi, India. Atmosphere 13(12):2090

Turkanović M, Brumen B, Hölbl M (2014) A novel user authentication and key agreement scheme for heterogeneous ad hoc wireless sensor networks, based on the Internet of Things notion. Ad Hoc Netw 20:96–112

# Chapter 16
# Climate Change and Desert Dust Storms Induced Public Health Impacts on Rural Inhabitants of Western India

**Rajyalakshmi Garaga, Arukonda Vaishnavi, and Pradeep Kumar Dammala**

## 16.1 Introduction

The arid regions of the world, known for their challenging landscapes, extreme climate conditions, and frequent dust storms, have attracted growing interest due to their potential environmental and public health consequences. Among these natural occurrences, desert dust storms emerge as significant factors that can modify atmospheric composition and impact human health over extensive distances. Inhaling fine dust particles during these occurrences has been linked to the development and worsening of various diseases, raising worries about the health of people living in these dry regions (Yadav et al. 2017; Middleton 2017; Münzel et al. 2019; Soleimani et al. 2020).

The Thar Desert, situated in the northwestern part of the Indian subcontinent, represents one of the most extensively studied desert regions globally (Rumana et al. 2014). While numerous studies have delved into the health impacts of desert dust storms in various geographical contexts, a noticeable research gap persists in comprehending the specific diseases prevalent in the Indian deserts, particularly the Thar Desert, and their direct association with dust storms (Singh et al. 2021). This research article aims to address these crucial gaps and shed light on the diseases linked to desert dust storms, with a special emphasis on the Thar Desert (Singh et al. 2022). Despite

R. Garaga (✉)
Center for Emerging Technologies for Sustainable Development (CETSD), Indian Institute of Technology Jodhpur, Jodhpur, India
e-mail: garaga@iitj.ac.in

A. Vaishnavi
Department of Mechanical Engineering, Indian Institute of Technology Jodhpur, Jodhpur, India

P. K. Dammala
Department of Civil & Infrastructure Engineering, Indian Institute of Technology Jodhpur, Jodhpur, India

© The Author(s), under exclusive license to Springer Nature Singapore Pte Ltd. 2024

G. Tripathi et al. (eds.), *Sustainability and Health Informatics*, Advances in Geographical and Environmental Sciences,
https://doi.org/10.1007/978-981-97-6706-9_16

the prevalence of studies examining the health effects of desert environments across Asia, a scarcity of research specifically concentrates on the diseases prevalent in the Indian deserts, including the Thar Desert (Meraj et al. 2021; Singh et al. 2021). Given the unique ecological conditions, anthropogenic influences, and diverse socio-economic factors, the health implications of dust storms in the Indian desert region may substantially differ from those in other arid areas. (Stanke et al. 2013; Dumka et al. 2019). Consequently, it is essential to conduct a thorough investigation into the health conditions affecting the residents of Indian desert regions to develop precise and effective public health strategies. Studies carried out on diseases associated with desert dust storms in arid areas of India often reveal shortcomings in conducting comprehensive literature reviews. This deficiency can impact the depth and accuracy of their findings. A robust literature review is not only crucial for grasping the current state of knowledge but also for identifying gaps and building on prior research to facilitate further investigations effectively (Farooq et al. 2022). By addressing this knowledge gap, our research aims to present a well-informed analysis of the health consequences of dust storms in the Thar Desert. While certain health aspects related to dust storms have been explored in studies conducted in the Thar Desert region, specific diseases resulting from the inhalation of desert dust have received insufficient attention (Gupta et al 2021; Goudie 2019; Schweitzer et al. 2018). Diseases like Pulmonary Tuberculosis, Valley fever, and Chronic Pulmonary disorders have been proposed to be associated with dust inhalation during desert storms. Nevertheless, there remains a need for more investigation to determine the extent of their occurrence and their direct connection with dust storms in the Thar Desert. Our research aims to address this gap by examining the influence of dust storms on these specific diseases and their potential consequences for public health (Vergadi et al. 2022; Trianti et al. 2017; Chatterjee 2017). The principal goal of this research article is to close the current gaps in knowledge by thoroughly exploring the diseases linked to desert dust storms, with a particular focus on the Thar Desert. Through a systematic examination of the connection between dust storms and diseases, our study intends to add to the expanding knowledge base regarding health concerns in arid environments (Meraj et al. 2022). Moreover, the findings derived from this study can establish a groundwork for devising preventive measures designed to protect the health and welfare of communities inhabiting dry regions, particularly those in the Thar Desert.

By conducting a comprehensive review of pertinent literature, empirical data analysis, and rigorous statistical scrutiny, this research article aims to offer valuable insights into the health consequences of desert dust storms and the diseases they are linked to. The goal is to generate knowledge that can serve as the foundation for evidence-based interventions and policies crafted to alleviate the health hazards associated with dust storms in the Thar Desert and arid regions worldwide. As we uncover the intricate connection between dust storms and public health, the importance of customized preventive measures becomes more evident, thus enhancing our overall comprehension of environmental health dynamics in arid landscapes.

## 16.2 Methodology

This study utilizes a systematic and exhaustive approach to investigate the health ramifications of desert dust storms in the Thar Desert region. The methodology encompasses a thorough literature review, data analysis, and statistical validation, ensuring a rigorous examination of the connection between dust storms and related diseases, as depicted in Fig. 16.1. The initial phase involves an extensive review of peer-reviewed articles, scientific reports, and pertinent publications from reputable databases, including PubMed, Web of Science, and Google Scholar. The search terms encompass a range of keywords, such as "dust storms," "public health," "Global Burden of Disease," "Thar desert," "pulmonary tuberculosis," "silicosis," and "environmental dynamics."

For inclusion in this study, the following criteria were applied: Studies that specifically investigate the health impacts of desert dust storms on human populations were considered. Studies conducted in arid regions, encompassing all deserts, with a particular focus on the Indian deserts, were included. Studies addressing the prevalence and relationship of diseases caused by desert dust storms were considered. Only original research articles, reviews, and systematic reviews/meta-analyses were considered (Ahmad et al. 2023). Moreover, only peer-reviewed articles published in the last 10 years were included. To establish correlations between dust storm occurrences, airborne particulate matter levels, and disease prevalence, statistical analysis was conducted. Descriptive statistics offer insights into disease distribution and dust storm frequency. Additionally, regression analysis was performed to quantify the association between dust storm events and health outcomes, taking potential confounding factors into consideration.

To enhance the scientific validity of the study, critical literature-based diseases identified in the literature review were cross-referenced with the Global Burden of Disease (GBD) data inventory. This step validates the relevance of diseases and their potential association with dust storms, strengthening the research findings. The study

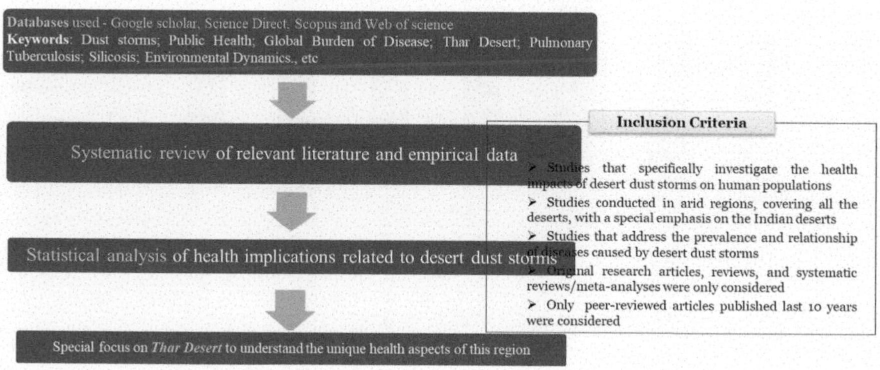


**Fig. 16.1** Strategic Literature Survey showing no. of studies chosen based on the inclusion criteria mentioned

acknowledges potential limitations, including data availability, variability in disease reporting, and the complexity of isolating dust storm impacts. These limitations were transparently addressed to ensure a comprehensive understanding of the research's scope and implications. In summary, this investigation employs a multidimensional methodology that combines literature review, data analysis, statistical validation, and mechanistic insights to comprehensively explore the health implications of desert dust storms in the Thar Desert region. Through this rigorous approach, the study aims to uncover the intricate relationship between dust storms and specific diseases while offering valuable insights for public health interventions and policies.

## 16.3 Results and Discussion

The impact of desert dust storms on human health has been extensively explored through various research articles, employing different analysis approaches and methodologies. Figure 16.2 depicts the studies that have contributed valuable insights into the health hazards associated with dust storms, their implications for different regions, and the potential strategies for mitigation (Santra et al. 2010).

Griffin et al. (2004) emphasized the significant role of desert dust in primary productivity, diversity flux, and climate change. Dust storms, when combined with pollutants, pose potential risks to both ocean and human health. It is encouraged that future research delves deeper into understanding the atmospheric dispersion of pollutants and disease-causing agents. Haldiya et al. (2005) found a gap in knowledge and practice among salt workers regarding protective measures. These results stress the need for increased awareness and protective measures in high-risk occupational environments. Meng et al. (2007) demonstrated a correlation between dust

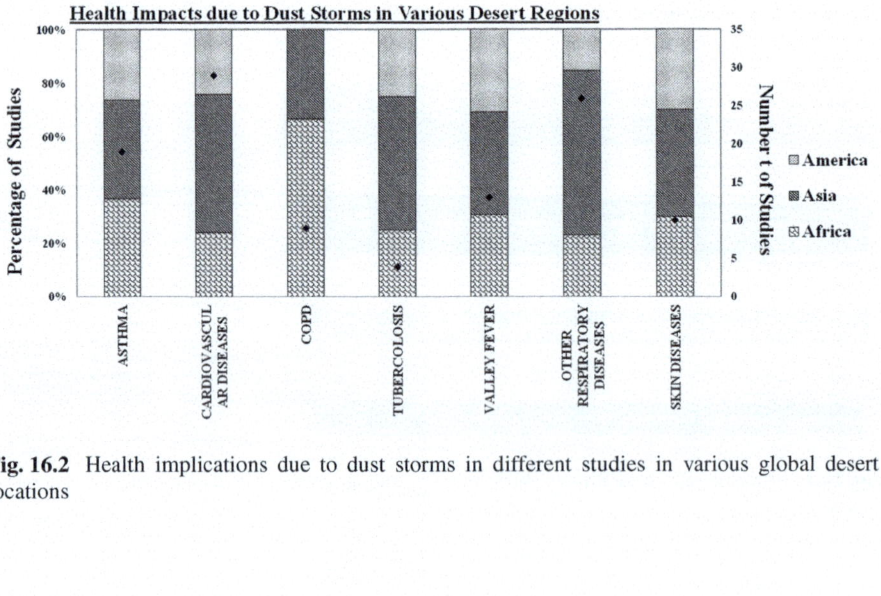


**Fig. 16.2** Health implications due to dust storms in different studies in various global desert locations

events and increased hospitalizations for respiratory and cardiovascular diseases, which aligns with recent data on the adverse effects of particulate matter on human health. Longueville et al. observed a prevalence of asthma and chronic obstructive pulmonary disease (COPD) in Rajasthan, despite frequent and intense dust events in West Africa, which remain largely unexplored in terms of their health effects. Singh et al. (2012) emphasized the potential exacerbation of respiratory diseases due to climate change, especially in high-altitude areas and deserts, calling for situation analysis and interventions. Goudie et al. (2014) underscored the diverse range of health disorders associated with dust storms, including respiratory, cardiovascular, skin, and infectious diseases, predicting increased dust storms due to land use pressures and climate change. Vodonos (2014) discussed the limitations of using outdoor air pollution measurements for studying indoor health impacts, highlighting chemical toxicity and the role of Sahara dust particles as carriers of pollutants. Wang et al. (2016) associated particulate matter from dust storms with increased clinic visits related to pulmonary tuberculosis, underscoring the importance of understanding the environmental factors influencing disease outbreaks. Aghababaeian et al. (2021) reviewed the significant impacts of dust storms on health, particularly cardiovascular and respiratory diseases, calling for more accurate exposure measurement and better epidemiological studies to inform policies. Lwin et al. (2023) conducted a scoping review, highlighting the main health outcomes associated with desert dust exposure, along with exposure measurement and study design challenges. They suggested standardized research protocols and more evidence for African populations. Broomandi et al. (2022) explored the correlation between dust storms and increased COVID-19 infection rates and noted the potential role of dust storms in the transmission and diffusion of infectious diseases. Khaniabadi et al. outlined the health risks associated with dust storms, emphasizing the importance of control measures and vegetation management to mitigate the impacts. Soltani et al. (2023) analyzed the factors contributing to dust storms using the DPSIR framework and emphasized the need for preventive measures to improve air quality. A review by Adams (2023) highlighted the global burden of premature deaths caused by particulate matter pollution, including dust and soot. It discussed both long-term and short-term health effects, along with the importance of mitigative strategies. Despite the multitude of studies on desert-related health impacts in Asia, there is a notable gap in research specific to Indian deserts, particularly the Thar Desert, with limited and insufficient literature available (Fig. 16.3).

These research articles collectively underscore the multidimensional impact of desert dust storms on human health, ranging from respiratory and cardiovascular diseases to infectious diseases and mental health issues (Issanova et al. 2015; Zhang et al. 2016; Sprigg et al. 2016; Lorentzou et al. 2019; Lloyd 2023; Querol et al. 2019; Adams 2023). Successful validation of critical literature-based diseases with GBD highlighted the likeliness of pulmonary diseases due to dust storms in the Thar Desert, India as depicted in Fig. 16.4.

While the findings reveal consistent patterns of adverse health effects, they also highlight the need for further research to improve exposure measurement accuracy and study design. The development of early warning systems, public health policies,

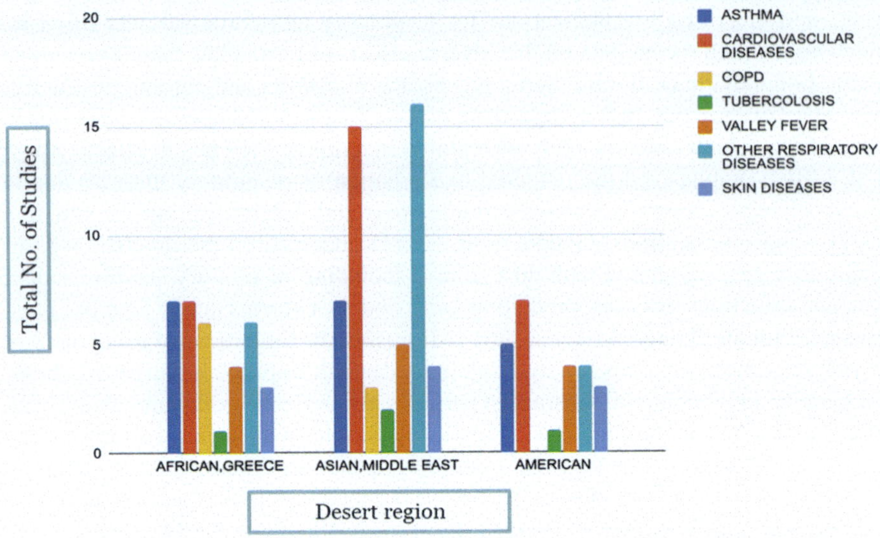


**Fig. 16.3** Distribution of studies conducted on different diseases across various regions globally

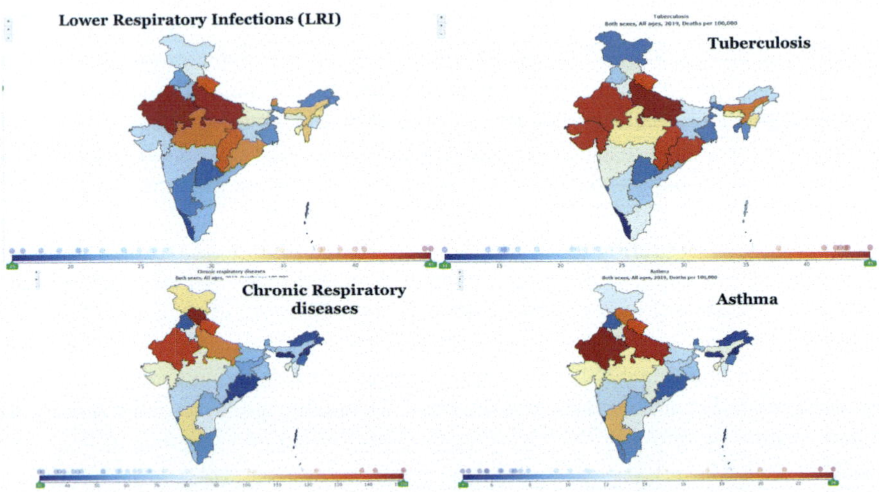


**Fig. 16.4** Diseases due to dust storms prevalent in Thar Desert, India highlighted by the Global Burden of Disease (GBD)

and mitigation strategies is crucial to address the increasing risks posed by dust storms, particularly in the context of climate change and urbanization.

## 16.4 Conclusion

The current review addresses a notable gap in knowledge by investigating the health effects of desert dust storms in the Thar Desert, India. The validation of disease associations through Global Burden of Disease data highlights the likelihood of pulmonary diseases being linked to dust storms in the region. The study underscores the importance of exploring diseases like Tuberculosis, Cardiovascular conditions, Asthma, and Lower Respiratory Infections caused by dust inhalation during storms. By the current study, attention is drawn to the health implications of dust storms, contributing to a better understanding of the interconnections between climate action and public health. Future research should focus on experimental investigations and understanding disease prevalence. The findings emphasize the urgency of tailored interventions to mitigate the health risks posed by dust storms in arid regions, particularly the Thar Desert, for the well-being of affected populations.

In summary, the study on the health effects of desert dust storms in the Thar Desert, India, is a relevant and integral part of the discourse surrounding sustainability, health informatics, and the systemic approach needed to confront the global challenges induced by climate change. The critical importance of considering the health dimension in the context of climate action and sustainability is underscored by this research.

## References


Adams BT (2023) Impact of dust on public health and strategies for mitigation. https://globalroadtechnology.com/impact-of-dust-on-public-health-and-strategies-for-mitigation/ last accessed on 21st Aug 2023.

Aghababaeian H, Ostadtaghizadeh A, Ardalan A, Asgary A, Akbary M, Yekaninejad MS, Stephens C (2021) Global health impacts of dust storms: a systematic review. Environ Health Insights. 15:11786302211018390

Ahmad FN et al. (2023) Unveiling Nature's Resilience: Exploring Vegetation Dynamics during the COVID-19 Era in Jharkhand, India, with the Google Earth Engine. Climate 11(9):187

Broomandi P, Crape B, Jahanbakhshi A, Janatian N, Nikfal A, Tamjidi M, Kim JR, Middleton N, Karaca F (2022) Assessment of the association between dust storms and COVID-19 infection rate in southwest Iran. Environ Sci Pollut Res 29(24):36392–36411

Chatterjee CM (2017) Desert dermatoses (Thar desert, India). Indian J Dermatol 62(1):52

De Longueville F, Ozer P, Doumbia S, Henry S (2013) Desert dust impacts on human health: an alarming worldwide reality and a need for studies in West Africa. Int J Biometeorol 57:1–9

Dumka UC, Kaskaoutis DG, Francis D, Chaboureau JP, Rashki A, Tiwari S, Singh S, Liakakou E, Mihalopoulos N (2019) The role of the Intertropical discontinuity region and the heat low in dust emission and transport over the Thar Desert, India: A Premonsoon case study. J Geophys Res: Atmospheres 124(23):13197–13219

Farooq et al. (2022). Slum Categorization for Efficient Development Plan—A Case Study of Udhampur City, Jammu and Kashmir Using Remote Sensing and GIS. Geospatial Technology for Landscape and Environmental Management: Sustainable Assessment and Planning, pp 283–299

Goudie AS (2014) Desert dust and human health disorders. Environ Int 1(63):101–113

Goudie AS. Dust storms and human health. InExtreme weather events and human health: International case studies (2019) 16. Springer International Publishing, Cham, pp 13–24

Griffin DW, Kellogg CA (2004) Dust storms and their impact on ocean and human health: dust in Earth's atmosphere. EcoHealth 1(3):284–295

Gupta L, Dev R, Zaidi K, Sunder Raman R, Habib G, Ghosh B (2021) Assessment of PM 10 and PM 2.5 over Ghaziabad, an industrial city in the Indo-Gangetic Plain: spatio-temporal variability and associated health effects. Environ Monit Assess. 193:1–21

Haldiya KR, Sachdev R, Mathur ML, Saiyed HN (2005) Knowledge, attitude and practices related to occupational health problems among salt workers working in the desert of Rajasthan. India. J Occup Health. 47(1):85–88

Issanova G, Abuduwaili J, Kaldybayev A, Semenov O, Dedova T (2015) Dust storms in Kazakhstan: frequency and division. J Geol Soc India 85:348–358

Khaniabadi YO, Bahrami P, Moulaei Birgani P, Rashidi R, Naqvi HR, Anbari K (2023) Risk assessment of exposure to the Middle Eastern dust storms in Iran. Hum Ecol Risk Assess Int J 29(3–4):743–756

Lloyd G (2023) Dust with a side of disease. https://ourenvironment.ac.nz/2021/05/22/dust-with-a-side-of-disease/ last accessed on 21st Aug 2023

Lorentzou C, Kouvarakis G, Kozyrakis GV, Kampanis NA, Trahanatzi I, Fraidakis O, Tzanakis N, Kanakidou M, Agouridakis P, Notas G (2019) Extreme desert dust storms and COPD morbidity on the island of Crete. Int J Chron Obstruct Pulmon Dis 6:1763–1768

Lwin KS, Tobias A, Chua PL, Yuan L, Thawonmas R, Ith S, Htay ZW, Yu LS, Yamasaki L, Roqué M, Querol X (2023) Effects of desert dust and sandstorms on human health: A scoping review. Geohealth. 7(3): e2022GH000728

Meng Z, Lu B (2007) Dust events as a risk factor for daily hospitalization for respiratory and cardiovascular diseases in Minqin. China Atmos Environ 41(33):7048–7058

Meraj FN: 2021: Meraj, G., Kanga, S., Kranjčić, N., Đurin, B. and Singh, S.K. et al. (2021) Role of natural capital economics for sustainable management of earth resources. Earth 2(3):622–634

Meraj FN et al. (2022) Modeling the sediment retention and ecosystem provisioning services in the Kashmir valley, India, Western Himalayas. Modeling Earth Systems and Environment 8(3):3859–3884

Middleton NJ (2017) Desert dust hazards: A global review. Aeol Res 1(24):53–63

Münzel T, Lelieveld J, Rajagopalan S, Daiber A (2019) Contribution of airborne desert dust to air quality and cardiopulmonary disease. Eur Heart J 40(28):2377–2378

Querol X, Tobías A, Pérez N, Karanasiou A, Amato F, Stafoggia M, García-Pando CP, Ginoux P, Forastiere F, Gumy S, Mudu P (2019) Monitoring the impact of desert dust outbreaks for air quality for health studies. Environ Int 1(130):104867

Rumana HS, Sharma RC, Beniwal V, Sharma AK (2014) A retrospective approach to assess human health risks associated with growing air pollution in urbanized area of Thar Desert, western Rajasthan, India. J Environ Health Sci Eng 12:1–9

Santra P, Mertia RS, Kushawa HL (2010) A new wind-erosion sampler for monitoring dust-storm events in the Indian Thar Desert. Curr Sci 25:1061–1067

Schweitzer MD, Calzadilla AS, Salamo O, Sharifi A, Kumar N, Holt G, Campos M, Mirsaeidi M (2018) Lung health in era of climate change and dust storms. Environ Res 1(163):36–42

Singh PK, Dhiman RC (2012) Climate change and human health: Indian context. J Vector Borne Dis 49(2):55

Singh FN et al. (2021) Impact of forest fires on air quality in Wolgan valley, New South Wales, Australia—A mapping and monitoring study using Google Earth Engine. Forests 13(1):4

Singh et al. (2021) Assessment of seasonal vegetation dynamics over parts of thar desert using geospatial techniques. ANGRAU, p 105

Singh FN (2022) Status of Air Pollution during COVID-19-Induced Lockdown in Delhi. India. Atmosphere 13(12):2090

Soleimani Z, Teymouri P, Boloorani AD, Mesdaghinia A, Middleton N, Griffin DW (2020) An overview of bioaerosol load and health impacts associated with dust storms: A focus on the Middle East. Atmos Environ 15(223):117187

Soltani MJ, Motamedvaziri B, Mosaffaei J, Noroozi AA, Ahmadi H (2023) Cause and effect analysis of the trend of dust storms using the DPSIR framework in the Hendijan region. Int J Environ Sci Technol 20(5):4919–4930

Sprigg WA (2016) Dust storms, human health and a global early warning system. Extreme Weather, Health, and Communities: Interdisciplinary Engagement Strategies. 59–87

Stanke C, Kerac M, Prudhomme C, Medlock J, Murray V (2013) Health effects of drought: a systematic review of the evidence. PLoS Currents. 5:5

Trianti SM, Samoli E, Rodopoulou S, Katsouyanni K, Papiris SA, Karakatsani A (2017) Desert dust outbreaks and respiratory morbidity in Athens. Greece. Environ Health. 16(1):1–9

Vergadi E, Rouva G, Angeli M, Galanakis E (2022) Infectious diseases associated with desert dust outbreaks: A systematic review. Int J Environ Res Public Health 19(11):6907

Vodonos A, Friger M, Katra I, Avnon L, Krasnov H, Koutrakis P, Schwartz J, Lior O, Novack V (2014) The impact of desert dust exposures on hospitalizations due to exacerbation of chronic obstructive pulmonary disease. Air Qual Atmos Health 7:433–439

Wang Y, Wang R, Ming J, Liu G, Chen T, Liu X, Liu H, Zhen Y, Cheng G (2016) Effects of dust storm events on weekly clinic visits related to pulmonary tuberculosis disease in Minqin. China Atmos Environ 1(127):205–212

Yadav R, Sahu LK, Beig G, Tripathi N, Jaaffrey SN (2017) Ambient particulate matter and carbon monoxide at an urban site of India: influence of anthropogenic emissions and dust storms. Environ Pollut 1(225):291–303

Zhang X, Zhao L, Tong DQ, Wu G, Dan M, Teng B (2016) A systematic review of global desert dust and associated human health effects. Atmosphere 7(12):158

# Chapter 17
# Therapeutic Potential of Natural Pharmacological Agents (Flavonoids and Other Phenolic Compounds) Found in Plants

**Kanchan Awasthi, Neeraj Jain, Madhu Prakash Srivastava, Sneha Verma, and Akshay Khanna**

## 17.1 Introduction

Medicinal plant use dates back to the earliest days of human civilization and has since been intertwined with its evolution. Traditional healers across the globe have utilized indigenous plants and extracts for extensive periods to cure a wide array of ailments, showcasing their pharmacological properties. Historically, herbal medicines were predominantly employed in forms such as tinctures, plasters, poultices, and teas, eventually progressing toward refined and isolated components. Knowledge of the use of medicinal plants has been preserved in the form of ancient folklore within families, lineages, and societies, passed down from generation to generation.

Plants of medicinal significance or their extracts have been utilized since ancient times for various conditions and have yielded important medicines such as anesthetics (morphine), antitussives (codeine), antihypertensives (reserpine), cardiotonics (digoxin), antineoplastics (vinblastine and taxol), and antimalarials (quinine and artemisinin). The discovery of medicinal plants and their therapeutic applications holds promising leads in curing or extending the life expectancy of individuals suffering from conditions such as cancer, malaria, cardiovascular issues, and neurological diseases. Plants have been utilized extensively due to their role as the primary source of bioactive natural compounds (Ahmad et al. 2023).

Naturally, throughout their resilience and development, plants have generated pharmacologically active secondary metabolites with a diverse array of structural characteristics. Over millions of years, they have evolved and adapted to fend off a wide range of biotic and abiotic challenges, including bacteria, insects, fungi, and

K. Awasthi (✉) · N. Jain · M. P. Srivastava · S. Verma · A. Khanna
Department of Science, Maharishi University of Information Technology, Lucknow, UP, India
e-mail: kanchan.awasthi@muit.in

© The Author(s), under exclusive license to Springer Nature Singapore Pte Ltd. 2024

G. Tripathi et al. (eds.), *Sustainability and Health Informatics*, Advances in Geographical and Environmental Sciences,
https://doi.org/10.1007/978-981-97-6706-9_17

environmental stressors. These secondary metabolites have been shown to possess active constituents for preventing major ailments. Nevertheless, plant-derived natural products represent a much more economical source of bioactive agents as they are non-toxic, readily available in nature, and cost-effective compared to synthetic compounds. These natural constituents and their derivatives might function differently from conventional medicines. Numerous natural products and synthetically modified derivatives have been successfully developed for clinical use across various therapeutic areas (Singh et al. 2022; Farooq et al. 2022).

The nineteenth century marked the isolation of multiple alkaloids from plants utilized as medicines, including atropine (*Atropa belladonna*), caffeine (*Coffea arabica*), cocaine (*Erythroxylum coca*), ephedrine (*Ephedra* species), morphine, codeine (*Papaver somniferum*), pilocarpine (*Pilocarpus jaborandi*), physostigmine (*Physostigma venenosum*), quinine (*Cinchona cordifolia*), salicin (*Salix* species), theobromine (*Theobroma cacao*), theophylline (*Camellia sinensis*), and ()-tubocurarine (*Chondodendron tomentosum*). Through these discoveries and subsequent advancements, bioactive secondary metabolites became recognized as bioactive agents, further establishing these plants' utilization as 'medicinal' or 'therapeutic' plants (Kumar et al. 2015).

The true nature of these compounds and agents is still under investigation, aiming to obtain these compounds through chemical synthesis. Presently, there are 125 clinically useful medicines of known constitution derived from approximately 100 species of advanced plants. Although an estimated 5000 plant species have been studied in detail as potential sources of new medicines, <10% of the world's biodiversity has been explored for potential natural applications. Numerous plant species remain of great importance and have the potential to cure or aid various conditions, calling for further exploration and discovery.

## 17.2 Conventional Healthcare Practices

Opinions and treatments for various conditions have always held paramount significance. Throughout history, conventional medical practices and traditional remedies have been utilized. Traditional medicines comprise natural ingredients, minerals, and medicinal plants. Only conventional treatments that primarily use processed plant products are regarded as herbal medicines. In developing countries, the traditional system of healing relies significantly on medicinal plants, which have been an integral part of their history and culture. Countries such as China and India have consistently relied on their traditional remedies for curing and treating numerous conditions (Ekor 2014).

## 17.3 Natural Products Derived from Medicinal Plants

Throughout history, secondary metabolites derived from natural sources and the chemicals derived from them have represented a wealth of novel compounds. Due to their wide array of chemical compositions and their capacity to impact various natural targets, natural products have historically been a rich resource for creating new compounds. The majority of naturally occurring substances originate from primary metabolites like amino acids, carbohydrates, and fatty acids, falling under the category of secondary metabolites. The creation and breakdown of proteins, fats, nucleic acids, and carbohydrates-essential for all life forms are the core processes of primary metabolism, where the substances involved in these pathways are termed "primary metabolites." On the flip side, secondary metabolites arise from primary metabolism and typically aren't involved in metabolic functions. These include alkaloids, phenolics, essential oils, terpenes, sterols, flavonoids, lignins, tannins, and various others. Secondary metabolites' biosynthesis originates from core processes like photosynthesis, glycolysis, and the Krebs cycle. These fundamental pathways produce biosynthetic intermediates, ultimately resulting in the creation of secondary metabolites known as natural products.

Intermediate compounds like Acetyl co-enzyme A (Acetyl-CoA), shikimic acid, mevalonic acid, and 1-deoxyxylulose-5-phosphate serve as essential foundational elements used in the synthesis of secondary metabolites in biological processes (Dias et al. 2012).

Some of the most common pharmacologically active chemicals are discussed herewith.

**Flavonoids**

Flavonoids, which are phytochemicals present in a wide array of plants, fruits, vegetables, and leaves, play implicit roles in medicinal chemistry. They offer a spectrum of medicinal benefits, such as anticancer, antioxidant, anti-inflammatory, and antiviral properties, along with neuroprotective and cardio-protective effects. The efficacy of these natural attributes relies on the specific type of flavonoid, its potential mode of action, and its bioavailability. These cost-effective medicinal properties bring significant natural benefits and have been proven effective for a variety of conditions. Ongoing research is concentrated on their isolation, synthesis of analogs, and their impact on human health through various methods and animal models (Sinha et al. 2022). A considerable number of flavonoids have been successfully isolated, and this number continues to grow steadily. As a result, endeavors have been undertaken to categorize the isolated flavonoids that possess beneficial properties to augment our understanding of their impact on health (Ullah et al. 2020).

**Flavonoids and Their Health Benefits**

The flavonoids show the following health benefits:

- Flavonoids exhibit the capacity to inhibit the growth of cancerous cells, functioning as potent anticancer agents.

- Hesperidin (Hsp), a crucial flavonoid, demonstrates remarkable anticancer activity.
- Aurone, a benzofuranone, and another significant flavonoid, has been widely utilized as an anticancer agent.
- Flavonoids found in *Emblica officinalis* showcase various pharmacological activities, including anticancer, anti-inflammatory, and immunomodulatory effects.
- They also function as antioxidants, proving to be more potent than both vitamin C and Vitamin E.
- *Bryonia alba* L., a medicinal plant from the Cucurbitaceae family, boasts a broad spectrum of biological activities.
- The consumption of dietary flavonoids demonstrates a positive correlation with reducing the risk of cardiovascular diseases.
- Foods rich in flavonoids have been shown to mitigate the adverse effects of neurogenetic diseases.
- Extracts containing flavonoids of *Stachys cretica* have exhibited relief from oxidative stress and a decrease in Alzheimer's disease, Hyperglycemia, and melisma. (Bahadori et al. 2019)
- Neuropathy, a condition characterized by nerve malfunction, has been alleviated using natural compounds containing flavonoids to relieve neuropathic pain.

## 17.4 Types of Flavonoids

(a) **Catechins**

Abundantly found in green tea, catechins potentially aid in preventing different forms of tumors.

(b) **Hesperidin**

Hesperidin, found in citrus fruits, is an antioxidant that reduces irritation and prevents chronic diseases.

(c) **Flavonols**

Notably, quercetin, a well-studied flavonol, is found in various foods such as apples, berries, kale, and onions.

(d) **Resveratrol**

Resveratrol, found in grapes, purple grape juice, and red wine, acts as an antioxidant and anti-inflammatory agent. Studies hint at its potential improvement of heart health and various tumors, as well as its possible contribution to longevity based on animal studies. However, further analysis on man is required to set up a clearer connection.

(e) **Glucosinolates**

Present in vegetables of the crucifer family like Brussels sprouts, cabbage, kale, and broccoli, glucosinolates provide them with distinct aroma and taste. During cooking and digestion, these compounds convert into other chemicals that may aid in controlling the formation and advancement of tumors.

(f) **Phytoestrogens**

Phytoestrogens can mimic or counteract estrogen effects. Isoflavones, a type of phytoestrogen found in soy, may decrease uterine wall tumor risk and bone loss in women, as suggested by some evidence.

## 17.5 Phenolic Compounds

Phenolic composites are secondary metabolites produced in plants through the shikimic acid pathway and pentose phosphate via phenylpropanoid metabolism. They comprise benzene rings with one or more hydroxyl substituents, varying from simple phenolic molecules to extensively polymerized compounds (Lin et al. 2016). Cardiac glycosides, which are actively being studied worldwide, are a unique group of secondary metabolites sourced from plants. Researchers are investigating their potential to minimize side effects in medicinal applications. Among these compounds, cardiac glycosides stand out as valuable remedies. Familiar instances encompass digitoxin, digoxin, and convallotoxin, which bolster cardiac vigor and control contractions, especially in instances of heart failure. Furthermore, these substances also exhibit diuretic properties, stimulating the production of urine and assisting in the elimination of surplus fluids from tissues and the circulatory system (Singh et al. 2021) (Table 17.1).

**Physiological Role of Glycosides**

(i) Cardiac glycosides are a class of steroids that possess the specific ability to exert a powerful action on the cardiac muscle. Even in very small amounts, they can beneficially stimulate a diseased heart. These compounds hold primary value in treating congestive heart failure.

**Table 17.1** Some common plants containing natural cardenolides (Morsy 2017)

| Family | Genera |
|---|---|
| Apocynaceae | *Adenium, Acokanthera, Strophanthus, Apocynum, Cerbera, Thevetia, Nerium, Carissa, Urechites* |
| Asclepiadaceae (subsumed in Apocynaceae) | *Gomphocarpus, Calotropis, Pachycarpus, Asclepias, Xysmalobium, Cryptostegia, Menabea, Periploca* |
| Moraceae | *Antiaris, Antiaropsis, Naucleopsis, Maquira, Castilla* |
| Leguminosae | *Coronilla* |
| Scrophulariaceae | *Digitalis, Isoplexis* |
| Cruciferae | *Erysimum, Cheiranthus* |
| Sterculiaceae | *Mansonia* |
| Tiliaceae | *Corchorus* |
| Celastraceae | *Euonymus, Lophopetalum* |

(ii) They effectively increase the force of heart contraction without a simultaneous increase in oxygen consumption. Consequently, the myocardium becomes a more efficient pump, enabling it to meet the demands of the circulatory system.

(iii) During cardiac failure or congestive heart failure, the heart faces difficulty in adequately pumping blood to meet the body's requirements. Each contraction of the heart involves sodium (Na+) influx and potassium (K+) outflow. To prepare for the next contraction, Na+, K+-ATPase is essential for re-establishing the concentration gradient by moving Na+ back into the cell against its gradient. This process, which demands energy, relies on the hydrolysis of ATP to ADP by Na+, K+-ATPase (Morsy 2017).

## 17.6 Minerals

Many plants have high mineral concentrations that extract them from the ground and transform them into simple forms for man. Constituent of minerals plays a vital role in a plant's efficacy as a medicinal remedy. For instance, horsetail is known for its high silica content, which is used in arthritis treatments due to its support for the formation of connective tissue.

**Pathophysiology**—It is well-established that certain inorganic trace elements such as vanadium, zinc, chromium, iron, potassium, sodium, and nickel play a crucial role in maintaining normoglycemia by stimulating the beta-cells of the pancreas. This study investigates the elemental composition present in the leaves of four traditional medicinal plants (*Murraya koenigii, Mentha piperita, Ocimum sanctum*, and *Aegle marmelos*) extensively utilized in treating diabetes-related metabolic conditions, using inductively coupled plasma atomic emission spectroscopy.

The concentrations of copper (Cu), nickel (Ni), zinc (Zn), potassium (K), and sodium (Na) were found to be in trace amounts, while iron (Fe), chromium (Cr), and vanadium (V) were observed in borderline concentrations. (Narendhirakannan et al. 2005).

**Proanthocyanins**—These compounds are pigments that impart red, purple, or blue hues to fruits and flowers and are closely related to tannins and flavonoids. They have been proven to be valuable in supporting various bodily functions, particularly in the protection of circulation, especially in the heart, eyes, and blood vessels. High levels of proanthocyanidins can be found in red grapes, blackberries, and hawthorn berries.

**Proanthocyanidins**, a polyphenol present in numerous fruits, such as cranberries, blueberries, and grape seeds (Fig. 17.1). These oligomeric flavonoids consist of oligomers of catechin and epicatechin, along with their gallic acid esters, and are categorized as such chemically. Composites of polyphenols with a similar polymeric structure make the category of tannins.

Jacques Masquelier first discovered oligomeric proanthocyanidins in 1947. He developed and patented methods for extracting them from pine bark and grape

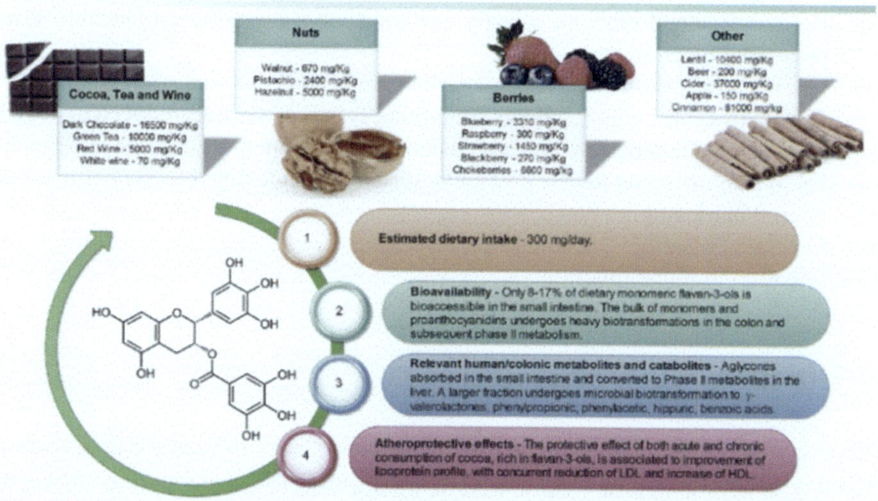


**Fig. 17.1** Dietary sources of Proanthocyanidins

seeds. While often associated with preventing urinary tract infections (UTIs) through the consumption of cranberries, grape seeds, or red wine, the effectiveness of proanthocyanidins for preventing or treating UTIs has not been conclusively proven.

**Uses of Proanthocyanidins**

i. It has been known to possess antioxidant capacities which can reduce the chance of cellular injury due to oxidative stress
ii. Cardiovascular diseases and associated risk factors, including high blood pressure and arterial hardening
iii. Cancer
iv. Compromised immune function
v. Diabetes
vi. Neurological disorders like Alzheimer's disease and dementia, manifest symptoms of impaired cognitive function, including memory issues and difficulty concentrating.

Manifestations of health issues include:

- Fatigue
- Slow recovery from exercise or physical activity
- Vision impairment
- Obesity

Several research studies have demonstrated the neuroprotective effects of proanthocyanidins in Alzheimer's disease (AD) by inhibiting amyloid-beta aggregation, reducing amyloid-beta production, amyloid plaques, and microgliosis. Additionally, they prevent amyloid-beta neurotoxicity in the brains of mice (Rauf et al. 2019).

(i) Boost lysosomal enzyme activity and enhance the phagocytic capability of peritoneal macrophages.
(ii) Obesity results from metabolic disturbance and can exacerbate the perturbation further. In a study cohort, tablets containing grape seed extract reduced oxidized LDL levels, thereby lowering the risk of arteriosclerosis.
(iii) Factors contributing to cardiovascular disease include high blood pressure and arterial hardening.
(iv) Cancer.
(v) Impaired immune function.

## 17.7 Saponins

This group of active compounds is named for their ability to produce lather when placed in water, resembling soap. There are two primary forms of saponins: steroidal and triterpenoid. Steroidal saponins closely resemble human hormones like estrogen and cortisol. Plants that contain saponins include agave, wild yam, and various lilies.

In instances of cardiac failure or congestive heart failure, the heart faces difficulty in adequately pumping blood to meet the body's demands. Each heartbeat involves an influx of Saponins have attracted considerable interest owing to their diverse properties, which can be both beneficial and detrimental (Fig. 17.2). Scientific research indicates that these compounds, known for promoting health, have the potential to enhance the immune system, combat cancer, and reduce cholesterol levels. They play a role in diminishing blood lipids, reducing cancer risk, and managing glycemic index. A diet high in saponins can aid in preventing dental caries and platelet aggregation, treating hypercalciuria in humans, and serving as a remedy for acute lead poisoning. Epidemiological studies have demonstrated an inverse relationship between saponins and the prevalence of kidney stones (Shi et al. 2004). Furthermore, several indigenous plants are utilized in the production of synthetic hormones for human use (Table 17.2).

## 17.8 Plant Pigments

Pigments are compounds known for their vivid colors and their application in coloring various materials. They are often referred to as Biological Pigments or Biochromes, representing true pigments. These biological pigments, insoluble in water, are utilized as solid ground particles mixed with liquids. They can be found in both plants and animals. Living organisms produce these pigments, resulting in a distinct color due to their absorbing skills.

There are two primary types of pigments classified based on their sources:

a. **Organic Pigments**: Derived from natural sources.

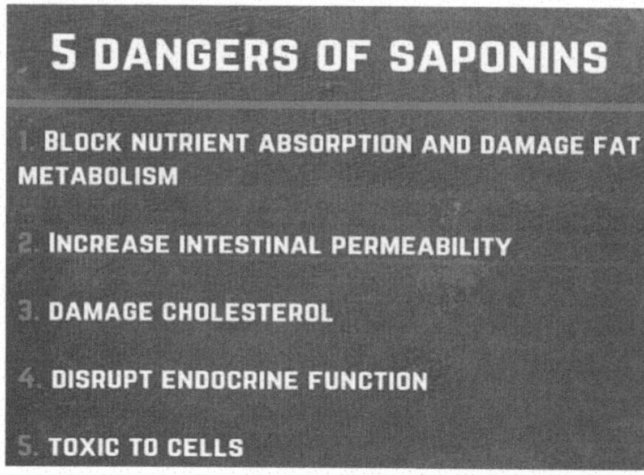


**Fig. 17.2** Dangers of Saponins

**Table 17.2** Plants in which Saponins are found

| Plant | Saponin content (g/kg) |
|---|---|
| Chick-pea (*Cicer arietinum*) | 2.3–6.0 |
| Soybean (*Glycine max*) | 5.6–56 |
| Navy bean (*Phaseolus vulgaris*) | 4.5–21 |
| Mung bean (*P. mungo*) | 0.5–5.7 |
| Mung bean shoots | 27 |
| Broad bean (*Vicia faba*) | 3.5 |
| Green pea (*Pisium sativum*) | 1.8–1.1 |
| Peanut (*Arachis hypogyea*) | 16 |
| Spinach (*Spinacea oleracea*) | 15 |
| Silver beet (*Beta vulgaris*) | 47 |
| Eggplant (*Solanum melongena*) | 58 |
| Yucca (Yucca schidigera) | 80 |

b. **Inorganic Pigments**: Derived from artificial sources like coal, and petroleum, and are sometimes created through simpler chemical reactions such as oxidation.

These are biochemical constituents that bounce back specific wavenumber of clear light, presenting a variety of appearances. Various organisms which include flora and fauna possess specific types of pigments that not only give them their distinctive appearance but also absorb particular wavelengths (Fig. 17.3).

**Types of Plant Pigments**

Plants possess a distinctive ability to capture photons and change them into carbohydrates by photosynthesis. This procedure initiates with the assimilation of photons by specific carbon-based particles known as pigments. In photosynthesis, a natural occurrence, green pigmentation referred to as chlorophyll is crucial, alongside other pigments in hues of unheroic and red. Other essential pigment varieties in plants

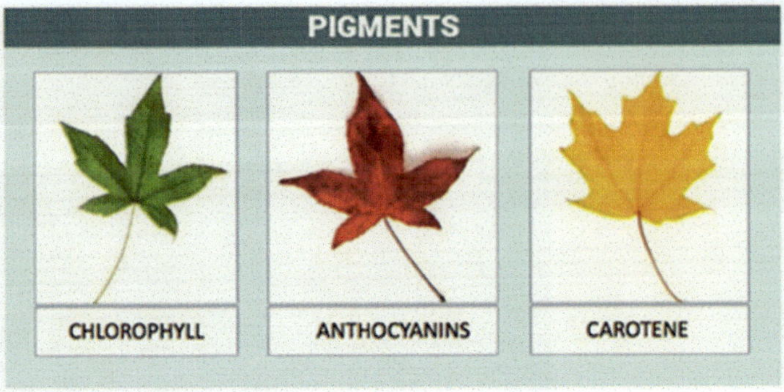


**Fig. 17.3** Different types of Pigments

include anthocyanins, betalains, carotenoids, porphyrins, etc. The pigments facilitate chemical reactions by interacting with and reflecting specific wavenumber.

The primary kinds of factory pigments are delineated as follows:

**Chlorophyll**: It stands as a fundamental pigment present within the cellular structure of all green plants. The green coloration of leaves and the soft parts of stems arises through chlorophyll. This pigment holds immense significance in photosynthesis.

Existence of chlorophyll pigment based on structure and function:

**Chlorophyll a**: Found in algae, cyanobacteria, and all higher plants.
**Chlorophyll b**: Present in green algae and higher plants exclusively.
**Chlorophyll c**: Discovered in certain photosynthetic Chromista and some marine algae.
**Chlorophyll d**: Only identified in red algae.
**Chlorophyll e**: Solely observed in algae.

Chlorophyll a and b are the chief pigments, essential for photosynthesis.

**Health Benefits of Chlorophyll**

- Chlorophyllin has shown potential benefits in reducing sensitivity and inhibiting the growth of bacteria in body injuries.
- A comprehensive study conducted in 2008 analyzed multiple studies regarding ointments containing papain, urea, and chlorophyllin for treating cracked skin.
- Although individual studies indicated the effectiveness of this ointment compared to other treatments, experts stress the need for larger, more controlled studies to authenticate these findings.
- Researchers have explored the effects of chlorophyll and chlorophyllin on cancer.
- A notable research conducted on trout demonstrated that chlorophyll lowered liver lumps by 29–63% and stomach lumps by 24–45%.

**Possible Side Effects**

Natural chlorophyll and chlorophyllin are not generally considered to be toxic. However, there might be potential side effects, such as:

- Digestive problems
- Diarrhea
- The excretion of green, tarry, or dark feces, could be misguided for gastrointestinal bleeding
- Topical application leads to an irritating or scorching sensation.

**Carotenoids**

Carotenoids, appearing in vibrant shades of orange, red, and yellow, are pigments insoluble in water and are closely associated with cell membrane structures. These biomolecules serve as antioxidants and play a pivotal role in maintaining optimal human vision.

**Health Benefits**

- Including carotenoid-rich foods in your diet promotes the health of eye cells and potentially inhibits the proliferation of cancerous cells.
- Carotenoids function as antioxidants, reducing bodily inflammation. While ongoing research delves into their precise mechanisms, these compounds have been associated with improving cardiovascular health by mitigating inflammation. Alleviating inflammation aids in safeguarding against heart conditions and averting arterial blockages.
- Enhancing your diet with more carotenoid-rich foods (Fig. 17.4) bolsters the immune system and overall health.
- While carotenoids can be obtained through vitamin A supplements, their consumption from natural sources amplifies their antioxidant effects.

Additionally, supplements carrying high levels of vitamin A can pose risks, potentially leading to toxicity when consumed excessively.

**Anthocyanins**

Anthocyanins represent a specific class of flavonoid pigments naturally occurring in the tissues of higher plants (Fig. 17.5). These pigments serve to impart color to the stems, leaves, roots, fruits, and flowers. Depending on the pH levels, these pigments manifest as red, blue, purple, and various other dark shades.

**Health Advantages**

- Reduction in heart problems, hypertension, and arterial stiffening.
- Potential cancer-fighting properties.
- Support for immune system function.
- Assistance in managing diabetes.
- Potential for aiding neurological disorders like Alzheimer's disease and dementia.

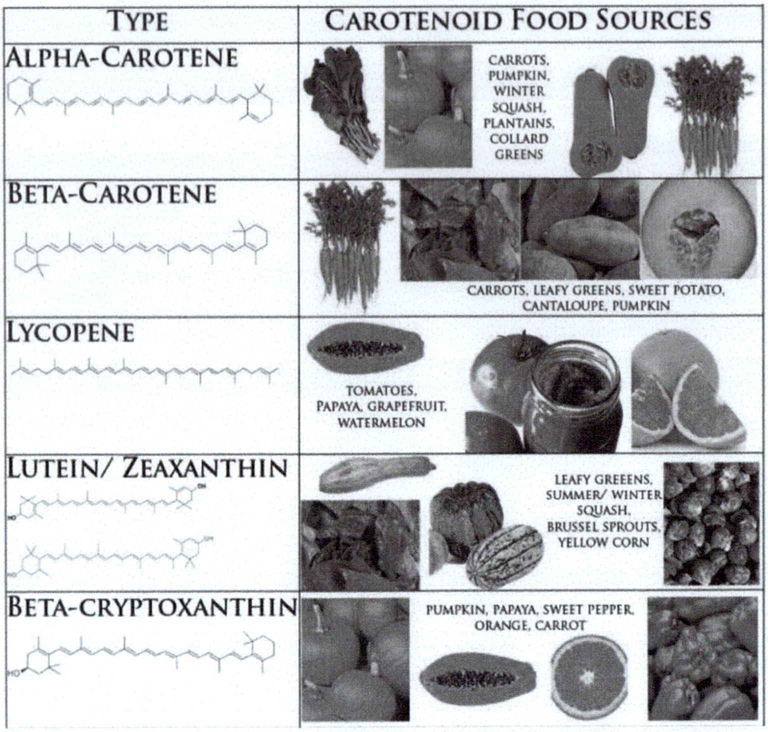


**Fig. 17.4** USDA database for flavonoid content of selected foods

- Alleviation of symptoms associated with poor cognitive function, such as memory issues and difficulty concentrating.
- Reduction of fatigue.
- Support for improved recovery following exercise or physical activity.
- Contribution to better vision health.
- Possible assistance in combating obesity.

### Essential Oils

Derived from fragrant medicinal plants, essential oils present a proven and age-old method for the management and prevention of various conditions. Their historical usage is rooted in the therapeutic effectiveness these oils offer. These natural, volatile compounds extracted from aromatic plants consist of a diverse array of substances, including esters of fatty acids, monoterpenes, sesquiterpenes, phenylpropanoids, aldehyde alcohols, and occasionally aliphatic hydrocarbons. For centuries, essential oils have served medicinal purposes, tracing their documented use back to Ancient Egypt. Renowned for their wide array of biological properties, these oils encompass antioxidants, anticancer elements, antiprotozoal, antifungal, antibacterial, and anti-inflammatory characteristics (Oliveira et al. 2018).

**Fig. 17.5** Different food sources that are rich in anthocyanin

## Health Benefits of Essential Oils (Fig. 17.6)

- Essential oils are characterized by a distinct blend of chemicals that facilitate their ability to permeate the cell walls of microorganisms, endowing them with robust antimicrobial attributes.
- They stand as a valuable reservoir of antioxidants, significantly enhancing their worth in the realms of food and nutrition industries.
- Thought to display properties that can counteract cancer, essential oils impede cell growth and the proliferation of cancerous cells. Derived from a variety of plants, these oils have displayed promise in combating several types of cancer, including those affecting the mouth, bones, lungs, prostate, liver, skin, colon, ovaries, pancreas, uterus, brain, leukemia, glioblastoma, and carcinoma.
- These oils demonstrate effectiveness in combating protozoan-related conditions, showcasing remarkable antiparasitic characteristics.
- Owing to their diverse chemical makeup in varying concentrations, essential oils find utility in medicinal formulations due to their potential anti-inflammatory effects.
- Recognized as a rich source of a wide array of compounds, essential oils not only aid in managing conditions but also hold the possibility to extend and enhance

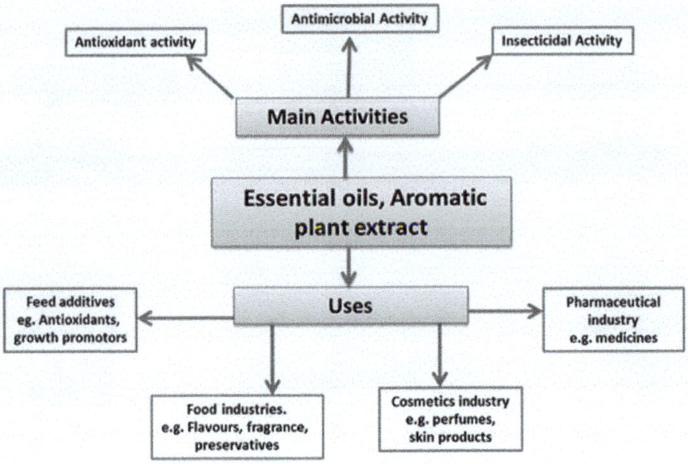


**Fig. 17.6** Activites of essential oils

life's quality for individuals. Their inherent properties present a genuine and natural alternative for addressing a variety of health concerns.

## 17.9 Conclusion

Plants play an integral role as life sources, providing essential elements such as oxygen, food, and fodder. Beyond these vital contributions, they offer an array of phytochemicals renowned for their diverse remedial properties. When consumed wisely and consistently in our diets, these compounds not only aid in alleviating ailments but also hold the potential, as suggested by various researchers, to effectively address a myriad of health concerns.

The utilization of plants for medicinal purposes is not a recent trend; many nations, including India, have long embraced this practice. Disciplines like Ayurveda have been wholly dedicated to leveraging plant-based remedies, asserting the ability to treat numerous conditions by harnessing the pharmacological elements and remedial qualities inherent in botanical sources.

Revisiting history prompts us to reconnect with our roots and re-establish a profound connection with nature. Throughout time, nature has bestowed countless benefits upon us, and it's crucial to approach its resources with utmost care, avoiding exploitation. While technological advancements have undoubtedly aided in certain health conditions, our current lifestyle trends have contributed to a decrease in the average lifespan. Therefore, by rekindling our understanding of nature's miraculous attributes and revisiting the potential of phytochemicals, we might pave the way for a healthier and more productive existence.

## References

Ahmad T, Gupta SK, Singh SK, Meraj G, Kumar P, Kanga S (2023) Unveiling nature's resilience: exploring vegetation dynamics during the COVID-19 era in Jharkhand, India, with the Google earth engine. Climate 11(9):187

Bahadori MB, Kirkan B, Sarikurkcu C (2019) Phenolic ingredients and therapeutic potential of Stachys cretica subsp. smyrnaea for the management of oxidative stress, Alzheimer's disease, hyperglycemia, and melasma. Indus Crops Prod 127(82–87). https://doi.org/10.1016/j.indcrop.2018.10.066

de Oliveira MS, Almeida MM, de Los Angeles Rodriguez Salazar M, Pires FCS, Bezerra FWF, Cunha VMB, … de Carvalho Junior RN (2018) Potential of medicinal use of essential oils from aromatic plants. InTech. https://doi.org/10.5772/intechopen.78002

Dias DA, Urban S, Roessner U (2012) A historical overview of natural products in drug discovery. Metabolites 2(2):303–336. https://doi.org/10.3390/metabo2020303

Ekor M (2014) The growing use of herbal medicines: issues relating to adverse reactions and challenges in monitoring safety. Front Pharmacol 4:177. https://doi.org/10.3389/fphar.2013.00177

Farooq M, Mushtaq F, Meraj G, Singh SK, Kanga S, Gupta A, Kumar P, Singh D, Avtar R (2022) Strategic slum upgrading and redevelopment action plan for Jammu city. Resources 11(12):120

Kumar S, Paul S, Walia YK, Kumar A, Singhal P (2015) Therapeutic potential of medicinal plants: a review. J Biol Chem Chron 1(1):46–54. https://doi.org/10.1016/j.indcrop.2018.10.066

Lin D, Xiao M, Zhao J, Li Z, Xing B, Li X, Kong M, Li L, Zhang Q, Liu Y, Chen H, Qin W, Wu H, Chen S (2016) An overview of plant phenolic compounds and their importance in human nutrition and management of type 2 diabetes. Molecules 21(10):1374. https://doi.org/10.3390/molecules21101374

Morsy N (2017) Cardiac glycosides in medicinal plants. INTECH. https://doi.org/10.5772/65963

Narendhirakannan RT, Subramanian S, Kandaswamy M (2005) Mineral content of some medicinal plants used in the treatment of diabetes mellitus. Biol Trace Elem Res 103:109–115. https://doi.org/10.1385/BTER:103:2:109

Rauf A, Imran M, Abu-Izneid T, Iahtisham-Ul-Haq, Patel S, Pan X, Naz S, Silva AS, Saeed F, Suleria HAR (2019) Proanthocyanidins: a comprehensive review. Biomed Pharmacother 116:108999. https://doi.org/10.1016/j.biopha.2019.108999

Shi J, Arunasalam K, Yeung D, Kakuda Y, Mittal G, Jiang Y (2004) Saponins from edible legumes: chemistry, processing, and health benefits. J Med Food 7(1):67–78. https://doi.org/10.1089/109662004322984734

Singh H, Meraj G, Singh S, Shrivastava V, Sharma V, Farooq M, Kanga S, Singh SK, Kumar P (2022) Status of air pollution during COVID-19-induced lockdown in Delhi, India. Atmosphere 13(12):2090

Singh S, Singh H, Sharma V, Shrivastava V, Kumar P, Kanga S, Sahu N, Meraj G, Farooq M, Singh SK (2021) Impact of forest fires on air quality in Wolgan valley, New South Wales, Australia—a mapping and monitoring study using Google earth engine. Forests 13(1):4

Sinha KK, Gupta MK, Banerjee MK, Meraj G, Singh SK, Kanga S, Farooq M, Kumar P, Sahu N (2022) Neural network-based modeling of water quality in Jodhpur, India. Hydrology 9(5):92

Ullah A, Munir S, Badshah SL, Khan N, Ghani L, Poulson BG, Emwas AH, Jaremko M (2020) Important flavonoids and their role as a therapeutic agent. Molecules 25(22):5243. https://doi.org/10.3390/molecules25225243

# Chapter 18
# Predicting Parkinson's Disease Using Analytical Algorithm: A Review

**Shashank Gaur and Sameer Awasthi**

## 18.1 Introduction

Parkinson's is a neurological disease that progresses as well as identified by motor indication and non-motor indication. Aside from a number of common indication, every individual will feel and show the condition in own unique way (Bind et al. 2015). Parkinson is a long-duration disorder of the brain that affects the motor system. Typically, indication seen gradually, and as the illness progress, non-motor effect increase in frequency. Shivers, inflexibility, slowness of motion, and poor mobility are the more noticeable initial signs. Although there is no known treatment for Parkinson's disease, medicines may significantly lessen your indication. Doctors may suggest rarely for surgery to treat your indication and control specific brain regions. Chronic Parkinson's disease affects both the neurological system and the portions of the body that the central system of nervous controls. The signs are seen gradually. A modest tremor in just one hand could be the initial indication. While shivers are so common with the condition, they can also cause you to move more slowly (Rana et al. 2022).

A brain ailment called Parkinson's disease causes stiffness, tremors, balance problems, and other accidental or unconscious movements. The disease, a synucleinopathy with a threshold of 4% in the over-80 population, is very prevalent in Western Europe with an incidence of 160/100,000 people. (Mutch et al. 1986). Typically, indications start out nonirritating and get bad over time. As the illness goes bad, people may be in difficult to talk and move around. Additionally, they may have mood swings, behavioral abnormalities, issues in sleep, melancholy, loss of memory, and tiredness. Despite the fact that almost everyone is at risk of getting Parkinson's, certain case studies indicate that men are more likely than women to suffer from this

S. Gaur (✉) · S. Awasthi
Bansal Institute of Engineering and Technology, Lucknow, India
e-mail: shshnk004@gmail.com

© The Author(s), under exclusive license to Springer Nature Singapore Pte Ltd. 2024

G. Tripathi et al. (eds.), *Sustainability and Health Informatics*, Advances in Geographical and Environmental Sciences,
https://doi.org/10.1007/978-981-97-6706-9_18

condition. Research is being done to identify any potential risk factors, despite the fact that the etiology is unclear.

Age clearly presents a risk. Despite the fact that 5% to 10% of patients experience indication before the age of 50, Parkinson's disease often first manifests in those over the age of 60. Early-onset Parkinson's disease is usually genetic, though not always, and certain types have been linked to specific gene abnormalities. Parkinson's disease is still predominantly diagnosed clinically, thus it's important to identify both early indications and those that suggest secondary causes of Parkinsonism. Therapy options for both the late and early stages of the illness have rapidly expanded along with a fuller knowledge of non-motor issues. For the detection and care of people with Parkinson's disease, the National Institute for Health and Clinical Excellence (NICE) in the guidelines issued by the UK (National Institute for Health and Clinical Excellence, 2006).

## 18.2 Related Work

Neurology offers a long range of methods that could be applied to support the clinical diagnosis of Parkinson's disease. These include imaging tests like SPECT and Cardiac MIBG, both of which show the physiological effects of dopaminergic denervation in an indirect manner.

### *18.2.1 Effect of Electroencephalography (EEG) Signals on Parkinson's Disease*

Compared to the previously described image-based exams, EEG testing is less expensive, non-invasive, and available at most medical facilities. This method, which captures the electrical activity of pyramidal neurons in the cerebral cortex, has been successful in treating other neurological conditions like epilepsy, developing biomarkers for Alzheimer's disease, and even experimentally detecting anomalies in schizophrenia (Gandal et al. 2012; Smailovic 2019; Kannathal et al., 2005;). Additionally, it can reproduce results regardless of outside factors thanks to its high temporal resolution and high test–retest ratio. However, the low signal-to-noise ratio and imaginary nature of the EEG signals, identify them and make their subsequent analysis challenging. Low signal-to-noise ratios are indicative of significant levels of noise in EEG signals, necessitating pre-processing to remove any distortions and filter out the noise before signal analysis and result gathering. The lack of a standardized cleaning methodology is the key issue when filtering signals.

Another issue is the possibility that important components of EEG signals can be destroyed when signal noise is removed, which could result in incorrect diagnosis.

The stochastic nature of an event means that its probability of occurrence is independent of earlier events. Therefore, for nonlinear dynamics study, which necessitates highly expensive computing approaches, advanced techniques must be utilized to extract the fundamental properties of the signals (Bigdely-Shamlo et al., 2015; Cole 2019), a medication that replaces the dopamine lost.

It is clinically determined whether someone has Parkinson's disease based on their motor indication, and this is done by evaluating their repetitive movements of the limb, spontaneous movements, resistance to passive mobilization, balance, and gait pattern. A highly suspicious attitude toward the disease and the evaluation of an experienced clinician is necessary for the diagnosis, and early-stage misdiagnosis is not uncommon (Beach, 2018) (Table 18.1).

Therefore, improvements in Parkinson's disease diagnostic techniques could aid in detecting the condition at an earlier stage, allowing for the anticipation of giving a drug containing dopaminergic, which may improve patients' quality of life for a longer period of time.

## 18.3 Proposed APPML Model

The Age Prediction in Parkinson's using ML (APPML) Model is a model that is used to predict the age of people who are going to be affected by Parkinson's disease based on the given datasets of a specific region. APPML model is inspired by the K Nearest Neighbor (KNN) algorithm in which the prediction is based on the K number of the nearest neighbor. In this model, three steps have to be followed that is the process the dataset, implementation of the APPML model, model analysis.

### *18.3.1 Process the Dataset*

The dataset is taken from Kaggle and this dataset is processed as training and testing dataset. The dataset is processed at different levels in which firstly, the label encoder works. The label encoder converts all the data into zero or one format. The process of label encoder is to standardize the dataset into zero or one category. Then, we have identified all the empty values from the dataset. In last, we have defined the training dataset and testing dataset. In machine learning, learning can be defined into three types that is supervised, unsupervised, and reinforcement learning. Reinforcement learning is based on reward and punishment that is if the prediction comes with less error then reward otherwise punishment. Now, if we look at unsupervised and supervised learning, every dataset is initially in an uncategourized format which is defined by unsupervised learning and if the dataset is processed and the dataset defined in a proper structured format then that is defined by supervised learning. So, in this way, we are converting an unsupervised learning-based dataset into a

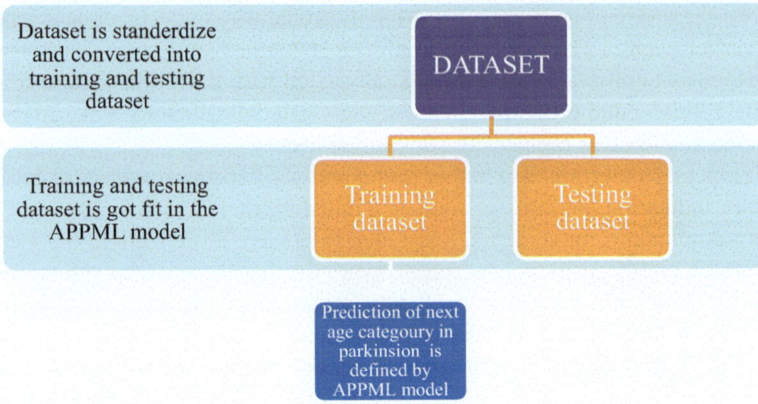


**Fig. 18.1** Dataflow diagram of APPML model

supervised learning dataset. The training and testing dataset after defining is used in the proposed model APPML model (Fig. 18.1).

### *18.3.2 APPML Model*

This model is initiated from the idea of KNN algorithm which predicts the targeted category of attribute by defining the similar category of attribute of targeted attribute in the dataset and from similar category attributes the attribute which has the highest percentage of appearance in the dataset will be the predicted valued of the targeted attribute.

### *18.3.3 APPML Model Analysis*

In this section, we will discuss the evaluation of APPML model because it is very important to verify that how model is working and for that we are using Root Mean Square Error (RSME) method the check the error rate of the APPML model. In this model we have taken 500 readings from the dataset and on the basis of which we predict the age of people who will be affected by Parkinson's disease in coming years but, ML-based model does not give full accuracy so RSME we can check that the prediction made by the APPML model has how much error rate that is from the actual value the predicted value is how much differ. This (Fig. 18.2) concludes that the model is not even overfitting and not even underfitting which shows that the model is working in proper format.

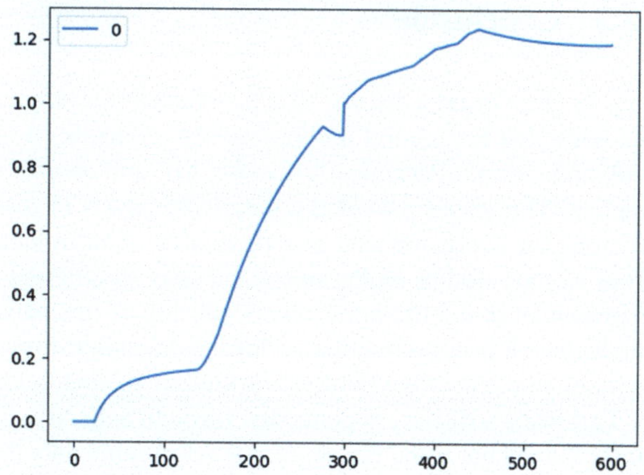


**Fig. 18.2** Root mean square error of APPML model

## 18.4 Result

In this section, we will discuss the confusion matrix that comes in the form of output for five different actual data and the prediction on those values. This APPML model is developed with the help of Google Colab in which a database is taken and this dataset is processed and converted into a training and testing dataset and then APPML model has been implemented (Fig. 18.3).

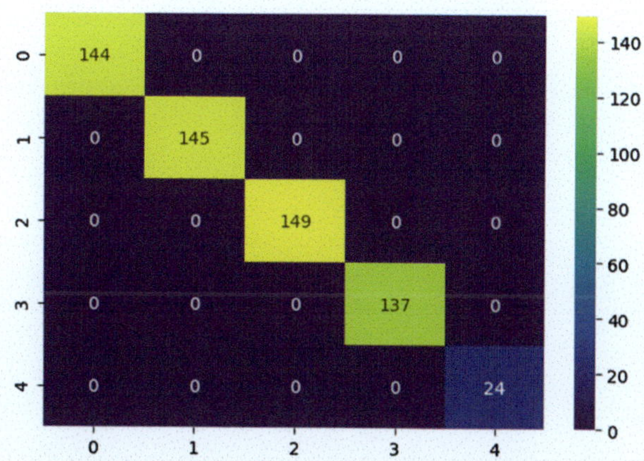


**Fig. 18.3** Confusion matrix of APPML model

## 18.5 Conclusion

Chronic Parkinson's disease affects both the neurological system and the portions of the body that the central nervous system controls. To differentiate between people with and without Parkinson's disease, ML and deep learning-related techniques are applied. This paper provided a detailed analysis of methods that use ML and deep learning for Parkinson's disease prediction. Additionally, a summary of the findings from various studies on Parkinson's disease prediction using ML and deep learning is presented in the literature review section of this paper. If we look at the APPML model then it is not showing overfitting and underfitting but, in this model we have to work on how the effect of other attributes can change the result of targeted attribute.

Parkinson's disease is generated through a disorder in the central nervous system of the human body and this disease is not curable only through medicines this disease can be controlled and this disease has now become a global challenge in the field of health. In this paper, a practical approach has been made in the area of health informatics to prevent from this disease. APPML model works in the field of health informatics in which prediction is done at which age a category of person has a higher probability of getting affected by this disease. This model can prepare people to take precautionary action before they become patients of Parkinson's disease.

**Table 18.1** Summary of recent related works

| Method | Performance | References |
|---|---|---|
| CNN + RNN | Accuracy: 96.9 | (Lee et al. 2019) |
| CNN + RNN, KNN, CNN, SVM, RF | Accuracy: 99.2 (CNN + RNN) | (Lee et al. 2021) |
| CNN | Accuracy: 88.25 | (Oh et al. 2020) |
| SVM, RF, DT, LR with LASSO, LR | Accuracy: 78 (RF) | (Chaturvedi et al. 2017) |
| RF | Accuracy: 92 | (Geraedts et al. 2021) |
| SVM | Accuracy: 64 | (Zhang et al., 2021) |
| SVM, KNN, LR, DT | Accuracy: 82 | (Anwar 2021) |
| SVM, KNN | Accuracy: 87 ± 3.5 (SVM)<br>Accuracy: 88 ± 2.8 (KNN) | (Betrouni et al., 2019) |
| CNN, SVM, MLP | Accuracy: 99.87 ± 0.03 (CNN) | (Chu et al. 2021) |
| VGG-16 | Accuracy: 99.62<br>F1-score: 0.996 | (Emamzadeh-Hashemi et al., 2022) |
| SVM | Accuracy: 87.54 ± 13.46 | (Guo et al., 2021) |
| MLP | Accuracy: 100 | (Jervis et al., 1994) |
| SVM, ANN, KNN, RF, LSSVM | LSSVM with:<br>Accuracy: 96.13 (HC vs Parkinson's disease OFF)<br>Accuracy: 97.65 (HC vs Parkinson's disease ON) | (Khare et al., 2021) |
| RF | Accuracy: 84.0 ± 4.2 | (Koch et al., 2019) |
| BNN | Accuracy: 86.8 | (Gilat et al. 2017) |
| BP-NN | Accuracy: 84.80 | (Handojoseno et al. 2016) |
| BN, NB, SVM, MLP, J48, RT, ELM, RF, mELM | Accuracy: 99.22 (RF with 50 tree) | (Oliveira et al. 2020) |
| CNN, RNN | Accuracy: 79 ± 1 (CNN)<br>Accuracy: 81 ± 1 (RNN) | (Ruffini et al. 2019) |
| RNN | Accuracy: 85 | (Ibañez et al. 2016) |
| MLP | Accuracy: 98.8 | (Saikia et al. 2019) |
| ANN | Accuracy: 98 | (Shaban et al., 2021) |
| EEGNet, HNet, DGHNet | Accuracy: 99.22 (HNet / DGHNet) | (Shah et al. 2020) |
| MLP | Accuracy: 97.5 | (Shreya Prabhu, Martis 2020) |
| DT, KNN, FKNN, NB, PNN, SVM | Accuracy: 99.62 ± 0.58 (Implementation of SVM with RBF) | (Yuvaraj et al. 2018) |
| BP-NN | Accuracy: 76.6 ± 3.4 | (Shine et al. 2012) |
| MLP, KNN | Accuracy: 73.47 (MLP)<br>Accuracy: 52.75 (KNN) | (Nguyen et al. 2013) |

(continued)

**Table 18.1** (continued)

| Method | Performance | References |
|---|---|---|
| MLP | Accuracy: 0.965<br>F1 score: 0.976 | (Rahman et al. 2013) |
| SVM | Accuracy: 93.34 ± 1.81 | (Vanneste et al., 2018) |
| Deep CNN | Accuracy 91.9% | (Yao et al. 2020) |
| ANN | Accuracy 98.3% | (Yurdakul et al., 2020) |
| SVM, logistic Regression | Accuracy 97% | (Yarnall et al. 2019) |
| Q-Backpropagated time delay neural network | Accuracy 91.49% | (Jane et al., 2016) |
| SVM, ANN | Accuracy 86.9% | (Perumal 2016) |
| ANN | Accuracy 96.28% | (Priya et al. 2021) |
| SVM | Accuracy 100% | (Andrei et al. 2019) |

## References

Andrei AG, Tăuțan AM, Ionescu B (2019) Parkinson's disease detection from gait patterns. In 2019 E-Health and Bioengineering Conference (EHB) (pp. 1–4). IEEE

Anwar T, Rehmat N, Naveed H (2021) A generic approach for classification of psychological disorders diagnosis using EEG. In Proceedings of the 43rd Annual International Conference of the IEEE Engineering in Medicine & Biology Society (EMBC), Guadalajara, Jalisco, Mexico, 1–5 pp. 2025–2029.

Beach TG, Adler CH (2018) Importance of low diagnostic Accuracy for early Parkinson's disease. Mov Disord off J Mov Disord Soc 33:1551–1554

Betrouni N, Delval A, Chaton L, Defebvre L, Duits A, Moonen A, Leentjens AFG, Dujardin K (2019) Electroencephalography- based machine learning for cognitive profiling in Parkinson's disease: Preliminary results. Mov Disord 34:210–217

Bigdely-Shamlo N, Mullen T, Kothe C, Su K-M, Robbins KA (2015) The PREP pipeline: Standardized preprocessing for large-scale EEG analysis. Front Neuroinform 9:16

Bind S, Tiwari AK, Sahani AK, Koulibaly P, Nobili F, Pagani M, Tatsch K (2015) A survey of machine learning based approaches for Parkinson disease prediction. Int J Comput Sci Inf Technol 6(2):1648–1655

Chaturvedi M, Hatz F, Gschwandtner U, Bogaarts JG, Meyer A, Fuhr P, Roth V (2017) Quantitative EEG (QEEG) measures differentiate Parkinson's disease (PD) patients from healthy controls (HC). Front Aging Neurosci. 9:3

Chu C, Zhang Z, Wang J, Liu S, Wang F, Sun Y, Han X, Li Z, Zhu X, Liu C (2021) Deep learning reveals personalized spatial spectral abnormalities of high delta and low alpha bands in EEG of patients with early Parkinson's disease. J Neural Eng 18:066036

Cole S, Voytek B (2019) Cycle-by-cycle analysis of neural oscillations. J Neurophysiol 122:849–861

Emamzadeh-Hashemi EA, Mahdizadeh A, Mirian MS, Lee S, McKeown MJ (2022) Deep transfer learning for parkinson's disease monitoring by image-based representation of resting-state EEG using directional connectivity. Algorithms 15:5

Gandal MJ, Edgar JC, Klook K, Siegel SJ (2012) Gamma synchrony: Towards a translational biomarker for the treatment-resistant symptoms of schizophrenia. Neuropharmacology 62:1504–1518

Geraedts VJ, Koch M, Contarino MF, Middelkoop HAM, Wang H, van Hilten JJ, Back THW, Tannemaat MR (2021) Machine learning for automated EEG-based biomarkers of cognitive impairment during Deep Brain Stimulation screening in patients with Parkinson's Disease. Clin Neurophysiol 132:1041–1048

Guo G, Wang S, Wang S, Zhou Z, Pei G, Yan T (2021) Diagnosing Parkinson's Disease Using Multimodal Physiological Signals. In Human Brain and Artificial Intelligence. HBAI 2021. Communications in Computer and Information Science, Yokohama, Japan, January 7, 2021; Wang, Y., Ed.; Springer: Singapore, 1369

Handojoseno AMA, Shine JM, Nguyen TN, Tran Y, Lewis SJG, Nguyen HT (2012) The detection of freezing of gait in parkinson's disease patients using EEG signals based on wavelet decomposition. In Proceedings of the Annual International Conference of the IEEE Engineering in Medicine and Biology Society, San Diego, CA, USA, August 28–September 1 2012; pp. 69–72

Handojoseno AMA, Shine JM, Nguyen TN, Tran Y, Lewis SJG, Nguyen HT (2013) Using EEG spatial correlation, cross frequency energy, and wavelet coefficients for the prediction of freezing of gait in parkinson's disease patients. In Proceedings of the 35th Annual International Conference of the IEEE Engineering in Medicine and Biology Society (EMBC), Osaka, Japan, 3–7 July 2013; pp 4263–4266

Jane YN, Nehemiah HK, Arputharaj K (2016) A Q-backpropagated time delay neural network for diagnosing severity of gait disturbances in Parkinson's disease. J Biomed Inform 60:169–176

Jervis BW, Saatchi MR, Lacey A, Roberts T, Allen EM, Hudson NR, Oke S, Grimsley M (1994) Artificial neural network and spectrum analysis methods for detecting brain diseases from the CNV response in the electroencephalogram. IEE Proc Sci Meas Technol 141:432–440

Kannathal N, Choo ML, Acharya UR, Sadasivan PK (2005) Entropies for detection of epilepsy in EEG. Comput Methods Programs Biomed 80:187–194

Khare SK, Bajaj V, Acharya UR (2021) Detection of Parkinson's disease using automated tunable Q wavelet transform technique with EEG signals. Biocybern. Biomed. Eng. 41:679–689

Koch M, Geraedts V, Wang H, Tannemaat M, Back T (2019) Automated machine learning for EEG-Based classification of parkinson's disease patients. In Proceedings of the 2019 IEEE International Conference on Big Data (Big Data), Los Angeles, CA, USA, 9–12 December 2019; pp. 4845–4852

Lee S, Hussein R, McKeown MJ (2019) A deep convolutional-recurrent neural network architecture for Parkinson's disease EEG classification. In Proceedings of the IEEE Global Conference on Signal and Information Processing (GlobalSIP), Ottawa, ON, Canada, 11–14 November 2019; pp. 1–4

Lee S, Hussein R, Ward R, Wang AJ, McKeown MJ (2021) A convolutional-recurrent neural network approach to resting-state EEG classification in Parkinson's disease. J Neurosci Methods 361:109282

Li B, Yao Z, Wang J, Wang S, Yang X, Sun Y (2020) Improved deep learning technique to detect freezing of gait in Parkinson's disease based on wearable sensors. Electronics 9(11):1919

Ly QT, Gilat M, Chai R, Martens KAE, Georgiades M, Naik GR, Tran Y, Lewis SJG, Nguyen HT (2017) Detection of turning freeze in Parkinson's disease based on S-transform decomposition of EEG signals. In Proceedings of the 2017 39th Annual International Conference of the IEEE Engineering in Medicine and Biology Society (EMBC), Jeju Island, Korea, 11–15 July 2017; pp. 3044–3047

Ly QT, Handojoseno, AMA, Gilat M, Nguyen N, Chai R, Tran Y, Lewis SJG, Nguyen HT (2016) Detection of Gait Initiation Failure in Parkinson's disease patients using EEG signals. In Proceedings of the 38th Annual International Conference of the IEEE Engineering in Medicine and Biology Society (EMBC), Orlando, FL, USA, 16–20 August 2016; pp. 1599–1602

Mutch WJ, Dingwall-Fordyce I, Downie AW et al (1986) Parkinson's disease in a Scottish City. BMJ 292:534–536

National Institute for Health and Clinical Excellence (2006) Parkinson's Disease: Diagnosis and Management in Primary and Secondary Care. London: NICE (http://guidance.nice.org. uk/ CG35)

Oh SL, Hagiwara Y, Raghavendra U, Yuvaraj R, Arunkumar N, Murugappan M, Acharya UR (2020) A deep learning approach for Parkinson's disease diagnosis from EEG signals. Neural Comput Appl 32:10927–10933

Oliveira APS, de Santana MA, Andrade MKS, Gomes JC, Rodrigues MCA, dos Santos WP (2020) Early diagnosis of Parkinson's disease using EEG, machine learning and partial directed coherence. Res. Biomed. Eng. 36:311–331

Perumal SV, Sankar R (2016) Gait and tremor assessment for patients with Parkinson's disease using wearable sensors. Ict Express 2(4):168–174

Priya SJ, Rani AJ, Subathra MSP, Mohammed MA, Damaševičius R, Ubendran N (2021) Local pattern transformation based feature extraction for recognition of Parkinson's disease based on gait signals. Diagnostics 11(8):1395

Rahman MA, Tutul AA, Islam ABMAA (2020) Solving the Maze of Diagnosing Parkinson's Disease based on Portable EEG Sensing to be Adaptable to Go In-The-Wild. In Proceedings of the 7th International Conference on Networking, Systems and Security, Dhaka, Bangladesh, 22–24 December 2020; pp. 65–73.

Rana A, Dumka A, Singh R, Panda MK, Priyadarshi N, Twala B (2022) Imperative role of machine learning algorithm for detection of parkinson's disease: review. Challenges and Recommendations. Diagnostics 12(8):2003

Rehman RZU, Del Din S, Guan Y, Yarnall AJ, Shi JQ, Rochester L (2019) Selecting clinically relevant gait characteristics for classification of early Parkinson's disease: a comprehensive machine learning approach. Sci Rep 9(1):1–12

Ruffini G, Ibañez D, Castellano M, Dubreuil-Vall L, Soria-Frisch A, Postuma R, Gagnon J-F, Montplaisir J (2019) Deep learning with EEG spectrograms in rapid eye movement behavior disorder. Front Neurol 10:806

Ruffini G, Ibañez D, Castellano M, Dunne S, Soria-Frisch A (2016) EEG-driven RNN classification for prognosis of neurodegenera- tion in at-risk patients. Lect. Notes Comput. Sci. (Incl. Subser. Lect. Notes Artif. In tell. Lect. Notes Bioinform.) 9886, 306–313

Saikia A, Hussain M, Barua AR, Paul S (2019) EEG-EMG correlation for Parkinson's disease. Int J Eng Adv Technol. 8:1179–1185

Shaban M (2021) Automated screening of parkinson's disease using deep learning based electroencephalography. In Proceedings of the 2021 10th International IEEE/EMBS Conference on Neural Engineering (NER), Virtually, 4–6 May 2021; pp. 158–161

Shah SAA, Zhang L, Bais A (2020) Dynamical system based compact deep hybrid network for classification of parkinson disease related EEG Signals. Neural Netw 130:75–84

Shreya Prabhu K, Martis RJ (2020) Diagnosis of parkinson's disease using computer aided tool based on EEG. In Proceedings of the IEEE 17th India Council International Conference INDICON, New Delhi, India, 10–13 December 2020; pp. 1–4

Smailovic U, Jelic V (2019) Neurophysiological markers of alzheimer's disease: quantitative EEG approach. Neurol. Ther. 8:37–55

Vanneste S, Song J-J, Ridder DD (2018) Thalamocortical dysrhythmia detected by machine learning. Nat Commun 9:1103

Yurdakul OC, Subathra MSP, George ST (2020) Detection of parkinson's disease from gait using neighborhood representation local binary patterns. Biomed Signal Process Control 62:102070

Yuvaraj R, Acharya UR, Hagiwara Y (2018) A novel Parkinson's Disease Diagnosis Index using higher-order spectra features in EEG signals. Neural Comput Appl 30:1225–1235

Zhang J, Gao Y, He X, Feng S, Hu J, Zhang Q, Zhao J, Huang Z, Wang L, Ma G et al (2021) Identifying Parkinson's disease with mild cognitive impairment by using combined MR imaging and electroencephalogram. Eur Radiol 31:7386–7394

SPRINGER NATURE

# GPSR Compliance

*The European Union's (EU) General Product Safety Regulation (GPSR) is a set of rules that requires consumer products to be safe and our obligations to ensure this.*

*If you have any concerns about our products, you can contact us on ProductSafety@springernature.com*

In case Publisher is established outside the EU, the EU authorized representative is:

Springer Nature Customer Service Center GmbH
Europaplatz 3
69115 Heidelberg, Germany

The manufacturer's authorised representative in the EU is Springer Nature Customer Service Centre GmbH, Europaplatz 3, 69115 Heidelberg, Germany. If you have any concerns regarding our products, please contact ProductSafety@springernature.com

Printed and bound by CPI Group (UK) Ltd, Croydon, CR0 4YY
30/09/2026
02225253-0010